Fluorescence Imaging and Biological Quantification

Fluorescence Imaging and Biological Quantification

Edited by

Raquel Seruca

Jasjit S. Suri

João M. Sanches

CRC Press is an imprint of the
Taylor & Francis Group, an **informa** business

CRC Press
Taylor & Francis Group
6000 Broken Sound Parkway NW, Suite 300
Boca Raton, FL 33487-2742

Printed on acid-free paper

International Standard Book Number-13: 978-1-4987-3704-3 (Hardback)

Library of Congress Cataloging-in-Publication Data

Names: Seruca, Raquel, editor. | Suri, Jasjit S., editor. | Sanches, João M., editor.
Title: Fluorescence imaging and biological quantification / [edited by] Raquel Seruca, Jasjit S. Suri, and João M. Sanches.
Description: Boca Raton : Taylor & Francis, 2017. | Includes bibliographical references.
Identifiers: LCCN 2017013420 | ISBN 9781498737043 (hardback : alk. paper)
Subjects: | MESH: Optical Imaging | Cytological Techniques | Molecular Imaging | Image Interpretation, Computer-Assisted
Classification: LCC R857.O6 | NLM WN 195 | DDC 616.07/54--dc23
LC record available at https://lccn.loc.gov/2017013420

Visit the Taylor & Francis Web site at
http://www.taylorandfrancis.com

and the CRC Press Web site at
http://www.crcpress.com

Printed and bound in the United States of America by Sheridan

The ideal scientist is enchanted by the scientific poetics of nature. As scientists, we dedicate this book to everyone who gets fascinated by Science and Art.

Contents

Preface

Huge numbers of biomedical images are generated each year in routine and research labs. In this book, we describe new developments and solutions to analyze and quantify fluorescence images, tagging DNA, RNA, and proteins in single cells as well as in cell populations. This book is a collaborative effort by large group of scientists working in complementary disciplines as biology, biochemistry, microscopy, physics, and engineering.

In Section I, we present different microscopic techniques that allow the production of high-quality 2D and 3D images as confocal microscopy to more quantitative methodologies, namely imaging flow cytometry and atomic force microscopy (AFM). This includes chapters that define strategies to circumvent limitations of fluorescent nanoparticles and include novel strategies to track, quantify, and map these signals. Multiple fluorochromes and fluorochrome dyes are currently available allowing single or multiple complex visualization of molecular events. In every chapter, the advantages and limitations of every microscopic approach will be discussed and the future technical developments in each scientific area will be addressed.

Section II compiles new imaging and computer-based technologies to access the inner machinery of living cells and shows how different methodologies contribute to advance on the understanding of highly dynamic biochemical processes occurring at cell, tissue, and organism level. We focused on a wide variety of biological questions related to signaling events and networks, formation of protein complexes, maintenance of cellular homeostasis by lysosomes, circadian rhythms, cell cycle, membrane trafficking, behavior of cancer-cell populations, and macrophages dynamics. Overall, we aim to demonstrate in this book how fluorescence microscopy and images can be mathematically processed to allow depiction of molecular events and pathways underlying cell function, tissue morphology and mechanics, and individual physiology.

Raquel Seruca, Jasjit S. Suri and João M. Sanches

Acknowledgments

We thank all coauthors for their contributions to the book and, in particular, to Paula Sampaio who provided an historical view of the topic.

Editors

Raquel Seruca (RS) received her PhD in Medicine from the faculty of Medicine from the University of Porto, Portugal in 1995. RS developed her PhD project in the field of genetics of gastric cancer.

Presently, RS is vice president of Institute of Molecular Pathology and Immunology at the University of Porto, Portugal (Ipatimup) and coordinator of the cancer research program at i3s, both located in Porto.

At i3s, Seruca is the group leader of Epithelial Interactions in Cancer (EPIC). As group leader, RS has been able to create a truly multidisciplinary environment, stimulating collaborations and interactions between surgeons, oncologists, pathologists, biologists, biochemists, and bioengineers.

The long-term goal of the group is to uncover how epithelial cell–cell and cell–matrix junctions, as well as the surrounding microenvironment, can influence gastric, breast, and colorectal cancer progression. The focus of the group is to unravel the role of E- and P-cadherin in epithelial homeostasis and cancer development.

Seruca's lab is a reference center of the International Gastric Cancer Linkage Consortium, responsible for the functional studies of E-cadherin mutations in hereditary gastric cancer. Using this disease model, the group has made significant contributions on the role of E-cadherin and associated signaling in cancer-cell migration, survival, and invasion.

Seruca has more than 200 publications in international peer-reviewed journals, with an h-index of 52.

Seruca collaborates with several research groups worldwide, serves as a reviewer for top journals in the area of cancer, and is invited to speak and chair sessions at major conferences dedicated to cancer research. Further, she is involved in the evaluation of several international grants and institutes.

In 2009, Seruca received the distinction Ordem do Infante D. Henrique from the Portuguese Presidency (Presidência da República) for her scientific merit. More recently, in 2014, she got the gold medal of the Porto City for her contribution for science internationalization.

Dr. Jasjit S. Suri, PhD, MBA, fellow of AIMBE is an innovator, visionary, scientist, and an internationally known world leader. Dr. Jasjit S. Suri received the Director General's gold medal in 1980 and the fellow of American Institute of Medical and Biological Engineering (AIMBE), awarded by National Academy of Sciences, Washington, DC in 2004. He has published more than 550, which includes journals, book chapters, and proceeding articles having an H-index 46, coauthored more than 40 books, 100 innovations, and trademarks. He is currently chairman of Global Biomedical Technologies, Inc., Roseville, California, company dedicated to cancer imaging and biomedical devices. Dr. Suri is also in the board of AtheroPoint, Roseville, California, a company dedicated to atherosclerosis imaging for early screening for stroke and cardiovascular monitoring. He has held positions as a chairman of IEEE Denver section and advisor board member to health care industries and several schools in the United States and abroad.

João M. Sanches (JS) received the EE, MSc, and PhD degrees from the Instituto Superior Técnico (IST) and Universidade Técnica de Lisboa (UTL), Lisbon, Portugal in 1991, 1996, and 2003, respectively and the habilitation (agregação) in 2013 by the Universidade de Lisboa (UL), Lisboa, Portugal in biomedical engineering. JS is associate professor at the Department of Bioengineering (DBE) at the IST and he is from the coordination board of the biomedical engineering master and doctoral programs. Before he was at the Department of Electrical and Computer Engineering (DEEC) where has taught in the area of signal processing, systems, and control.

JS is a senior researcher and member of the board of directors of the Institute for Systems and Robotics (ISR–IST). His work has been focused in biomedical engineering (BME), namely, in biological and medical image processing and statistical signal processing of physiological and behavioral data. Currently, JS aims to develop new tools and methodological strategies to quantify and map molecular and morphometric pathologic cancer biomarkers.

As a group leader, Professor Sanches has been able to create a truly multidisciplinary and stimulating environment where researchers from top institutions from the biology and medicine areas work close together with engineers and computer scientists. Approximately 150 international publications (ORCID and Google Scholar) and several patents were already produced in the scope of this collaborative work.

He is senior member of the IEEE Engineering in Medicine and Biology Society (EMBS) since 2011 and member of the bioimaging or bio-imaging and Signal Processing Technical Committee (BISP-TC) of the IEEE Signal Processing Society.

Contributors

Mónica Abreu
Institute for Theoretical Biology (ITB)
Charité-Universitätsmedizin Berlin and Humboldt-Universität zu Berlin
and
Medical Department of Hematology, Oncology, and Tumor Immunology, and Molekulares Krebsforschungszentrum (MKFZ)
Charité-Universitätsmedizin Berlin
Berlin, Germany

Catarina R. Almeida
i3S - Instituto de Investigação e Inovação em Saúde da Universidade do Porto
and
INEB - Instituto de Engenharia Biomédica
Universidade do Porto
Porto, Portugal

Linda Arngården
Science for Life Laboratory
Department of Immunology, Genetics and Pathology
Biomedical Center
Uppsala University
Uppsala, Sweden

Miguel Aroso
iBiMED - Instituto de Biomedicina
Departamento de Ciências Médicas
Universidade de Aveiro
Aveiro, Portugal

Patrícia Carneiro
Instituto de Investigação e Inovação em Saúde (i3S)
and
Institute of Molecular Pathology and Immunology of the University of Porto (IPATIMUP)
and
Instituto de Investigação e Inovação em Saúde/Institute for Research and Innovation in Health (i3S)
University of Porto
Porto, Portugal

Filomena A. Carvalho
Faculdade de Medicina
Instituto de Medicina Molecular
Universidade de Lisboa
Lisboa, Portugal

Patrícia Carvalho
i3S - Instituto de Investigação e Inovação em Saúde
Universidade do Porto
and
IPATIMUP - Instituto de Patologia e Imunologia Molecular da Universidade do Porto
Porto, Portugal

Patrícia Castro
Instituto de Investigação e Inovação em Saúde (i3S)
and
Institute of Molecular Pathology and Immunology of the University of Porto (IPATIMUP)
and
Department of Pathology and Oncology
Medical Faculty of the University of Porto
Porto, Portugal

Angela Margarida Costa
i3S - Instituto de Investigação e Inovação em Saúde
Universidade do Porto
and
INEB - Instituto de Engenharia Biomédica
Universidade do Porto
Porto, Portugal

Rukeia El-Athman
Institute for Theoretical Biology (ITB)
Charité-Universitätsmedizin Berlin and Humboldt-Universität zu Berlin
Berlin, Germany

Sofia Esménio
Institute for Systems and Robotics, Instituto Superior Técnico
Lisboa, Portugal

Tiago Esteves
i3S - Instituto de Investigação e Inovação em Saúde
Universidade do Porto
and
Faculdade de Engenharia
Universidade do Porto
Porto, Portugal

Maria Sofia Fernandes
Instituto de Investigação e Inovação em Saúde/Institute for Research and Innovation in Health (i3S)
University of Porto
and
Institute of Molecular Pathology and Immunology of the University of Porto (IPATIMUP)
Porto, Portugal

Julia Fernandez-Rodriguez
Centre for Cellular Imaging at Sahlgrenska Academy
University of Gothenburg
Gothenburg, Sweden

Anabela Ferro
Instituto de Investigação e Inovação em Saúde (i3S)
and
IPATIMUP, Institute of Molecular Pathology and Immunology
Universidade do Porto
and
Instituto de Investigação e Inovação em Saúde/Institute for Research and Innovation in Health (i3S)
University of Porto
Porto, Portugal

Joana Figueiredo
Instituto de Investigação e Inovação em Saúde (i3S)
and
Institute of Molecular Pathology and Immunology of the University of Porto (IPATIMUP)
Porto, Portugal

Martina Fonseca
Institute for Systems and Robotics, Instituto Superior Técnico
Lisboa, Portugal

A. Freitas
i3S - Instituto de Investigação e Inovação em Saúde
Universidade do Porto
and
INEB - Instituto de Engenharia Biomédica
Universidade do Porto
and
FMUP, Faculdade de Medicina da Universidade do Porto
Porto, Portugal

Luise Fuhr
Institute for Theoretical Biology (ITB)
Charité-Universitätsmedizin Berlin and Humboldt-Universität zu Berlin
and
Medical Department of Hematology, Oncology, and Tumor Immunology, and Molekulares Krebsforschungszentrum (MKFZ)
Charité-Universitätsmedizin Berlin
Berlin, Germany

Nikolai Genov
Institute for Theoretical Biology (ITB)
Charité-Universitätsmedizin Berlin and Humboldt-Universität zu Berlin
and
Medical Department of Hematology, Oncology, and Tumor Immunology, and Molekulares Krebsforschungszentrum (MKFZ)
Charité-Universitätsmedizin Berlin
Berlin, Germany

M. Gomez-Lazaro
i3S - Instituto de Investigação e Inovação em Saúde
Universidade do Porto
and
INEB - Instituto de Engenharia Biomédica
Universidade do Porto
Porto, Portugal

Johan Heldin
Science for Life Laboratory
Department of Immunology, Genetics and Pathology
Biomedical Center
Uppsala University
Uppsala, Sweden

Axel Klaesson
Science for Life Laboratory
Department of Immunology, Genetics and Pathology
Biomedical Center
Uppsala University
Uppsala, Sweden

Benjamin König
Institute of Chemistry and Biochemistry
Freie Universität Berlin
Berlin, Germany

Catarina S. Lopes
Faculdade de Medicina
Instituto de Medicina Molecular
Universidade de Lisboa
Lisboa, Portugal

Jeannine Mazuch
Institute for Theoretical Biology (ITB)
Charité-Universitätsmedizin Berlin and Humboldt-Universität zu Berlin
Berlin, Germany

Tânia Mestre
Institute for Systems and Robotics (ISR), Evolutionary Systems and Biomedical Engineering Laboratory (LaSEEB), Instituto Superior Técnico (IST)
and
Institute for Systems and Robotics, Instituto Superior Técnico
Lisboa, Portugal

Tommy Nilsson
The Research Institute of the McGill University Health Centre
McGill University
Montreal, Québec, Canada

Maria José Oliveira
i3S - Instituto de Investigação e Inovação em Saúde
Universidade do Porto
and
INEB - Instituto de Engenharia Biomédica
Universidade do Porto
and
Department of Pathology and Oncology
Faculty of Medicine
Porto, Portugal

Joana Paredes
Instituto de Investigação e Inovação em Saúde (i3S)
and
Institute of Molecular Pathology and Immunology of the University of Porto (IPATIMUP)
and
Department of Pathology and Oncology
Medical Faculty of the University of Porto
Porto, Portugal

Camila C. Portugal
Glial Cell Biology Laboratory (GCB)
Instituto de Investigação e Inovação em Saúde (i3S)
Porto, Portugal

Pedro Quelhas
i3S - Instituto de Investigação e Inovação em Saúde
Universidade do Porto
Porto, Portugal

Doroteya Raykova
Science for Life Laboratory
Department of Immunology, Genetics and Pathology
Biomedical Center
Uppsala University
Uppsala, Sweden

Angela Relógio
Institute for Theoretical Biology (ITB)
Charité-Universitätsmedizin Berlin and Humboldt-Universität zu Berlin
and
Medical Department of Hematology, Oncology, and Tumor Immunology, and Molekulares Krebsforschungszentrum (MKFZ)
Charité-Universitätsmedizin Berlin
Berlin, Germany

João B. Relvas
Glial Cell Biology Laboratory (GCB)
Instituto de Investigação e Inovação em Saúde (i3S)
Porto, Portugal

C.C. Ribeiro
i3S - Instituto de Investigação e Inovação em Saúde
Universidade do Porto
and
INEB - Instituto de Engenharia Biomédica
Universidade do Porto
and
ISEP - Instituto Superior de Engenharia do Porto
Instituto Politécnico do Porto
Porto, Portugal

Ana Sofia Ribeiro
Instituto de Investigação e Inovação em Saúde (i3S)
and
Institute of Molecular Pathology and Immunology of the University of Porto (IPATIMUP)
Porto, Portugal

André Roma
i3S - Instituto de Investigação e Inovação em Saúde
Universidade do Porto
and
IPATIMUP - Instituto de Patologia e Imunologia Molecular da Universidade do Porto
Porto, Portugal

and

Faculdade de Medicina da Universidade de Coimbra
Coimbra, Portugal

Ivan Sahumbaiev
Institute for Systems and Robotics (ISR)
Evolutionary Systems and Biomedical Engineering Laboratory (LaSEEB), Instituto Superior Técnico (IST)
Lisboa, Portugal

Paula Sampaio
Advanced Light Microscopy unit (ALM), Instituto de Investigação e Inovação em Saúde (i3S) Instituto de Biologia Molecular e Celular (IBMC)
Universidade do Porto
and
Instituto de Investigação e Inovação em Saúde (i3S)
Porto, Portugal

João M. Sanches
Institute for Systems and Robotics (ISR)
Evolutionary Systems and Biomedical Engineering Laboratory (LaSEEB), Instituto Superior Técnico (IST)
Lisboa, Portugal

Nuno C. Santos
Faculdade de Medicina
Instituto de Medicina Molecular
Universidade de Lisboa
Lisboa, Portugal

Raquel Seruca
Instituto de Investigação e Inovação em Saúde (i3S)
and
IPATIMUP, Institute of Molecular Pathology and Immunology
Universidade do Porto
and
Department of Pathology and Oncology
Faculdade de Medicina da Universidade do Porto
Porto, Portugal

Renato Socodato
Glial Cell Biology Laboratory (GCB)
Instituto de Investigação e Inovação em Saúde (i3S)
Porto, Portugal

Ola Söderberg
Science for Life Laboratory
Department of Immunology, Genetics and Pathology
Biomedical Center
Uppsala University
Uppsala, Sweden

Tobias Stauber
Institute of Chemistry and Biochemistry
Freie Universität Berlin
Berlin, Germany

Sérgia Velho
i3S - Instituto de Investigação e Inovação em Saúde
Universidade do Porto
and
IPATIMUP - Instituto de Patologia e Imunologia Molecular da Universidade do Porto
Porto, Portugal

Lisa von Kleist
Institute of Chemistry and Biochemistry
Freie Universität Berlin
Berlin, Germany

Introduction

Brief Historical view

Microscopy, fluorescence, and imaging

Two millenniums ago, Romans discovered that glass could be used to enlarge objects, but the first instruments capable to make visible structures not visible by the naked eye, only appeared in the past 400 years. Microscopes made possible to observe microorganisms, blood cells, sperm, and small details in plants and animals for the first time. Anton van Leeuwenhoek and Robert Hooke represented their pioneer observations by hand drawings and quantified some of their observations. As an example, Leeuwenhoek estimated the number of *animalcules* (bacteria and protozoa) in drops of water and determined the size and shape of *red corpuscles* (red blood cells).

Improvements in optics and theoretical studies of image formation and optics carried out in the second half of nineteenth century founded the grounds for the modern optical microscopy we experience today. Nowadays, scientists have a broad range of imaging methods available to study biological systems. The most classical transmission microscopy contrast techniques such as brightfield, phase-contrast, differential interference contrast, polarization, and dark-field are based on the alterations of light induced by interaction with the specimen. Transmission microscopy is a routinely choice approach for morphological analysis and for live-cell imaging to study cell shape, cell cycle, or migration of cell in culture. However, it is limited, as it does not allow to differential and simultaneously label molecules with high sensitivity and in a quantitative way. Fluorescence microscopy surpasses these problems and allows exploring the cells and tissues at more molecular and subcellular levels. In fluorescence microscopy, the specimens are stained by fluorochrome dyes, quantum dots, or express chimeric fluorescently tagged proteins. These fluorophores get into an excited state after absorbing energy, as a photon, at specific wavelengths. This high-energy state is transient and first, the excited state electrons go to lower energy levels by vibrational relaxation, after which the molecule returns to ground state by emission of a photon of lower energy than excitation. Multiple fluorochromes with very sensitive emission profiles are commercially available to label different cellular components allowing their independent visualization with high sensitivity as well as their spatial and temporal correlation.

The main constrain of fluorescence is blurring as fluorophores are self-luminous. So, out-of-focus light from different focal planes mix with the in-focus signals originating blurred images in widefield fluorescence microscopy (WFM). Confocal fluorescence microscopy overcomes this problem, by using a spatial filter, known as pinhole, at the detection level that suppresses the light from out-of-focus areas. This leads to the formation of a high contrasted image, optical section, that have reduced contribution of out-of-focus light. A specimen can then be optically sectioned and those images can be used to build a 3D reconstruction of the specimen.

Imaging cell in culture is still the most common approach to study the biology of the cells; however, higher eukaryotes are multicellular organism where the cells are integrated in a 3D community, tissues. So, the ability to study the cells within its natural environment is essential to have an integrate vision of the biological processes, and confocal microscopy ability to generate optical sections is still limited to few dozens of micrometers deep into tissues due to scattering of the light by matter. Multiphoton microscopy use pulsed IR light that is less scattered by tissues allowing to image deep. Since, IR photon have less energy, fluorophores must absorb energy of two photons to get excited. The high probability of excitation occurs only at focal plane, with no out-of-focus light being generated. Emitted light can be only detected close to the objective and this increases the detection sensitivity experimented in multiphoton microscopy.

In the past decade, a new microscopy technique to image live embryos in toto had a great development. In light sheet microscopy, a plane of the sample is illuminated by a sheet of light and emitted light is detected on an objective at 90°. Combining the acquisition at multiple planes and angles permits to make a total 3D reconstruction of each sample. Due to light sheet microscopy low phototoxicity, it is instrumental for live imaging of embryo development. Actually, live-cell imaging is essential to know the rules of biological processes or cell fate decisions and had a great expansion with development of new probes, including fluorescent proteins, which could *illuminate* targeted cell components. This enabled observing and measuring dynamic cellular events at molecular level with high spatial and temporal resolution.

The chemical distribution within an unstained specimen can be also evaluated by Raman confocal microscopy, a spectroscopic approach that provides a specific fingerprint of molecules.

Recent developments let to create techniques based in switch on/off of fluorescents molecules (generically known as *super-resolution microscopy*) that overcome the limit of resolution (~200 nm) of optical microscopy and approaching it to nanoscopy with resolution limits reaching already 10 nm. These techniques are *super-resolution* microscopy plus the combination of light microscopy with electron microscopy and atomic force microscopy will open a new vision of cell biology in future years.

As the pioneers already demonstrated, imaging is not only getting a nice picture of fine details. This is particularly evident with digital imaging that generates images, which are 2D matrices. So, the digital images can be analyzed for extraction of quantitative information in order to perform accurate evaluation of data and science of excellence.

Paula Sampaio

Section I

Advanced methods of analysis

Fluorescence Imaging and Biological Quantification

Edited by
Raquel Seruca
Jasjit S. Suri
João M. Sanches

CRC Press is an imprint of the
Taylor & Francis Group, an **informa** business

CRC Press
Taylor & Francis Group
6000 Broken Sound Parkway NW, Suite 300
Boca Raton, FL 33487-2742

Printed on acid-free paper

International Standard Book Number-13: 978-1-4987-3704-3 (Hardback)

Library of Congress Cataloging-in-Publication Data

Names: Seruca, Raquel, editor. | Suri, Jasjit S., editor. | Sanches, João M., editor.
Title: Fluorescence imaging and biological quantification / [edited by] Raquel Seruca, Jasjit S. Suri, and João M. Sanches.
Description: Boca Raton : Taylor & Francis, 2017. | Includes bibliographical references.
Identifiers: LCCN 2017013420 | ISBN 9781498737043 (hardback : alk. paper)
Subjects: | MESH: Optical Imaging | Cytological Techniques | Molecular Imaging | Image Interpretation, Computer-Assisted
Classification: LCC R857.O6 | NLM WN 195 | DDC 616.07/54--dc23
LC record available at https://lccn.loc.gov/2017013420

Visit the Taylor & Francis Web site at
http://www.taylorandfrancis.com

and the CRC Press Web site at
http://www.crcpress.com

Printed and bound in the United States of America by Sheridan

The ideal scientist is enchanted by the scientific poetics of nature. As scientists, we dedicate this book to everyone who gets fascinated by Science and Art.

Contents

Preface

Huge numbers of biomedical images are generated each year in routine and research labs. In this book, we describe new developments and solutions to analyze and quantify fluorescence images, tagging DNA, RNA, and proteins in single cells as well as in cell populations. This book is a collaborative effort by large group of scientists working in complementary disciplines as biology, biochemistry, microscopy, physics, and engineering.

In Section I, we present different microscopic techniques that allow the production of high-quality 2D and 3D images as confocal microscopy to more quantitative methodologies, namely imaging flow cytometry and atomic force microscopy (AFM). This includes chapters that define strategies to circumvent limitations of fluorescent nanoparticles and include novel strategies to track, quantify, and map these signals. Multiple fluorochromes and fluorochrome dyes are currently available allowing single or multiple complex visualization of molecular events. In every chapter, the advantages and limitations of every microscopic approach will be discussed and the future technical developments in each scientific area will be addressed.

Section II compiles new imaging and computer-based technologies to access the inner machinery of living cells and shows how different methodologies contribute to advance on the understanding of highly dynamic biochemical processes occurring at cell, tissue, and organism level. We focused on a wide variety of biological questions related to signaling events and networks, formation of protein complexes, maintenance of cellular homeostasis by lysosomes, circadian rhythms, cell cycle, membrane trafficking, behavior of cancer-cell populations, and macrophages dynamics. Overall, we aim to demonstrate in this book how fluorescence microscopy and images can be mathematically processed to allow depiction of molecular events and pathways underlying cell function, tissue morphology and mechanics, and individual physiology.

Raquel Seruca, Jasjit S. Suri and João M. Sanches

Acknowledgments

We thank all coauthors for their contributions to the book and, in particular, to Paula Sampaio who provided an historical view of the topic.

Editors

Raquel Seruca (RS) received her PhD in Medicine from the faculty of Medicine from the University of Porto, Portugal in 1995. RS developed her PhD project in the field of genetics of gastric cancer.

Presently, RS is vice president of Institute of Molecular Pathology and Immunology at the University of Porto, Portugal (Ipatimup) and coordinator of the cancer research program at i3s, both located in Porto.

At i3s, Seruca is the group leader of Epithelial Interactions in Cancer (EPIC). As group leader, RS has been able to create a truly multidisciplinary environment, stimulating collaborations and interactions between surgeons, oncologists, pathologists, biologists, biochemists, and bioengineers.

The long-term goal of the group is to uncover how epithelial cell–cell and cell–matrix junctions, as well as the surrounding microenvironment, can influence gastric, breast, and colorectal cancer progression. The focus of the group is to unravel the role of E- and P-cadherin in epithelial homeostasis and cancer development.

Seruca's lab is a reference center of the International Gastric Cancer Linkage Consortium, responsible for the functional studies of E-cadherin mutations in hereditary gastric cancer. Using this disease model, the group has made significant contributions on the role of E-cadherin and associated signaling in cancer-cell migration, survival, and invasion.

Seruca has more than 200 publications in international peer-reviewed journals, with an h-index of 52.

Seruca collaborates with several research groups worldwide, serves as a reviewer for top journals in the area of cancer, and is invited to speak and chair sessions at major conferences dedicated to cancer research. Further, she is involved in the evaluation of several international grants and institutes.

In 2009, Seruca received the distinction Ordem do Infante D. Henrique from the Portuguese Presidency (Presidência da República) for her scientific merit. More recently, in 2014, she got the gold medal of the Porto City for her contribution for science internationalization.

Dr. Jasjit S. Suri, PhD, MBA, fellow of AIMBE is an innovator, visionary, scientist, and an internationally known world leader. Dr. Jasjit S. Suri received the Director General's gold medal in 1980 and the fellow of American Institute of Medical and Biological Engineering (AIMBE), awarded by National Academy of Sciences, Washington, DC in 2004. He has published more than 550, which includes journals, book chapters, and proceeding articles having an H-index 46, coauthored more than 40 books, 100 innovations, and trademarks. He is currently chairman of Global Biomedical Technologies, Inc., Roseville, California, company dedicated to cancer imaging and biomedical devices. Dr. Suri is also in the board of AtheroPoint, Roseville, California, a company dedicated to atherosclerosis imaging for early screening for stroke and cardiovascular monitoring. He has held positions as a chairman of IEEE Denver section and advisor board member to health care industries and several schools in the United States and abroad.

João M. Sanches (JS) received the EE, MSc, and PhD degrees from the Instituto Superior Técnico (IST) and Universidade Técnica de Lisboa (UTL), Lisbon, Portugal in 1991, 1996, and 2003, respectively and the habilitation (agregação) in 2013 by the Universidade de Lisboa (UL), Lisboa, Portugal in biomedical engineering. JS is associate professor at the Department of Bioengineering (DBE) at the IST and he is from the coordination board of the biomedical engineering master and doctoral programs. Before he was at the Department of Electrical and Computer Engineering (DEEC) where has taught in the area of signal processing, systems, and control.

JS is a senior researcher and member of the board of directors of the Institute for Systems and Robotics (ISR–IST). His work has been focused in biomedical engineering (BME), namely, in biological and medical image processing and statistical signal processing of physiological and behavioral data. Currently, JS aims to develop new tools and methodological strategies to quantify and map molecular and morphometric pathologic cancer biomarkers.

As a group leader, Professor Sanches has been able to create a truly multidisciplinary and stimulating environment where researchers from top institutions from the biology and medicine areas work close together with engineers and computer scientists. Approximately 150 international publications (ORCID and Google Scholar) and several patents were already produced in the scope of this collaborative work.

He is senior member of the IEEE Engineering in Medicine and Biology Society (EMBS) since 2011 and member of the bioimaging or bio-imaging and Signal Processing Technical Committee (BISP-TC) of the IEEE Signal Processing Society.

Contributors

Mónica Abreu
Institute for Theoretical Biology (ITB)
Charité-Universitätsmedizin Berlin and Humboldt-Universität zu Berlin
and
Medical Department of Hematology, Oncology, and Tumor Immunology, and Molekulares Krebsforschungszentrum (MKFZ)
Charité-Universitätsmedizin Berlin
Berlin, Germany

Catarina R. Almeida
i3S - Instituto de Investigação e Inovação em Saúde da Universidade do Porto
and
INEB - Instituto de Engenharia Biomédica
Universidade do Porto
Porto, Portugal

Linda Arngården
Science for Life Laboratory
Department of Immunology, Genetics and Pathology
Biomedical Center
Uppsala University
Uppsala, Sweden

Miguel Aroso
iBiMED - Instituto de Biomedicina
Departamento de Ciências Médicas
Universidade de Aveiro
Aveiro, Portugal

Patrícia Carneiro
Instituto de Investigação e Inovação em Saúde (i3S)
and
Institute of Molecular Pathology and Immunology of the University of Porto (IPATIMUP)
and
Instituto de Investigação e Inovação em Saúde/Institute for Research and Innovation in Health (i3S)
University of Porto
Porto, Portugal

Filomena A. Carvalho
Faculdade de Medicina
Instituto de Medicina Molecular
Universidade de Lisboa
Lisboa, Portugal

Patrícia Carvalho
i3S - Instituto de Investigação e Inovação em Saúde
Universidade do Porto
and
IPATIMUP - Instituto de Patologia e Imunologia Molecular da Universidade do Porto
Porto, Portugal

Patrícia Castro
Instituto de Investigação e Inovação em Saúde (i3S)
and
Institute of Molecular Pathology and Immunology of the University of Porto (IPATIMUP)
and
Department of Pathology and Oncology
Medical Faculty of the University of Porto
Porto, Portugal

Angela Margarida Costa
i3S - Instituto de Investigação e Inovação em Saúde
Universidade do Porto
and
INEB - Instituto de Engenharia Biomédica
Universidade do Porto
Porto, Portugal

Rukeia El-Athman
Institute for Theoretical Biology (ITB)
Charité-Universitätsmedizin Berlin and Humboldt-Universität zu Berlin
Berlin, Germany

Sofia Esménio
Institute for Systems and Robotics, Instituto Superior Técnico
Lisboa, Portugal

Tiago Esteves
i3S - Instituto de Investigação e Inovação em Saúde
Universidade do Porto
and
Faculdade de Engenharia
Universidade do Porto
Porto, Portugal

Maria Sofia Fernandes
Instituto de Investigação e Inovação em Saúde/Institute for Research and Innovation in Health (i3S)
University of Porto
and
Institute of Molecular Pathology and Immunology of the University of Porto (IPATIMUP)
Porto, Portugal

Julia Fernandez-Rodriguez
Centre for Cellular Imaging at Sahlgrenska Academy
University of Gothenburg
Gothenburg, Sweden

Anabela Ferro
Instituto de Investigação e Inovação em Saúde (i3S)
and
IPATIMUP, Institute of Molecular Pathology and Immunology
Universidade do Porto
and
Instituto de Investigação e Inovação em Saúde/Institute for Research and Innovation in Health (i3S)
University of Porto
Porto, Portugal

Joana Figueiredo
Instituto de Investigação e Inovação em Saúde (i3S)
and
Institute of Molecular Pathology and Immunology of the University of Porto (IPATIMUP)
Porto, Portugal

Martina Fonseca
Institute for Systems and Robotics, Instituto Superior Técnico
Lisboa, Portugal

A. Freitas
i3S - Instituto de Investigação e Inovação em Saúde
Universidade do Porto
and
INEB - Instituto de Engenharia Biomédica
Universidade do Porto
and
FMUP, Faculdade de Medicina da Universidade do Porto
Porto, Portugal

Luise Fuhr
Institute for Theoretical Biology (ITB)
Charité-Universitätsmedizin Berlin and Humboldt-Universität zu Berlin
and
Medical Department of Hematology, Oncology, and Tumor Immunology, and Molekulares Krebsforschungszentrum (MKFZ)
Charité-Universitätsmedizin Berlin
Berlin, Germany

Nikolai Genov
Institute for Theoretical Biology (ITB)
Charité-Universitätsmedizin Berlin and Humboldt-Universität zu Berlin
and
Medical Department of Hematology, Oncology, and Tumor Immunology, and Molekulares Krebsforschungszentrum (MKFZ)
Charité-Universitätsmedizin Berlin
Berlin, Germany

M. Gomez-Lazaro
i3S - Instituto de Investigação e Inovação em Saúde
Universidade do Porto
and
INEB - Instituto de Engenharia Biomédica
Universidade do Porto
Porto, Portugal

Johan Heldin
Science for Life Laboratory
Department of Immunology, Genetics and Pathology
Biomedical Center
Uppsala University
Uppsala, Sweden

Axel Klaesson
Science for Life Laboratory
Department of Immunology, Genetics and Pathology
Biomedical Center
Uppsala University
Uppsala, Sweden

Benjamin König
Institute of Chemistry and Biochemistry
Freie Universität Berlin
Berlin, Germany

Catarina S. Lopes
Faculdade de Medicina
Instituto de Medicina Molecular
Universidade de Lisboa
Lisboa, Portugal

Jeannine Mazuch
Institute for Theoretical Biology (ITB)
Charité-Universitätsmedizin Berlin and Humboldt-Universität zu Berlin
Berlin, Germany

Tânia Mestre
Institute for Systems and Robotics (ISR), Evolutionary Systems and Biomedical Engineering Laboratory (LaSEEB), Instituto Superior Técnico (IST)
and
Institute for Systems and Robotics, Instituto Superior Técnico
Lisboa, Portugal

Tommy Nilsson
The Research Institute of the McGill University Health Centre
McGill University
Montreal, Québec, Canada

Maria José Oliveira
i3S - Instituto de Investigação e Inovação em Saúde
Universidade do Porto
and
INEB - Instituto de Engenharia Biomédica
Universidade do Porto
and
Department of Pathology and Oncology
Faculty of Medicine
Porto, Portugal

Joana Paredes
Instituto de Investigação e Inovação em Saúde (i3S)
and
Institute of Molecular Pathology and Immunology of the University of Porto (IPATIMUP)
and
Department of Pathology and Oncology
Medical Faculty of the University of Porto
Porto, Portugal

Camila C. Portugal
Glial Cell Biology Laboratory (GCB)
Instituto de Investigação e Inovação em Saúde (i3S)
Porto, Portugal

Pedro Quelhas
i3S - Instituto de Investigação e Inovação em Saúde
Universidade do Porto
Porto, Portugal

Doroteya Raykova
Science for Life Laboratory
Department of Immunology, Genetics and Pathology
Biomedical Center
Uppsala University
Uppsala, Sweden

Angela Relógio
Institute for Theoretical Biology (ITB)
Charité-Universitätsmedizin Berlin and Humboldt-Universität zu Berlin
and
Medical Department of Hematology, Oncology, and Tumor Immunology, and Molekulares Krebsforschungszentrum (MKFZ)
Charité-Universitätsmedizin Berlin
Berlin, Germany

João B. Relvas
Glial Cell Biology Laboratory (GCB)
Instituto de Investigação e Inovação em Saúde (i3S)
Porto, Portugal

C.C. Ribeiro
i3S - Instituto de Investigação e Inovação em Saúde
Universidade do Porto
and
INEB - Instituto de Engenharia Biomédica
Universidade do Porto
and
ISEP - Instituto Superior de Engenharia do Porto
Instituto Politécnico do Porto
Porto, Portugal

Ana Sofia Ribeiro
Instituto de Investigação e Inovação em Saúde (i3S)
and
Institute of Molecular Pathology and Immunology of the University of Porto (IPATIMUP)
Porto, Portugal

André Roma
i3S - Instituto de Investigação e Inovação em Saúde
Universidade do Porto
and
IPATIMUP - Instituto de Patologia e Imunologia Molecular da Universidade do Porto
Porto, Portugal

and

Faculdade de Medicina da Universidade de Coimbra
Coimbra, Portugal

Ivan Sahumbaiev
Institute for Systems and Robotics (ISR)
Evolutionary Systems and Biomedical Engineering Laboratory (LaSEEB), Instituto Superior Técnico (IST)
Lisboa, Portugal

Paula Sampaio
Advanced Light Microscopy unit (ALM), Instituto de Investigação e Inovação em Saúde (i3S) Instituto de Biologia Molecular e Celular (IBMC)
Universidade do Porto
and
Instituto de Investigação e Inovação em Saúde (i3S)
Porto, Portugal

João M. Sanches
Institute for Systems and Robotics (ISR)
Evolutionary Systems and Biomedical Engineering Laboratory (LaSEEB), Instituto Superior Técnico (IST)
Lisboa, Portugal

Nuno C. Santos
Faculdade de Medicina
Instituto de Medicina Molecular
Universidade de Lisboa
Lisboa, Portugal

Raquel Seruca
Instituto de Investigação e Inovação em Saúde (i3S)
and
IPATIMUP, Institute of Molecular Pathology and Immunology
Universidade do Porto
and
Department of Pathology and Oncology
Faculdade de Medicina da Universidade do Porto
Porto, Portugal

Renato Socodato
Glial Cell Biology Laboratory (GCB)
Instituto de Investigação e Inovação em Saúde (i3S)
Porto, Portugal

Ola Söderberg
Science for Life Laboratory
Department of Immunology, Genetics and Pathology
Biomedical Center
Uppsala University
Uppsala, Sweden

Tobias Stauber
Institute of Chemistry and Biochemistry
Freie Universität Berlin
Berlin, Germany

Sérgia Velho
i3S - Instituto de Investigação e Inovação em Saúde
Universidade do Porto
and
IPATIMUP - Instituto de Patologia e Imunologia Molecular da Universidade do Porto
Porto, Portugal

Lisa von Kleist
Institute of Chemistry and Biochemistry
Freie Universität Berlin
Berlin, Germany

Introduction

Brief Historical view

Microscopy, fluorescence, and imaging

Two millenniums ago, Romans discovered that glass could be used to enlarge objects, but the first instruments capable to make visible structures not visible by the naked eye, only appeared in the past 400 years. Microscopes made possible to observe microorganisms, blood cells, sperm, and small details in plants and animals for the first time. Anton van Leeuwenhoek and Robert Hooke represented their pioneer observations by hand drawings and quantified some of their observations. As an example, Leeuwenhoek estimated the number of *animalcules* (bacteria and protozoa) in drops of water and determined the size and shape of *red corpuscles* (red blood cells).

Improvements in optics and theoretical studies of image formation and optics carried out in the second half of nineteenth century founded the grounds for the modern optical microscopy we experience today. Nowadays, scientists have a broad range of imaging methods available to study biological systems. The most classical transmission microscopy contrast techniques such as brightfield, phase-contrast, differential interference contrast, polarization, and dark-field are based on the alterations of light induced by interaction with the specimen. Transmission microscopy is a routinely choice approach for morphological analysis and for live-cell imaging to study cell shape, cell cycle, or migration of cell in culture. However, it is limited, as it does not allow to differential and simultaneously label molecules with high sensitivity and in a quantitative way. Fluorescence microscopy surpasses these problems and allows exploring the cells and tissues at more molecular and subcellular levels. In fluorescence microscopy, the specimens are stained by fluorochrome dyes, quantum dots, or express chimeric fluorescently tagged proteins. These fluorophores get into an excited state after absorbing energy, as a photon, at specific wavelengths. This high-energy state is transient and first, the excited state electrons go to lower energy levels by vibrational relaxation, after which the molecule returns to ground state by emission of a photon of lower energy than excitation. Multiple fluorochromes with very sensitive emission profiles are commercially available to label different cellular components allowing their independent visualization with high sensitivity as well as their spatial and temporal correlation.

The main constrain of fluorescence is blurring as fluorophores are self-luminous. So, out-of-focus light from different focal planes mix with the in-focus signals originating blurred images in widefield fluorescence microscopy (WFM). Confocal fluorescence microscopy overcomes this problem, by using a spatial filter, known as pinhole, at the detection level that suppresses the light from out-of-focus areas. This leads to the formation of a high contrasted image, optical section, that have reduced contribution of out-of-focus light. A specimen can then be optically sectioned and those images can be used to build a 3D reconstruction of the specimen.

Imaging cell in culture is still the most common approach to study the biology of the cells; however, higher eukaryotes are multicellular organism where the cells are integrated in a 3D community, tissues. So, the ability to study the cells within its natural environment is essential to have an integrate vision of the biological processes, and confocal microscopy ability to generate optical sections is still limited to few dozens of micrometers deep into tissues due to scattering of the light by matter. Multiphoton microscopy use pulsed IR light that is less scattered by tissues allowing to image deep. Since, IR photon have less energy, fluorophores must absorb energy of two photons to get excited. The high probability of excitation occurs only at focal plane, with no out-of-focus light being generated. Emitted light can be only detected close to the objective and this increases the detection sensitivity experimented in multiphoton microscopy.

In the past decade, a new microscopy technique to image live embryos in toto had a great development. In light sheet microscopy, a plane of the sample is illuminated by a sheet of light and emitted light is detected on an objective at 90°. Combining the acquisition at multiple planes and angles permits to make a total 3D reconstruction of each sample. Due to light sheet microscopy low phototoxicity, it is instrumental for live imaging of embryo development. Actually, live-cell imaging is essential to know the rules of biological processes or cell fate decisions and had a great expansion with development of new probes, including fluorescent proteins, which could *illuminate* targeted cell components. This enabled observing and measuring dynamic cellular events at molecular level with high spatial and temporal resolution.

The chemical distribution within an unstained specimen can be also evaluated by Raman confocal microscopy, a spectroscopic approach that provides a specific fingerprint of molecules.

Recent developments let to create techniques based in switch on/off of fluorescents molecules (generically known as *super-resolution microscopy*) that overcome the limit of resolution (~200 nm) of optical microscopy and approaching it to nanoscopy with resolution limits reaching already 10 nm. These techniques are *super-resolution* microscopy plus the combination of light microscopy with electron microscopy and atomic force microscopy will open a new vision of cell biology in future years.

As the pioneers already demonstrated, imaging is not only getting a nice picture of fine details. This is particularly evident with digital imaging that generates images, which are 2D matrices. So, the digital images can be analyzed for extraction of quantitative information in order to perform accurate evaluation of data and science of excellence.

Paula Sampaio

Chapter 1 Confocal microscopy in the life sciences

Miguel Aroso and
M. Gomez-Lazaro

Contents

Introduction

Microscopy imaging is, in general, achieved by reflecting light off the specimen or by illumination of fluorescently labeled molecules (e.g., proteins). One of the main advantages of fluorescence microscopy is the increase in signal of the fluorophores against a dark background [1]. In widefield microscopy, the brightest and highest intensity of the incident light, from an incoherent mercury or xenon arc-discharge lamp, is at the focal point of the objective but there is illumination of other parts of the sample and as a result, different focal planes emits light resulting in high background, which might compromise the quality of the image [2]. This effect is more pronounced in thicker specimens (>2 μm), where out-of-focus fluorescence contributes to a higher background and to a degradation of most of the fine details. In this respect, the development of the laser scanning confocal microscopy (referred as confocal microscopy in this chapter) revolutionized the field of life sciences, since this technology allows the generation of sharper images with significant lower background. The basis of confocal microscopy was developed by Marvin Minsky in 1955 and patented in 1957 [3]. However, further developments of Minsky's prototype were hampered by limitations in the illumination and in the imaging system. The first commercial confocal microscope arrived at the market 30 years later—the Bio-Rad MRC-500.

Modern confocal microscopes can be considered as completely integrated electronic systems [4], where the optical microscope plays a central role in a configuration that consists of one or more electronic detectors, a computer (for image display, processing, output, and storage), and several laser systems combined with wavelength selection devices and a beam-scanning assembly [5]. One of the most important components of the scanning unit is the pinhole aperture, which acts as a spatial filter and is positioned directly in front of the detector [6] (Figure 1.1).

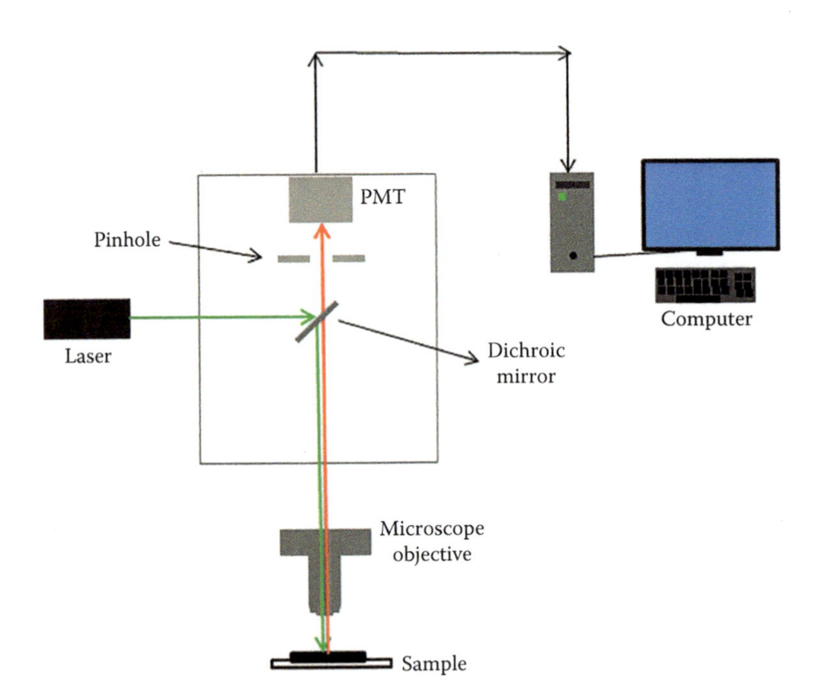

Figure 1.1 Schematic representation of a confocal microscope. A typical confocal microscope is composed of a laser as a source of excitation light (which can be different lasers with different laser lines or even a unique white laser), which will be used to scan the focused laser beam across the sample. The light is reflected off a dichroic mirror to direct the light to the sample. The objective of the microscope is used to focus the point illumination in the sample that will be scanned. The emitted light passes through the mirror and will be detected by the detector (usually a photomultiplier tube [PMT]) after passing through a pinhole that removes the out-of-focus light. The smaller the opening of the pinhole the higher amount of out-of-focus light is rejected. The photons arriving at the detector are processed by a computer for image display.

It is possible to adjust the pinhole aperture to exclude fluorescent signals from out-of-focus features positioned above and below the focal plane and control the optical section thickness [5,7]. Thus, the image obtained has less haze and better contrast and represents a thin cross section of the specimen [8–13]. It is also possible to acquire several optical sections from the specimen that later can be used to create 3D representations. Nevertheless, the reduction of the pinhole size leads to a reduction of the image intensity, as fewer photons can be captured. Thus, there is a need to have a bright and coherent excitation source (e.g., laser) and very sensitive photon detectors [14]. Those detectors should be highly sensitive and respond very quickly to a continuous flux of varying light intensity. The most common choice in many commercial confocal microscopes are the photomultiplier tubes (PMT) [15–17], which convert the fluorescent signals that pass through the pinhole into an analog electrical signal with a continuously varying voltage that corresponds to the intensity of the signal. Then, the analog signal is converted into pixels and the image information is displayed in the computer's monitor. The confocal image of a specimen is reconstructed, point-by-point, from emission photon signals and does not exist as a real image that can be observed through the microscope eyepieces [7].

Within this chapter, you will find advice on sample preparation, image acquisition, and preprocessing. It also includes the description of two common applications of confocal microscopy within the life science field (colocalization and fluorescence recovery after photobleaching) and a summary of commonly used fluorophores. Although confocal microscopy represents a popular technology, it has some limitations that will be revealed together with some advanced technological developments.

Experimental procedures

Sample processing: Needs and troubles

The observation of biological samples by confocal microscopy should, ideally, be carried out in living specimens. However, most of the times, it is not possible and previous sampling preparation is required. Common sample preparation for widefield and confocal microscopy relies on four main steps: (1) fixation to preserve cellular morphology and adherence of the specimen to the coverslip, (2) permeabilization to grant access of the labeling reagents to intracellular components, (3) labeling of the desired structures, and (4) mounting of the sample with addition of appropriate antifading reagent.

The fixation step must preserve the cellular organization, 3D structure, and antigenicity of the target when performing immunofluorescence. However, no fixation protocol is perfect and it should be chosen accordingly with the main objectives of the sample visualization. For example, if the samples will be analyzed by confocal microscopy with the objective of 3D reconstruction extra care should be taken to avoid cell shrinkage in the *z*-direction. Most often, the sample dimensions are easier to preserve in the *x*–*y*-direction due to the adherence of the cells to the glass coverslip. Chemical solvents as ethanol, methanol, or acetone are fast precipitating fixatives but they might lead to cell shrinkage. However, these solvents have the advantage of being fast, reproductive, good at preserving the antigenicity of the target, and good as permeabilizing agents with no need for further permeabilization step. Glutaraldehyde and paraformaldehyde are cross-linking fixatives that are good at preserving the 3D cellular structure. However, glutaraldehyde can lead to a high degree of cross-linking and the destruction of the antigen-binding sites. Paraformaldehyde is a more equilibrated reagent in the preservation of the cellular structures and the antigenicity of the targets, but it should be kept in mind that some commercial preparations might contain methanol that can induce the shrinkage artifacts mentioned previously.

The permeabilization of the sample has the objective of granting access of labeling agents to intracellular molecules or components. When solvents are used as fixatives, there is no need of permeabilization because the solvents are able to extract membrane lipids. In the case of cross-linking fixatives, the cellular membrane remains intact and thus detergents are commonly used (i.e., Triton X-100 and digitonin) to extract membrane lipids and permeabilize cell membranes.

One of the most important aspects of the mounting step is the addition of an antifading agent to decrease photobleaching. Those agents are free radical scavengers that typically include n-propyl gallate, p-phenylenediamine, and 1,4-diazobicyclo-(2,2,2)-octane (DABCO).

Image acquisition

After adequate sample preparation appropriate image acquisition begins with a clear idea of the purpose of the image and the purpose of showing the most representative images of the sample. Several aspects should be considered before image acquisition: the capacity of the system to excite and collect the emission of the fluorophores and the size of the resolved feature under study (Is it a group of cells, a cell, or an organelle?). Another important feature is the choice of the objective with special attention to numerical aperture (NA) and magnification. NA is defined as the half of the angular aperture of the lens and influences not only the resolution of the image but also the ability of the objective to collect light that is translated into signal intensity. Higher resolution can be achieved with higher NA objectives. However, those objectives display commonly shorter working distances, which can be undesirable when imaging thick samples. As the objective NA is proportional to the refractive index of the immersion medium, higher NA are displayed by immersion oil objectives (NA oil > water > air). The use of proper immersion media for each objective reduces the problems of spherical and chromatic aberrations. A mismatch between the diffractive indexes of the sample and the immersion medium might cause: (1) elongated-like images when performing z-stack acquisition because the travel distances of the focal plane will differ from the mechanical focus, (2) chromatic errors that will lead to different axial registration in multichannel acquisition, and (3) spherical aberrations since the optical resolution is associated with the diameter of the diffraction-limited-focus. For a thorough explanation on objectives in confocal microscopy, see Reference 18.

By changing the objective magnification, the size of the pixel changes accordingly. A pixel can be defined as the geometrical subunit in an image, where light intensity is stored as a single number. The area of each pixel is defined by the magnification, the NA of the objective, and the adjustment of the zoom magnification. By zooming in an area of the image, we are also decreasing the size of the pixel and with smaller pixels, it should be possible to image smaller objects. The diffraction limit constraints the smallest size possible to acquire.

At last, the acquisition of the fluorescent signal requires special attention. This signal can be amplified by changing the accelerating voltage of the PMT by regulation of the gain and offset. The gain controls the brightness and contrast of the image, and the offset adjusts the background displayed. The adjustment of these two parameters aims at displaying the full gray-scale range in the image (8, 12, or 16 bits that correspond to 256, 4096, and 65536 gray levels, respectively).

In summary, good practices to improve the image quality (increase signal-to-noise ratio) include: an increase in the laser power, however, care should be taken to not to damage the sample since high-intensity light might bleach the fluorophore or damage the cells; an increase in the acquisition time, this can be done by decreasing the scanning speed and using frame or line averaging during acquisition, again long scanning periods might damage your fluorophore or cells; use of objectives with high NA since their lenses have greater light collecting capacity; an increase in the pinhole size, it will reduce the resolution in Z, however will increase the amount of photons arriving at the detector; and an increase in the detector gain, this will increase both signals and noise [19]. For a detailed explanation of fundamentals on concepts of image acquisition, see Reference 20.

Image processing and analysis

The selected picture for publication should be representative of the sample and should contain the appropriate controls—a picture is worth a thousand words. Pictures must be presented according to general guidelines for ethical data processing [21–24]. For comparative purposes, images must be acquired under the same conditions and image acquisition parameters and processing must be consistent across sampling. After acquisition, some adjustments are allowed when preparing the picture for publication, such as small modifications of brightness, contrast, and gamma levels. However, these adjustments have to aid in representative purposes but not change the data [25]. For a better representation of the sample, it is also accepted to crop an image as long as the purpose is just to improve the picture and not exclude any data. For a thorough explanation of the ethical guidelines for image manipulation in science, see References [21,23,25].

The information contained in the digital images can be used for representation but also for quantification of the fluorescent signal but it should be noted that the quantified intensity, although proportional to the number of photons emitted by the sample in the focal plane, is not a direct measurement of the photons. For comparison purposes, the same settings for acquisition of the images should be taken into consideration. In this chapter, we show two examples for image analysis: quantification of the fluorescence intensity recovered after photobleaching and quantification of colocalization between two fluorescent probes.

Applications

Labeling tools in confocal microscopy

The use of fluorophores is a routine procedure in confocal microscopy within the life sciences field. They are usually organic compounds that contain aromatic groups or several π bonds. Furthermore, some proteins present intrinsic fluorescent properties. More recently, fluorescent nanocrystals or quantum dots (QD) are emerging molecules with fluorescent properties that are used, as well, in fluorescent microscopy. The QDs present very attractive advantages such as their nanometer size, continuous excitation spectra and narrow emission band, and their high resistance to photobleaching [26–29].

The use of fluorophores is vastly disseminated due to the high sensitivity of the emitted fluorescence and the large array of different fluorophores with differentiated features that allows for multicolor fluorescence microscopy with the detection of multiple targets at the same time (Table 1.1). Additional advantages are the possibility of targeting specific components or even dynamic processes in cells [30–33]. However, unstained samples can also be imaged by using the light reflected by the sample after irradiation with a laser source and additional information can be extracted from these images.

In the past decades, we have assisted to a huge increase in the availability of different fluorophores and fluorescent probes, which can be specific for different cellular compartments (Tables 1.1 through 1.4). The difference between fluorescent probes and fluorophores is their ability to provide a specific staining. Fluorescent probes are designed to:

Table 1.1 Examples of Fluorophores. λ_{Ex} Represents the Peak Excitation Wavelength in nm, whereas λ_{Em} Represents the Peak Emission Wavelength in nm

Fluorophore	λ_{Ex} (nm)	λ_{Em} (nm)	Fluorophore	λ_{Ex} (nm)	λ_{Em} (nm)
Hydroxycoumarin	325	386	Cal Fluor Gold 540	522	544
Hoechst 33342	343	483	Alexa fluor 532	530	555
DAPI	345	455	HEX	535	556
Hoechst 33258	345	478	Propidium iodide	536	617
Aminocoumarin	350	445	Cal Fluor Orange 560	538	559
Biosearch Blue	350	447	VIC	538	554
Methoxycoumarin	360	410	5 (6) TAMRA	540	565
Acridine	362	462	5 TAMRA	542	568
Alexa fluor 430	430	545	6 TAMRA	542	564
SYTOX blue	431	480	NED	546	575
Coumarin	432	472	SYTOX orange	547	570
Chromomycin A3	445	575	TRITC	547	572
Mithramycin	445	575	Quasar 570	548	566
Pulsar 650	460	650	Cy3	550	570
Red 613	480;565	613	Alexa fluor 546	556	573
R-phycoerythrin (PE)	480;565	578	Alexa fluor 555	556	573
Cy2	490	510	Rhodamine red	560	580
Fluorescein	490	514	Tamara	565	580
TruRed	490;675	695	Cal Fluor Red 590	569	591
5 (6) FAM	492	517	Rox	575	602
5 FAM	492	518	Alexa fluor 568	578	603
6 FAM	492	515	Cy3.5	581	596
Ethidium bromide	493	620	Texas Red	583	603
Alexa fluor 488	494	517	Alexa fluor 594	590	617
5 (6) FAM SE	495	519	Cal Fluor Red 610	590	610
5 FAM SE	494	520	Cal Fluor Red 635	618	637
6 FAM SE	496	516	Alexa fluor 633	621	639
Rhodamine Green	503	528	LIZ	638	650
Dichlorofluorescein	504	529	DRAQ5	646	681
5(6)Carboxy Dichlorofluorescein	504	529	DRAQ7	646	681
5Carboxy Dichlorofluorescein	504	529	Pulsar 647	647	667
6Carboxy Dichlorofluorescein	504	529	Alexa fluor 647	650	668
5(6)Carboxy Dichlorofluorescein SE	504	529	Allophycocyanin	650	660
5Carboxy Dichlorofluorescein SE	504	529	Cy5	650	670
6Carboxy Dichlorofluorescein SE	504	529	Alexa fluor 660	663	690
SYTOX green	504	533	Cy5.5	675	694
TOTO 1, TO-PRO-1	509	533	Alexa fluor 680	679	702
JOE	520	548	Pulsar 705	690	705
TET	521	536	Cy7	743	770

Table 1.2 Examples of Fluorescent Proteins Used in Confocal Microscopy

Fluorescent Protein	λ_{Ex} (nm)/λ_{Em} (nm)	Fluorescent Protein	λ_{Ex} (nm)/ λ_{Em} (nm)
Y66H	382/459	mCitrine	516/529
EBFP	383/445	YPet	517/530
EBP2	383/448	PhiYFP	525/537
Azurite	384/450	ZsYellow	529/539
GFP (uv)	395/509	mBanana	540/553
Stemmer	395/509	Kusabira–Orange1	548/561
mTagBFP	399/456	mKusabira–Orange1	548/559
T-Sapphire	399/511	mOrange	548/562
ECFP	433/475	mOrange2	549/565
Cerulean	433/475	mKO2 (Kusabira Orange2)	551/565
CyPet	435/477	Tomato	554/581
dKeima-Tandem	440/620	tdTomato (tandem dimer)	554/581
AmCyan1	458/489	DsRed-Express	555/584
TagCFP	458/480	TagRFP	555/584
mTFP1 (Teal)	462/492	TagRFP-T	555/584
Midori–ishi Cyan	472/495	DsRed-Monomer	556/586
GFP (wt)	475/509	DsRed/RFP	558/583
AcGFP	480/505	mRuby	558/605
TurboGFP	482/502	DsRed2	563/582
TagGFP	482/505	mApple	568/592
EGFP (S65T/F64L)	484/507	mTangerine	568/585
sfGFP (Superfolder GFP)	485/510	mStrawberry	574/596
Emerald	487/509	AsRed2	576/592
GFP–S65T	488/509	JRed	584/610
Azami–Green	492/505	mRFP1	584/607
mAzami–Green1	492/505	mCherry	587/610
ZsGreen1	493/505	HcRed1	588/618
mWasabi	493/509	mPlum	590/649
TagYFP	508/524	tHcRed (tandem)	590/637
EYFP	514/527	AQ143	595/655
Topaz	514/527	mRaspberry	598/625
Venus	515/528		

(1) bind to specific macromolecules; (2) localize in specific cellular regions (organelles, cellular components, etc.); (3) monitor dynamic processes (endocytosis, exocytosis, membrane fluidity, enzymatic activity, and cell viability); and (4) examine environmental variables (ion concentration [31], pH, reactive oxygen species [ROS], and membrane potential [34]) [32]. The possibility to add fluorophores to specific biomolecules (such as antibodies or specific proteins) expands the

Table 1.3 Examples of Phototransformable Fluorescent Proteins Used in Confocal Microscopy

Fluorescent Protein	λ_{Ex} (nm)/λ_{Em} (nm)	λ_{Ex} (nm)/λ_{Em} (nm)	Reference
Photoconvertible Proteins			
PSmOrange	548/565	636/662	[35]
PSmOrange2	546/561	619/651	[36]
Dendra 2	490/507	553/573	[37]
Photoactivable Proteins			
PA–GFP	488/507	× 100 in emission	[38]
PAmCherry	564/595	× 100 in emission	[39]
PATagRFP	562/595	× 540 in emission	[40]
PAmKate	586/628	× 100 in emission	[41]
PAiRFP1	690/717	× 18 in emission	[42]
PAiRFP2	690/717	× 26 in emission	[42]
Photoswitchable proteins			
Dronpa	Switch off 488 nm/ on 405 nm	503/518	[43]
Padron	Switch off 405 nm/ on 488 nm	503/522	[44]

Note: PA, photoactivable.

Table 1.4 Examples of Vital Dyes Used in Confocal Microscopy to Measure Dynamic Properties of Living Cells

Fluorescent Dye	Specificity	Fluorescent Dye	Specificity
Mitotracker Red–CMX–ROS	Mitochondria	FMR 1–43FX and FMR 4–64FX	Plasma membrane
ER Tracker	ER	CellTrace™ BODIPYRTR methyl ester	Intracellular membranes
Hoechst 33342	Nucleus	Calcein AM	Cytoplasm
Propidium iodide	Nucleus	INDO-1	Ca^{2+}
DAPI	Nucleus	QUIN-2	Ca^{2+}
TO-PRO.1	Nucleus	FLUO-3	Ca^{2+}
YO-PRO-1	Nucleus	FLUO-4	Ca^{2+}
TOTO-3	Nucleus	FURA-2	Ca^{2+}
TO–PRO-3	Nucleus	SNARF-1	H^{+}
DRAQ5	Nucleus	BCECF	H^{+}
DRAQ7	Nucleus	DCFH-DA	Peroxynitrite
SYBR Green I	Nucleus	DHR123	Reactive oxygen species
FITC–phalloidin	Actin	Rhodamine 123	mp Mitochondria
DiOC6	ER–mitochondria	TMRM	mp Mitochondria
Lyso Tracker Red DND-99	Lysosomes	TMRE	mp Mitochondria
NBDC6–Ceramide	Golgi	Oxonol V	mp Cytoplasm
Cell Tracker	Cytoplasm	Oxonol VI	mp Cytoplasm
TubulinTracker™ Green Probe	Cytoskeleton	DiBAC	mp Cytoplasm
SYTORRNASelect™ Green Statin	Nucleolus		

Note: ER: Endoplasmic reticulum; mp: Membrane potential.

available possibilities. The use of fluorescently conjugated antibodies is a widespread practice for the labeling of specific macromolecules to perform immunofluorescent assays.

DNA probes for nonspecific and stable labeling of the cell nucleus are widely used fluorescent probes due to their easy handling and quick staining. Numerous bisintercalating agents are available such as 4′,6-diamidino-2-phenylindole (DAPI), Hoechst, Propidium iodide, TOTO, YOYO, DRAQ5, and DRAQ7 among others [33,45,46] (Table 1.4). They allow not only for nuclear staining and visualization of healthy *versus* apoptotic nuclei in fixed samples but also due to the cell nonpermeant features that some dyes might be used to assess cell viability because these specific nonpermeant dyes only cross the nuclear envelope when its integrity is affected, such as in necrotic and apoptotic cells (i.e., propidium iodide and DRAQ7) [47]. In addition, the use of fluorescent probes that change their fluorescent properties in response to specific changes in the microenvironment are very useful to investigate physiological states, cellular communication, and protein trafficking within the endolysosomal pathway, among others [48,49].

The use of fluorescent proteins has grown exponentially since the discovery of the green fluorescent protein (GFP). As the recognition of its relevance in the life science field, in 2008, Osamu Shimomura, Martin Chalfie, and Roger Tsien received the Nobel Prize in Chemistry for their work on the GFP [50]. Nowadays, many fluorescent proteins are available, which can be used to be expressed alone in cells or as tags for other proteins (Table 1.2). The main advantage of fluorescent proteins is their suitability for live-cell imaging [51,52]; however, when antibodies are not available for immunofluorescence, the cloning of a specific protein tagged with a fluorescent protein might aid in the study of its localization and function [32,53]. For example, the DsRed2-mito construct is a nice example that allows the labeling of mitochondria assisting in the investigation of mitochondrial dynamics with confocal microscopy [52]. This construct takes advantage of the mitochondrial targeting signal of the subunit VIII of human cytochrome c oxidase protein coupled with the fluorescent protein DsRed2 (Clontech, Mountain View, CA).

Recently, great advances are performed in the design of phototransformable fluorescent proteins whose properties can be controlled and/or changed by light [54,55]. These proteins include photoactivatable, photoconvertible, and photoswitchable fluorescent proteins. A transition from nonfluorescent to a fluorescent state is induced by light in the photoactivatable proteins, whereas photoconvertible fluorescent proteins can change between two fluorescent states on stimulation. Both processes are irreversible since they are induced by a covalent modification in the protein. In the case of reversible photoswitching, the conversion between fluorescent states is reversible [55,56] (Table 1.2). The use of these kinds of phototransformable proteins expand the utilization of fluorescent proteins to the study of diverse cellular processes, such as the measurement of protein half-life within the cell [57] and cell tracking among others [37].

Rational choices should be made when choosing the specific fluorescent probes necessary for the experiment, taking into account the targeting specificity, preservation of cell physiology, and minimization of cross talk and bleed-through artifacts.

Selected applications

Fluorescence recovery after photobleaching

Fluorescence recovery after photobleaching (FRAP) is a fluorescence microscopy technique that takes advantage of the irreversible photodestruction of fluorophores to gain insight into the intracellular macromolecular dynamics [58]. In brief, a typical FRAP assay comprises three main acquisition steps. In the first set of images (prebleaching), low laser power is used to image the fluorescent probe. The second step (bleaching) involves the application of an intense laser beam in a small region-of-interest (ROI) that will lead to the irreversible photodamage of the fluorophore in the specific ROI. The last step (postbleach) involves the time acquisition of several images at low laser power. The recovery of the fluorescence intensity in the ROI is the result of the diffusion of nonbleached fluorescent molecules into the bleached area (Figure 1.2). The macromolecules diffusion properties within the cell depend on the macromolecule spatial location, interactions, binding properties, existence of membrane barriers, and so on [59,60]. Mobile proteins can exchange their location with their unbleached counterparts, leading to the recovery of fluorescence in the bleached area at rates proportional to protein mobility (Figure 1.2). However, the repetitive illumination, even at low power, can cause unintended photobleaching. This has to be taken into consideration for the quantification of the fluorescence recovery.

For the representation of the fluorescence intensity *versus* time in the ROI, we can use the following formula:

$$I_t = \frac{\frac{Ib_t}{Ic_t} - \frac{Ib_{\min}}{Ic_{\min}}}{\left(\frac{Ib_t}{Ic_t} - \frac{Ib_{\min}}{Ic_{\min}}\right)_{\max}} \times 100$$

where:

Ib is the fluorescence intensity of the photobleached region

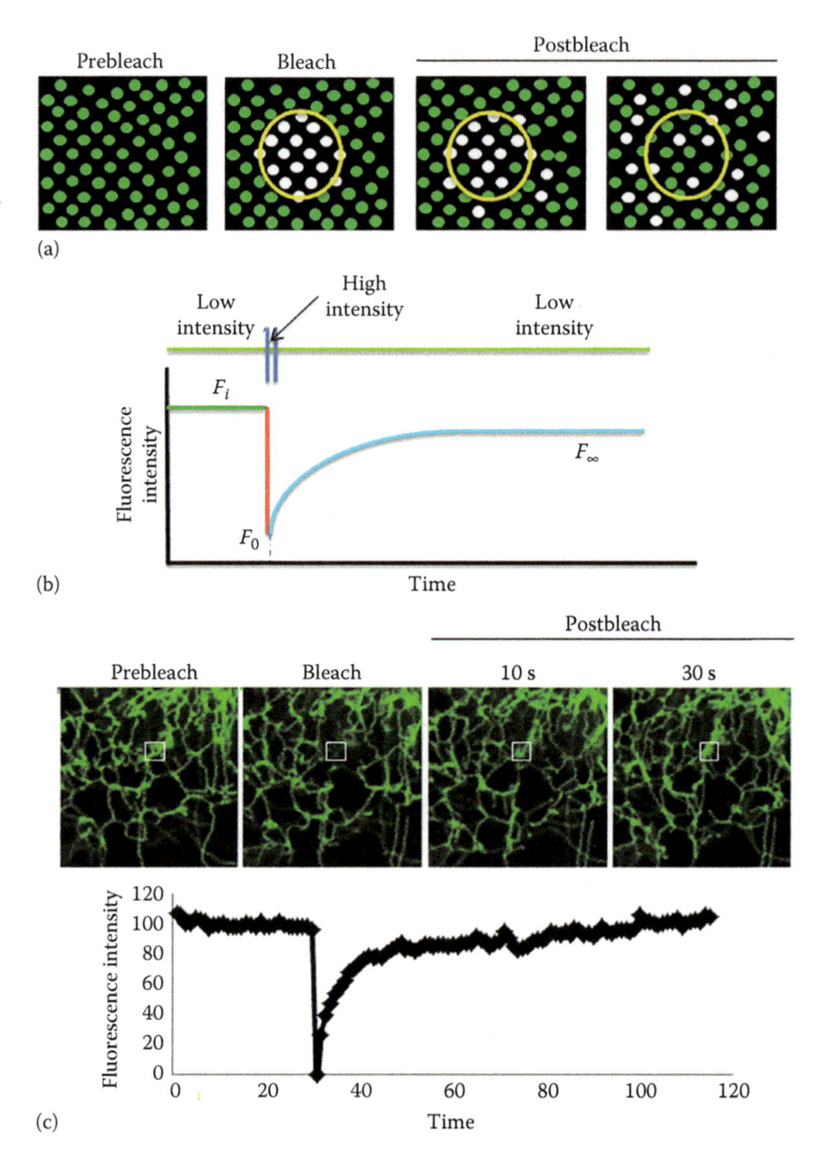

Figure 1.2 Fluorescence recovery after photobleaching (FRAP). (a) Representative scheme of a FRAP experiment. When a selected ROI is subjected to high-intensity laser, the fluorophores present in that region are irreversibly damaged; however, with time and depending on the diffusion properties of the molecules nonbleached fluorophores might enter into the previously bleached area and, as well, bleached-fluorophores might exit leading to the recovery in the fluorescence. From this kind of experiment, a typical graphical representation of the fluorescence intensity can be seen in (b). In (c), COS7 cells expressing the GFP–Sec61β that localizes to the endoplasmic reticulum was subjected to a FRAP experiment. Pictures of the prebleached, bleached, and postbleached steps are shown. In addition, a graph showing the fluorescence intensity over time is displayed.

Ic is the fluorescence intensity of a nonbleach region

Being min, the minimal intensity (after the bleach) and max, the mean of the prebleach intensities with background subtraction in all the intensity values.

Then the mobile fraction (M_f) can be calculated from the recovery graphs (Figure 1.2b and c) by applying the formula

$$M_f = \frac{F_\infty - F_0}{F_i - F_0}$$

where:

F_∞ is the fluorescence intensity after recovery
F_i is the fluorescence intensity before bleaching
F_0 is the value just after bleaching

Other photobleaching techniques include the following: inverse FRAP (iFRAP) where the ROI is excluded, whereas the surrounding region suffers bleaching and fluorescence loss in photobleaching (FLIP) where there is a repeated bleaching in a selected region during all monitorization period together with the fluorescence measurement in the nonbleached region. The reduction of fluorescence is due to the movement of the bleached fluorophores that leave the bleached region, increasing the population of total bleached fluorophores. In addition, in fluorescence localization after photobleaching (FLAP) two proteins with different fluorescence features are expressed at the same time and in the same proportion. Only one of the two fluorophores suffers bleaching and the other is used as the reference signal.

Colocalization

When the function of a protein is unknown, identifying its spatial location within the cell might be a first approach to unravel its function. This can be done by the quantification of the overlapping or colocalization of the signal with known specific marker proteins. The quantification of signal colocalization measures, in an image, the degree of association of two different labels in the same location. The interpretation of the data has to be done carefully because colocalization does not mean interaction. Although colocalization measurements are widely used to quantify the subcellular association of different fluorescent probes, its application is constrained by the resolution limit of the system. For comprehensive reviews on colocalization topics in microscopy, see Reference 61–65.

For this kind of studies some considerations should be fulfilled when acquiring the images, such as the signal of each channel can be distinguished from the background and is representative of the specific fluorescent probes used; there is no autofluorescence; there is neither bleed-through nor cross talk between the different probe signals; the use of proper controls; adequate image quality and pixel size for the cellular structure in consideration;

the adjustment of parameters for acquisition such as illumination, detector gain, and offset in order to collect the different signals in the linear range of the detector. Of note, the objectives should be corrected for chromatic and spherical aberrations for the wavelengths in use. Without these corrections, it would neither be possible to bring both signals in the same focus nor the whole field in the same focal plane.

For the quantification of the degree of colocalization, several parameters can be used. The most common are the Pearson's correlation coefficient (PCC) and the Mander's overlap coefficient (MOC) [62].

The PCC for images with two channels can be calculated by the following formula:

$$\text{PCC} = \frac{\sum_i (R_i - \bar{R}) \times (G_i - \bar{G})}{\sqrt{\sum_i (R_i - \bar{R})^2 \times (G_i - \bar{G})^2}}$$

where R_i and G_i represent the intensity values of each channel of pixel i and $\bar{R}$ and $\bar{G}$ represent the mean intensities of each channel in the whole image [61]. The values obtained can range between 1 and −1. PCC = 1 implies that there is a perfect correlation between the intensity distribution in both channels, whereas PCC = −1 indicates its complete absence. When we get values ranging from 0 to 1, it denotes the degree of overlap between the channels, whereas values from 0 to −1 indicate that there is no overlap or the channels have extreme opposite degrees of intensity.

The PCC coefficient measures the pixel-by-pixel covariance in the signal levels and is independent of signal levels and background. However, artifacts in the values of the coefficient can be obtained when the nonpositive pixels in regions without labeling are taken into consideration for the calculation of the coefficient. Setting of a threshold value for the positive signal constitutes a good practice, and if the signal is restricted to a specific region in the cell, then calculate the coefficient just for that region or ROI.

The Mander's coefficient can be quantified by the following formula:

$$\text{MOC} = \frac{\sum_i (R_i - G_i)}{\sqrt{\sum_i R_i^2 \times \sum_i G_i^2}}$$

In the case of this coefficient, the values are never negative, and its value is independent of signal levels of both probes and is sensitive to occurrence in the same pixel. MOC is 0 when both probes are completely exclusive. The main relevance of this coefficient is that it is possible to obtain the fraction of one probe overlapping with the other and *vice versa*.

An example of colocalization analysis is shown in Figure 1.3. Prior to the quantification of any colocalization coefficient, the signal threshold

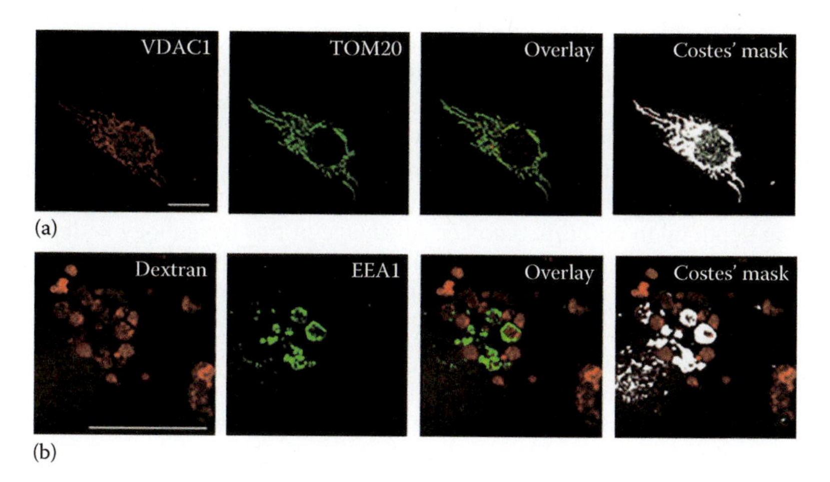

Figure 1.3 Colocalization analysis using the plug-in JACoP from ImageJ. Multichannel images were acquired. (a) Immunofluorescence using antibodies against the mitochondrial proteins TOM20 (green) and VDAC1 (red) was performed in COS-7 cells. The overlay image and the Costes' mask of colocalization are also displayed. (b) COS-7 cells were incubated with Dextran–TRITC for 24 h and immunofluorescence against the early endosomal protein EEA1 was performed following cell fixation with 4% PFA. The overlay image and the Costes' mask of colocalization are also displayed. Scale bar 20 μm.

was adjusted for both channels removing the background contribution. By using the JACoP plug-in from ImageJ [62], from the images in Figure 1.3a, we obtained a PCC value of 0.747, whereas the Mander's coefficient for the fraction of VDAC1 colocalizing with TOM20 is 0.769 (M1) and the coefficient for TOM20 colocalizing with VDAC1 is 0.792 (M2). Although in the case of Figure 1.3b, PCC is 0.17, whereas the Mander's coefficient for the fraction of DEXTRAN colocalizing with EEA1 is 0.096 (M1) and the coefficient for EEA1 colocalizing with DEXTRAN is 0.415 (M2). The differences in the PCC coefficients from both examples Figure 1.3a and b are expected since both VDAC1 and TOM20 are known mitochondrial proteins and a high colocalization degree was expected. However, in the case of EEA1 and dextran, only a partial overlap of the signal occurs since the dextran is not only present in the early endosomes (EEA1) but also present in the lysosomes.

Limitations of confocal microscopy and advanced technologies

Despite the advantages offered by using a confocal microscope, this technology has several limitations that we have been refereeing along this chapter, including the resolution limit, imaging depth, and acquisition speed. Several techniques have been developed to overcome such limitations.

For example, with total internal reflection fluorescence (TIRF) microscopy, it is possible to image structures less than 200 nm. However, it presents a major limitation since the major penetration depth is just 100 nm [66]. Still this technology is very useful when analyzing mechanisms that occur at the cell membrane level. To reduce the imaging depth limitation, two-photon laser scanning microscopy increases the imaging depth up to 600 μm. Another advantage of this technology is that it causes less photobleaching and phototoxicity, due to the reduced absorption and scattering of the excitation light and the use of lasers with longer wavelengths, making this technique highly valuable for live-cell experiments, imaging of thicker samples, such as tissue slices and scaffolds, and *in vivo* studies. However, due to the use of longer wavelength lasers, the resolution obtained in two-photon microscopy is slightly worse than in confocal microscopy [67].

The acquisition speed is a key fundamental feature mainly when acquiring images on live-cell experiments for high-speed events, such as organelle movement. In this regard, spinning disk confocal microscopes present an advantage over confocal microscopes since the specimen is scanned by thousands of points of light at the same time speeding up the acquisition rate, which is achieved by using a rotating disk with thousands of pinholes instead of a single pinhole [68,69]. A different approach to improve the acquisition speed is the use of resonant scanners that can function at 8 or 12 Hz [70]. The time-consuming acquisition of sufficient amount of images to obtain statistically relevant data is another problem arising from the acquisition speed in confocal microscopy. In this regard, high-content screening confocal systems have arisen [71]. With this technology, it is possible to increase the automation and speed of the image acquisition. However, these systems need a huge storage capacity for the series of images acquired and significant computing power along with high-end software for image analysis.

Another limitation of the confocal microscope relies on the specific setup of the equipment regarding the different lasers available, which will ultimately limit the number of fluorophores allowed to use in the experiment. Recently, a new type of laser called *white laser* became available. With this technology, it is possible to excite the sample within the 470–670 nm range with a unique laser [72] and some confocal microscopes are able to select eight excitation lines from 3 trillion unique combinations for simultaneous imaging. In addition, due to the reduced light intensity on the sample, photobleaching and phototoxicity are reduced.

The detectors used in confocal microscopy might constitute another limitation because of the quantum efficiency and the noise level are important features in image acquisition. As explained in the previous section (Introduction), the PMTs are the most common used detectors in confocal microscopes despite that they display a quantum efficiency of around 30% [73]. Emergent technologies are applied to improve the efficiency that translates in the use of lower laser powers for imaging.

Among them are the GaAsP, Hybrid, and Airyscan detectors, which also allow for higher acquisition speed. Furthermore, when imaging multiple fluorophores, the fluorescent signal might bleed through different channels. Optimization of the respective detection range together with low-laser intensity will reduce this effect. When possible (i.e., when there are no limitations in acquisition speed), sequential scan is the best choice. However, in cases when this does not work, mathematical restoration of the dyes into different channels can be performed [74].

Future outlook

A hot technological improvement in the field is the increase in resolution. Nowadays, with the advent of super-resolution microscopy, the light diffraction limit is reduced allowing to image at the molecular level (down to tens of nm). Different approaches exist: (1) stimulated emission depletion (STED) microscopy [75]; (2) structured illumination microscopy (SIM) [76]; and (3) localization microscopy (photoactivated localization microscopy [PALM], fluorescence photoactivation localization microscopy [FPALM], stochastic optical reconstruction microscopy [STORM], fPALM, and STORM) [77]. However, they still present phototoxicity problems when working with live cells and acquisition speed limitation [78]. With STED, a maximum resolution of 20–50 nm can be achieved. By selective deactivation of fluorophores, sharper images are obtained because no fluorescence from neighboring fluorophores occurs [79]. SIM is a wide-field approach to break the light diffraction limit that requires considerable expertise in acquisition and processing, since artifacts are easily obtained when working with suboptimal conditions [80]. For image acquisition in SIM, the sample is illuminated with spatially structured excitation light. The high-resolution information resides in the form of moiré fringes, which are processed to reconstruct images with improved resolution [76]. Localization microscopy approaches use sequential activation and time-resolved localization of photoswitchable fluorophores. After repeated cycle of photoactivation–imaging–bleaching, the merging of all the single-molecule positions produces the final image, allowing for resolving structures and biomolecular interactions [81]. Ultimately, correlative light electron microscopy (CLEM) combines light microscopy with the resolution of electron microscopy allowing an increase in resolution down to the nanometer level [82].

Other advanced technology that is worth to mention is fluorescence lifetime imaging microscopy (FLIM). With this technology, images are obtained based on the differences of the exponential decay of the fluorescence of a fluorophore. The fluorescence lifetime is the time that a given fluorophore stays in the excitation state. This value is characteristic of every fluorophore and depends on the microenvironment [83]. One of the most popular applications of this technique is the quantification of föster resonance energy transfer (FRET) (Chapter 10 from this book). Since with this equipment, there are no artifacts due to differences in excitation power, photobleaching, spectral cross talk, or differences in fluorophore concentrations, the use of FLIM offers a real quantitative analysis of the molecular interactions [84].

There is also a growing tendency to use approaches directed to high throughput and automated image acquisition to overcome the time constrains of acquiring and processing high-quality images or screen through large datasets. In addition, this approach generates a huge amount of data that is pushing the development of workstations with massive storage capacities and computing power to be able to work with the large collection of files created. For example, the amount of data generated in a limited time can reach the terabyte order in a high-content screening confocal microscope setup [85]. These setups not only are capable of automatically acquiring a vast amount of data but also with high quality. Along with the implementation of these screening systems is the need to develop high-end software for automated analysis of the data generated with capacity to extract the maximum of information with high confidence or consistency. As an example, the software developed for *intelligent screening* where an initial scan of the sample is performed and then localizes the areas with the feature of interest, extracting the spatial location within the sample for further high-resolution imaging [86].

With all these in mind, ideally the field should evolve toward affordable systems with increased resolution (toward super-resolution), acquisition speed and quantitative data, displaying less photodamage, and working in a high-content screening mode. The increasing quality and amount of data must be accompanied with the development of powerful automated software and hardware and will certainly have a pronounced impact in the knowledge evolution of the life sciences.

Acknowledgments

This work was supported by the Bioimaging Center for Biomaterials and Regenerative Therapies (b.IMAGE), and financed by Portuguese funds through *FCT—Fundação para a Ciência e a Tecnologia* in the framework of project UID/BIM/04293/2013.

References

1. Ishikawa-Ankerhold, H.C., R. Ankerhold, and G.P. Drummen, Advanced fluorescence microscopy techniques—FRAP, FLIP, FLAP, FRET and FLIM. *Molecules*, 2012. **17**(4): 4047–4132.
2. Nwaneshiudu, A. et al., Introduction to confocal microscopy. *J Invest Dermatol*, 2012. **132**(12): e3.

3. Minsky, M., Memoir on inventing the confocal scanning microscope. *Scanning*, 1988. **10**(4): 128–138.
4. Inoué, S. and K.R. Spring, (Eds.) *Video Microscopy: The Fundamentals*, New York: Plenum Press, 1997.
5. Rai, V. and N. Dey., *The Basics of Confocal Microscopy, Laser Scanning, Theory and Applications*, P.C.-C. Wang, (Ed.) 2011.
6. Wilson, T. and A.R. Carlini, Three-dimensional imaging in confocal imaging systems with finite sized detectors. *J Microsc*, 1988. **149**(1): 51–66.
7. Claxton, N.S., Fellers, T.J., and Davidson M.W., Laser scanning confocal microscopy. Department of Optical Microscopy and Digital Imaging, National High Magnetic Field Laboratory, Florida State University, (http://www.aptechnologies.co.uk/images/Data/Vertilon/PP6207.pdf), 2005.
8. Diaspro, A.E., *Confocal and Two-Photon Microscopy: Foundations, Applications, and Advances*, 2002, New York: Wiley-Liss.
9. Hibbs, A.R.E., *Confocal Microscopy for Biologists*, 2004, New York: Kluwer Press.
10. Matsumoto, B.E., *Methods in Cell Biology: Cell Biology Applications of Confocal Microscopy*, 2002, New York: Academic Press.
11. Muller, W.E., *Introduction to Confocal Fluorescence Microscopy*, 2002, Maastrich, the Netherlands: Shaker Publishing.
12. Paddock, S.W., *Confocal Microscopy: Methods and Protocols*, 1999, New York: Humana Press Springer.
13. Pawley, J.B.E., *Handbook of Biological Confocal Microscopy*, 1995, New York: Springer.
14. Sanderson, M.J. et al., Fluorescence microscopy. *Cold Spring Harb Protoc*, 2014. **2014**(10): pdb.top071795.
15. Spring, K., Detectors for Fluorescence Microscopy, in *Methods in Cellular Imaging*, A. Periasamy, (Ed.) 2001, New York: Springer. pp. 40–52.
16. Art, J., Photon detectors for confocal microscopy, in *Handbook of Biological Confocal Microscopy*, J.B. Pawley, (Ed.) 2006, New York: Springer. pp. 251–264.
17. Amos, W.B., Instruments for fluorescence imaging, in *Protein Localization by Fluorescence Microscopy: A Practical Approach*, V.J. Allan, (Ed.), 2000, New York: Oxford University Press. pp. 67–108.
18. Keller, H.E., Objective lenses for confocal microscopy, in *Handbook of Biological Confocal Microscopy*, J.E. Pawley, (Ed.) 2006, New York: Springer. pp. 145–161.
19. Tsien, R.Y., L. Ernst, and A. Waggoner, Fluorophores for confocal microscopy: Photophysics and photochemistry, in *Handbook of Biological Confocal Microscopy*, J. B. Pawley, (Ed.) 2006, New York: Springer. pp. 338–352.
20. Pawley, J.B., Points, pixels, and gray levels: Digitizing image data, in *Handbook of Biological Confocal Microscopy*, J.E. Pawley, (Ed.) 2006, New York: Springer.
21. Johnson, J., Not seeing is not believing: Improving the visibility of your fluorescence images. *Mol Biol Cell*, 2012. **23**(5): 754–757.
22. Rossner, M. and R. O'Donnell, The JCB will let your data shine in RGB. *J Cell Biol*, 2004. **164**(1): 11–13.
23. Cromey, D.W., Digital images are data: And should be treated as such. *Methods in Molecular Biology*, 2013. **931**: 1–27.
24. Rossner, M. and K.M. Yamada, What's in a picture? The temptation of image manipulation. *J Cell Biol*, 2004. **166**(1): 11–15.
25. Cromey, D.W., Avoiding twisted pixels: Ethical guidelines for the appropriate use and manipulation of scientific digital images. *Sci Eng Ethics*, 2010. **16**(4): 639–667.

26. Wu, X. et al., Immunofluorescent labeling of cancer marker Her2 and other cellular targets with semiconductor quantum dots. *Nat Biotechnol*, 2003. **21**(1): 41–46.
27. Jaiswal, J.K. et al., Long-term multiple color imaging of live cells using quantum dot bioconjugates. *Nat Biotechnol*, 2003. **21**(1): 47–51.
28. Texier, I. and V. Josser, In vivo imaging of quantum dots. *Methods Mol Biol*, 2009. **544**: 393–406.
29. Deerinck, T.J., The application of fluorescent quantum dots to confocal, multiphoton, and electron microscopic imaging. *Toxicologic Pathology*, 2008. **36**(1): 112–116.
30. Gomez-Lazaro, M. et al., 6-Hydroxydopamine activates the mitochondrial apoptosis pathway through p38 MAPK-mediated, p53-independent activation of Bax and PUMA. *J Neurochem*, 2008. **104**(6): 1599–1612.
31. Kao, J.P., A.T. Harootunian, and R.Y. Tsien, Photochemically generated cytosolic calcium pulses and their detection by fluo-3. *J Biol Chem*, 1989. **264**(14): 8179–8184.
32. Zorov, D.B. et al., Examining intracellular organelle function using fluorescent probes: From animalcules to quantum dots. *Circ Res*, 2004. **95**(3): 239–252.
33. Akagi, J. et al., Real-time cell viability assays using a new anthracycline derivative DRAQ7(R). *Cytometry A*, 2013. **83**(2): 227–234.
34. Gross, D. and L.M. Loew, Fluorescent indicators of membrane potential: Microspectrofluorometry and imaging. *Methods Cell Biol*, 1989. **30**: 193–218.
35. Subach, O.M. et al., A photoswitchable orange-to-far-red fluorescent protein, PSmOrange. *Nat Meth*, 2011. **8**(9): 771–777.
36. Subach, O.M. et al., A FRET-facilitated photoswitching using an orange fluorescent protein with the fast photoconversion kinetics. *J Am Chem Soc*, 2012. **134**(36): 14789–14799.
37. Caires, H.R. et al., Finding and tracing human MSC in 3D microenvironments with the photoconvertible protein Dendra2. *Sci Rep*, 2015. **5**: 10079.
38. Patterson, G.H. and J. Lippincott-Schwartz, A photoactivatable GFP for selective photolabeling of proteins and cells. *Science*, 2002. **297**(5588): 1873–1877.
39. Subach, F.V. et al., Photoactivatable mCherry for high-resolution two-color fluorescence microscopy. *Nat. Methods*, 2009. **6**(2): 153–159.
40. Subach, F.V. et al., Bright monomeric photoactivatable red fluorescent protein for two-color super-resolution sptPALM of live cells. *J Am Chem Soc*, 2010. **132**(18): 6481–6491.
41. Gunewardene, M.S. et al., Superresolution imaging of multiple fluorescent proteins with highly overlapping emission spectra in living cells. *Biophys J*, 2011. **101**(6): 1522–1528.
42. Piatkevich, K.D., F.V. Subach, and V.V. Verkhusha, Far-red light photoactivatable near-infrared fluorescent proteins engineered from a bacterial phytochrome. *Nat Commun*, 2013. **4**: 2123.
43. Warren, M.M. et al., Ground-state proton transfer in the photoswitching reactions of the fluorescent protein Dronpa. *Nat Commun*, 2013. **4**: 1461.
44. Fron, E. et al., Excited state dynamics of photoswitchable fluorescent protein padron. *J Phys Chem B*, 2013. **117**(51): 16422–16427.
45. Rye, H.S. et al., Stable fluorescent complexes of double-stranded DNA with bis-intercalating asymmetric cyanine dyes: Properties and applications. *Nucleic Acids Res*, 1992. **20**(11): 2803–2812.
46. Errington, R.J. et al., Advanced microscopy solutions for monitoring the kinetics and dynamics of drug-DNA targeting in living cells. *Adv Drug Deliv Rev*, 2005. **57**(1): 153–167.
47. Suzuki, T. et al., DNA staining for fluorescence and laser confocal microscopy. *J Histochem Cytochem*, 1997. **45**(1): 49–53.

48. Pérez-Alvarez, A., A. Araque, and E.D. Martín, Confocal microscopy for astrocyte in vivo imaging: Recycle and reuse in microscopy. *Front Cell Neurosci*, 2013. **7**: 51.
49. Humphries, W.H.I.V., C.J. Szymanski, and C.K. Payne, Endo-lysosomal vesicles positive for Rab7 and LAMP1 are terminal vesicles for the transport of dextran. *PLoS One*, 2011. **6**(10): e26626.
50. *The Nobel Prize in Chemistry 2008*, 2015, Nobelprize.org. Nobel Media AB 2014.
51. Wang, Y., J.Y. Shyy, and S. Chien, Fluorescence proteins, live-cell imaging, and mechanobiology: Seeing is believing. *Annu Rev Biomed Eng*, 2008. **10**: 1–38.
52. Gomez-Lazaro, M. et al., 6-Hydroxydopamine (6-OHDA) induces Drp1-dependent mitochondrial fragmentation in SH-SY5Y cells. *Free Radic Biol Med*, 2008. **44**(11): 1960–1969.
53. Borta, H. et al., Analysis of low abundance membrane-associated proteins from rat pancreatic zymogen granules. *J Proteome Res*, 2010. **9**(10): 4927–4939.
54. Zhou, X.X. and M.Z. Lin, Photoswitchable fluorescent proteins: Ten years of colorful chemistry and exciting applications. *Curr Opin Chem Biol*, 2013. **17**(4): 682–690.
55. Adam, V. et al., Phototransformable fluorescent proteins: Future challenges. *Curr Opin Chem Biol*, 2014. **20**: 92–102.
56. Bourgeois, D. and V. Adam, Reversible photoswitching in fluorescent proteins: A mechanistic view. *IUBMB Life*, 2012. **64**(6): 482–491.
57. Flores, A. and I. Arinze, Use of Dendra2 to monitor degradation of the transcription factor Nrf2 in promyelocytic leukemia-nuclear bodies in living cells (594.1). *The FASEB J*, 2014. **28**(1 Suppl): 594.
58. Reits, E.A. and J.J. Neefjes, From fixed to FRAP: Measuring protein mobility and activity in living cells. *Nat Cell Biol*, 2001. **3**(6): E145–E147.
59. Nouar, R. et al., FRET and FRAP imaging: approaches to characterise tau and stathmin interactions with microtubules in cells. *Biol Cell*, 2013. **105**(4): 149–161.
60. Ribrault, C. et al., Syntaxin1A lateral diffusion reveals transient and local SNARE interactions. *J Neurosci*, 2011. **31**(48): 17590–17602.
61. Dunn, K.W., M.M. Kamocka, and J.H. McDonald, A practical guide to evaluating colocalization in biological microscopy. *Am J Physiol–Cell Physio*, 2011. **300**(4): C723–C742.
62. Bolte, S. and F.P. CordelièRes, A guided tour into subcellular colocalization analysis in light microscopy. *J Microsc*, 2006. **224**(3): 213–232.
63. Scriven, D.R., R.M. Lynch, and E.D. Moore, Image acquisition for colocalization using optical microscopy. *Am J Physiol Cell Physiol*, 2008. **294**(5): C1119–C1122.
64. Oheim, M. and D. Li, Quantitative colocalisation imaging: Concepts, measurements, and pitfalls, in *Imaging Cellular and Molecular Biological Functions*, S. Shorte and F. Frischknecht, (Eds.) 2007, Berlin, Germany: Springer. pp. 117–155.
65. Demandolx, D. and J. Davoust, Multicolour analysis and local image correlation in confocal microscopy. *J Microsc*, 1997. **185**(1): 21–36.
66. Axelrod, D., Total internal reflection fluorescence microscopy in cell biology. *Traffic*, 2001. **2**(11): 764–774.
67. Benninger, R.K.P. and D.W. Piston, Two-photon excitation microscopy for the study of living cells and tissues. *Curr Protoc Cell Biol*, 2013. **4**: 4–11.
68. Stehbens, S. et al., Imaging intracellular protein dynamics by spinning disk confocal microscopy. *Methods Enzymol*, 2012. **504**: 293–313.
69. Graf, R., J. Rietdorf, and T. Zimmermann, Live cell spinning disk microscopy. *Adv Biochem Eng Biotechnol*, 2005. **95**: 57–75.

70. Borlinghaus, R.T., MRT letter: High speed scanning has the potential to increase fluorescence yield and to reduce photobleaching. *Microsc Res Tech*, 2006. **69**(9): 689–692.
71. Buchser, W., M. Collins, T. Garyantes, R. Guha, S. Haney, V. Lemmon, Z. Li, and O.J. Trask, Assay development guidelines for image-based high content screening, high content analysis and high content imaging, in *Assay Guidance Manual*, G.S. Sittampalam, N.P. Coussens, H. Nelson et al., (Eds.) 2014, Bethesda, MD: Eli Lilly & Company and the National Center for Advancing Translational Sciences. pp. 1–71.
72. Borlinghaus R.T and Kuschel. L.R., The White Confocal—Spectral Gaps Closed in *Current Microscopy Contributions to Advances in Science and Technology*, A. Mendez-Vilas, (Ed.) 2012, Badajoz, Spain: Formatex Research Centre.
73. Pawley, J., Fundamental limits in confocal microscopy, in *Handbook Of Biological Confocal Microscopy*, J.B. Pawley, (Ed.) 2006, New York: Springer. pp. 20–42.
74. Zimmermann, T., Spectral imaging and linear unmixing in light microscopy, in *Microscopy Techniques*, J. Rietdorf, (Ed.) 2005, Berlin Heidelberg: Springer. pp. 245–265.
75. Klar, T.A. et al., Fluorescence microscopy with diffraction resolution barrier broken by stimulated emission. *Proc Natl Acad Sci USA*, 2000. **97**(15): 8206–8210.
76. Gustafsson, M.G.L., Surpassing the lateral resolution limit by a factor of two using structured illumination microscopy. *J Microsc*, 2000. **198**(2): 82–87.
77. Betzig, E. et al., Imaging intracellular fluorescent proteins at nanometer resolution. *Science*, 2006. **313**(5793): 1642–1645.
78. Cox, S., Super-resolution imaging in live cells. *Dev Biol*, 2015. **401**(1): 175–181.
79. Westphal, V. et al., Video-rate far-field optical nanoscopy dissects synaptic vesicle movement. *Science*, 2008. **320**(5873): 246–249.
80. Ball, G. et al., SIMcheck: A toolbox for successful super-resolution structured illumination microscopy. *Sci Rep*, 2015. **5**: 15915.
81. Vangindertael, J. et al., Super-resolution mapping of glutamate receptors in C. elegans by confocal correlated PALM. *Sci Rep*, 2015. **5**: 13532.
82. Heath, J. and Verkade. P., *Correlative Light and Electron Microscopy*. 2nd ed. Essential Knowledge Briefing, D.J. Heath, (Ed.) 2015, Chichester, West Sussex: John Wiley & Sons.
83. van Munster, E. and T.J. Gadella, Fluorescence lifetime imaging microscopy (FLIM), in *Microscopy Techniques*, J. Rietdorf, (Ed.) 2005, Berlin Heidelberg: Springer. pp. 143–175.
84. Wang, X. et al., Gamma-secretase modulators and inhibitors induce different conformational changes of presenilin 1 revealed by FLIM and FRET. *J Alzheimers Dis*, 2015. **47**(4): 927–937.
85. DeBernardi, M., S. Hewitt, and A. Kriete, Automated confocal imaging and high-content screening for cytomics, in *Handbook Of Biological Confocal Microscopy*, J.B. Pawley, (Ed.) 2006, New York: Springer. pp. 809–817.
86. Sedic, M. et al., Haploinsufficiency for BRCA1 leads to cell-type-specific genomic instability and premature senescence. *Nat Commun*, 2015. **6**: 7505.

Chapter 2 Imaging flow cytometry for quantification of cellular parameters

Catarina R. Almeida

Contents

Introduction

Anyone who has worked with microscopy images knows the difficulty in objectively quantifying the data from a large enough number of events for robust statistics. Flow cytometry allows acquisition of data from thousands of cells in a short period of time but lacks the imaging component. Imaging flow cytometry (IFC) results from a combination of conventional cytometry with microscopy: cells are run in flow and excited with lasers, while charge-coupled device (CCD) cameras with time delay integration (TDI) technology acquire images from the flowing events. Analysis of these images, which include both brightfield and fluorescence images, allows for high-content analysis of a large number of events in a very short time.

The first commercial equipment became available in 2005 and since then the number of manuscripts using this technology has been increasing. Most publications use the commercial instruments but there has been a recent effort in developing new custom-made alternatives [1–8]. Currently Merck Millipore (EMD Millipore in the United States and Canada) commercializes two Amnis imaging cytometers, the ImageStream®X Mark II and the low-end FlowSight®, the main difference between the two systems being their resolution and number of available lasers for excitation. The ImageStream®X Mark II system is a high-resolution flow cytometer that can be equipped with 20, 40, or 60 X objectives, with each pixel corresponding to 1.0, 0.25, or 0.1 μm^2, respectively. It can simultaneously acquire images in up to 12 channels, for brightfield, darkfield (corresponding to scatter), and fluorescence images, at rates of up to 5000 objects per second. Currently available lasers can excite at 375, 405, 561, 592, 642, 730 and 785 nm, with a 375 nm laser announced to become available soon. This system can also include the extended depth of field (EDF™) option, which increases the depth of cells in focus without loss of fluorescence sensitivity. This allows visualization of different focus planes in a single image, similar to a maximum projection image obtained by confocal microscopy.

Performing an imaging flow cytometry experiment

Before performing an imaging flow cytometry experiment, one should first of all think whether and how will quantification of a large number of imaging parameters be useful and also whether the events to analyze can be run in flow. If imaging is not essential and there is only a need to know the intensity of fluorescence, then conventional flow cytometry is a better option. On the other hand, if cells are adherent and cannot be run in flow without significantly altering their properties, it is best to use imaging cytometry.

Every imaging flow cytometry analysis should then start with a careful planning of the experiment. First, it is necessary to decide what to label and search which dyes and antibodies are available. Colors must be chosen depending on the system offered, paying special attention to the existing excitation lasers. At the same time, it is essential to think in advance how data will be analyzed, as this will determine whether any extra markers are necessary. This planning will also reveal which positive controls will be important to establish gates for certain applications, such as apoptosis and cell-cycle analysis. As with any cytometry experiment, negative controls, preferentially isotype controls, are important. In addition, when using more than one color, it will be necessary to compensate, and thus one should think about strategies to ensure positive staining for each fluorophore in single-stained controls. When studying rare events using calibration beads might be a good option. After designing the experiment, all antibodies and dyes have to be carefully titrated. This is particularly important when using different labels that will be excited by the same laser, as the laser power will have to be sufficient to excite all fluorophores without reaching saturation of any. As with conventional fluorescence imaging, it is essential for quantification that the fluorescence signal does not reach saturation. Thus, this titration is critical to ensure acquisition of good quality data for posterior analysis. Broadly speaking, the staining protocol to be performed is equivalent to what is done for conventional fluorescence imaging: (1) cells can be fixed and permeabilized if intracellular staining is being performed; (2) depending on the antibody of choice it may be necessary to block for unspecific binding; and (3) cells should then be incubated with the primary or conjugated antibodies at a low temperature, thoroughly washed and if required incubated with the secondary antibodies before washing again. Once everything has been optimized, one can proceed with labeling the samples and its analysis. Samples should be filtered immediately prior to acquisition; if cells are clustered in clumps, the system will run unstable and will not acquire nice focused images of a good cell number. Samples are run at the same time as calibration beads, crucial to maintain the cells in focus.

Data obtained with the ImageStream®X Mark II system can be analyzed with the software IDEAS®. After some training, this is an intuitive, versatile, and easy-to-use software. The typical analysis will be straightforward to anyone who is familiar with conventional cytometry analysis and image treatment. As with flow cytometry, one can have dot plots and histograms and it is possible to draw gates in populations of interest, but the parameters to be analyzed are not limited to fluorescence intensity. As with fluorescence microscopy, it is possible to observe the cells, change the brightness and contrast, and draw masks to limit areas of interest in a cell, such as the nucleus. But here, analysis is performed in many different events simultaneously. A very interesting feature of imaging flow cytometry is the possibility of visualizing an event corresponding to a specific dot in a dot plot, and vice versa. This is very useful when drawing gates for a certain population. An analysis flow typically involves the following: (1) compensation—compensation is performed pixel-by-pixel; it can be done either automatically or manually; (2) selection of single cells on a plot with the aspect ratio versus area (Figure 2.1a); (3) selection of events in focus by using the gradient feature, which reflects the

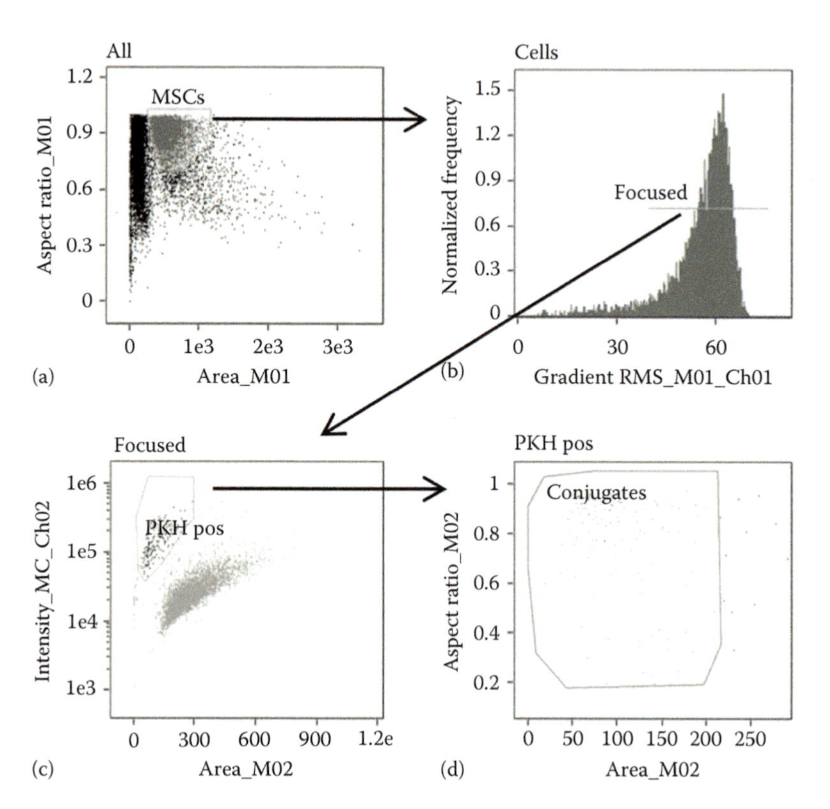

Figure 2.1 Example of analysis performed with imaging flow cytometry data. (a) First, a gate is drawn to select cells of interest (in this case, MSC) on an aspect ratio versus area dot plot. When in doubt where to draw the gate, it is possible to select a dot and visualize the corresponding event. (b) Second, cells in focus are selected by defining a gate on a gradient histogram, to include sharp images. These first two steps can be performed in the opposite order. Most often, the brightfield channel is the one used to select these gates. Then, cells positive for the marker(s) of interest are selected to proceed with analysis in the subsets of interest (c). Different masks as well as features can then be calculated. In this example, a gate was drawn on an aspect ratio versus area plot for the fluorescence channel of our marker, in order to quantify the percentage of conjugated cells (d). Imaging flow cytometry was essential to ensure that these gates were indeed including cell conjugates.

sharpness of each image (Figure 2.1b); (4) selection of cells positive for the markers of interest (Figure 2.1c); (5) design masks of interest, for example, for the cell membrane or for stained spots (Figure 2.2); and (6) design and calculus of different features (Figure 2.1d).

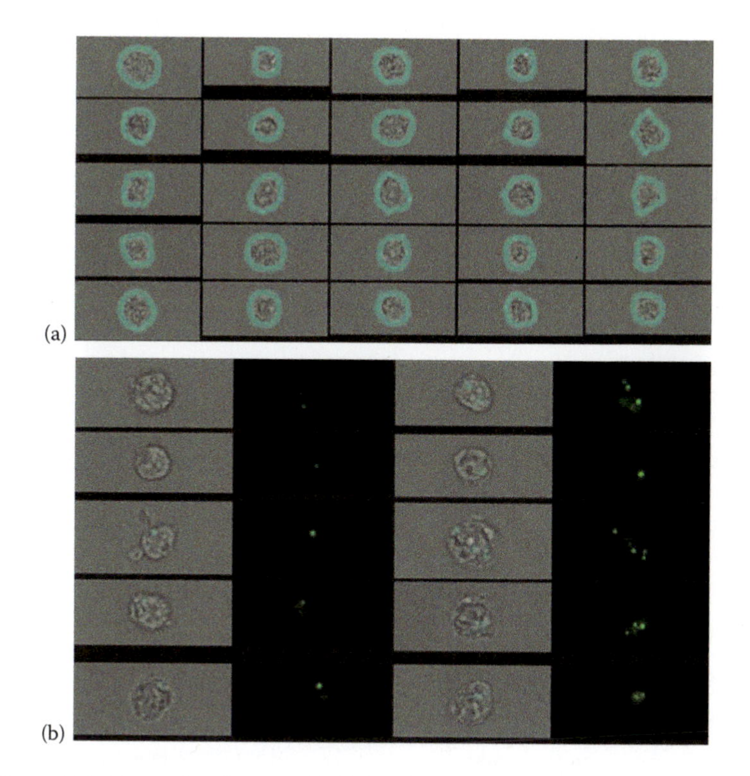

Figure 2.2 Example of masks (in light blue) applied to identify the cell membrane (a) or fluorescence spots (b). After defining a mask, features such as the intensity of a fluorophore within a certain area of the cell can be easily quantified.

Applications

The statistical rigor, throughput, and objectivity of imaging flow cytometry have led many researchers in cell biology to adopt this technology. Here we will discuss some studies that have used imaging flow cytometry to study morphology, protein location (including trafficking, internalization, and protein colocalization), and cell–cell interactions. Other applications not explored in this article include analysis of cell death and apoptosis, autophagy, cell cycle, DNA damage, and detection of rare events. The possibility of analyzing rare events in heterogeneous populations has been nicely reviewed by others [9]. All of these applications can be combined and the number of different analysis that can be performed is enormous.

Morphology analysis

One of the most obvious applications of this technology is the study of cell morphology. Many works have analyzed the morphology of cells, including mammalian cells, yeast [10,11], and bacteria [12]. In addition, analysis of morphological and texture features has allowed the detection of extracellular microparticles down to 0.2 μm [13,14]. Morphological analyses can be combined with markers to identify specific cell subsets or to analyze other cellular events, such as nuclear translocation of signaling proteins.

Some of the best examples on morphological analysis with imaging flow cytometry come from studies with erythroid cells [15]. Indeed, by combining staining of specific markers with morphological analysis, it is possible to study the different stages in erythroid maturation [16]. This has allowed identification of erythrocyte precursors [17] and contributed to reveal molecular mechanisms regulating maturation [18–20]. Moreover, tools to detect shape changes in sickle cell disease have been developed [21]. Similarly, it is possible to identify the different stages in megakaryopoiesis, based on cell's ploidy, cell and nucleus morphology, and staining for the markers CD117 and CD41 [22,23]. Megakaryocytes are rare cells with an amorphous shape that would otherwise be difficult to distinguish from clumps of cells [22,23].

Lymphocytes and monocytes can also adopt a different morphology depending on their phenotype or activation status. For example, T-lymphocytes stimulated with chemokines suffer shape alterations. Thus, imaging flow cytometry has been applied to correlate morphological changes induced by CCL19 with polarization of actin and phosphorylated Ezrin–Radixin–Moesin (ERM) [24] and to show that binding of oversulfated chondroitin sulfate (OSCS) to SDF-1α inhibited chemokine-induced morphological changes [25].

As another example, cells freshly isolated with different methods from bovine intervertebral disc, more specifically from the central, nucleus pulposus (NP) region, have been characterized. Imaging flow cytometry analysis allowed quantification of the cell's diameter, area, nuclear-to-cytoplasmic ratio, and also the presence of vesicles [26]. Interestingly, analysis of the area calculated from the brightfield images and from fluorescence images after actin staining revealed the presence of a pericellular matrix surrounding the cells, which was further confirmed by staining against collagen type VI [26].

Morphological analysis has also contributed to identify and characterize rare cell populations. For example, imaging flow cytometry has been instrumental to, in combination with flow cytometry and confocal microscopy, describe very small embryonic-like stem cells (VSELs) [27]. This is a rare population of cells that has been proposed to be pluripotent [28], but that whose existence and role is still the subject of intense debate [29,30].

Protein location

One of the most used applications of imaging flow cytometry is the study of protein subcellular location [31–33]. Studies on protein trafficking have been performed by determining colocalization with specific markers, for early and late endosomes, lysosomes, and so on. As an example, it is possible to analyze trafficking of major histocompatibility complex (MHC) class II (MHCII) molecules on specific subsets of primary mouse or human dendritic cells (DCs) [34]. DCs are a heterogeneous population of cells that uptake antigens (Ag) that are processed, loaded in MHCII, and presented on the cell's surface, to prime naïve CD4 T cells. Imaging flow cytometry will contribute to dissect in detail the mechanisms regulating the proteostasis of MHCII-peptide complexes on DC, as well as their surface expression.

Specific location at the cell membrane can be determined from either brightfield or fluorescence images, being then easy to analyze internalization. Internalization studies have been performed with membrane receptors [35], viral proteins [36], exosomes [37], microparticles [38,39], apoptotic cells [40], and bacteria [41,42]. Thus, there is a growing interest from the nanomaterials community in applying this tool [43]. Indeed, some works have tried to tailor nanoparticles to affect its internalization by cell lines [42,44,45] or primary cell macrophages [46]. For systems including the EDF option, it is also possible to quantify the number of fluorescent events that can be found within the cells. The actual number of nanoparticles cannot be determined due to their small size but quantifying the number of nanoparticle-loaded vesicles can contribute to nanoparticle design for therapy [47]. In addition, measuring uptake of antibodies by cancer cells can be an interesting tool when testing monoclonal antibodies designed for cancer therapy [48].

Imaging flow cytometry can be used to quantify protein colocalization [49–52]. However, caution must be taken in interpreting colocalization between two molecules, as image resolution is too low for colocalization to assume direct protein interaction. This would have to be confirmed by other biochemical techniques. In a recent study, dysregulation of N-glycosylation of the T-cell receptor (TCR) in ulcerative colitis patients has been studied in detail [53]. Corroborating data obtained by immunoprecipitation, flow cytometry, and immunohistochemistry, imaging flow cytometry analysis confirmed colocalization of N-glycans with the TCR and also the impaired glycosylation at the membrane of lamina propria T-lymphocytes. A detailed protocol on this imaging flow cytometry analysis has been established [54].

The activity of some signaling proteins, particularly transcription factors such as NF-κB, nuclear factor of activated T cells (NFAT), and others, can be inferred from their location. The study of NF-κB translocation is one of the best examples on the utility of imaging flow cytometry. Indeed, by simply staining for NF-κB and the nucleus, it is possible to determine how much of this molecule is in the nucleus and how much is in the cytoplasm and thus to quantify its nuclear translocation on stimulation. For example, it was shown that resveratrol treatment decreases the NF-κB translocation that is seen when stimulating primary human DC with TNF-α [55]. Most importantly, simultaneous analysis of other cellular events allows more complex analysis and spatial and temporal dissection of signaling pathways. For example, NFAT translocation has been analyzed in combination with Ca^{2+} measurements based on DiH–Rhod2 and MitoDR staining in T cells [56].

Intercellular communication

Another interesting possibility is the analysis of intercellular communication. Intercellular communication occurs when two cells come together, and involves different steps, which may include cell adhesion, protein reorganization/accumulation at the intercellular contact, activation of signaling pathways, secretion, and signaling termination. Cell–cell contact is mostly studied by imaging, but quantification and automated image analysis is extremely hard as it will be difficult to detect two cells in close contact when many cells are together in the same slide or culture dish. The possibility of running thousands of cells in flow increases the chances of finding cell–cell conjugates that can be further analyzed.

Figure 2.3 presents an example of how imaging flow cytometry can contribute to our knowledge on intercellular communication. In our quest to understand the impact of immune responses on tissue repair/regeneration, we have been studying interactions between natural killer (NK) lymphocytes and mesenchymal stem cells (MSC). MSC are being widely studied due to their capacity to differentiate into different lineages, such as osteoblasts and chondrocytes but also due to their paracrine and immunomodulatory properties [57]. Depending on the cells microenvironment, they can either stimulate or suppress the response of different immune cells, including NK cells [58]. On the other hand, NK cells are capable of lyzing target cells and of cytokine secretion, thus regulating the function of other cell types [59]. Different groups have been studying the bidirectional interactions between these two cell types [60–72]. We have found that freshly isolated human NK cells can be activated by MSC but on the other hand, those NK cells are capable of stimulating MSC recruitment [73]. In order to further dissect the cell–cell interactions, imaging flow cytometry was performed to analyze whether these two cell types form conjugates (Figure 2.3). For that, NK cells were labeled with PKH67, coincubated with bone marrow MSC, at different ratios, fixed at defined time points, and run on ImageStream^{X}. It was then possible to quantify the number of NK cells (labeled in green) that were conjugated to MSC (with a bigger size). Indeed NK cells formed conjugates with MSC albeit at a very low percentage. In addition, not surprisingly, the percentage of conjugated NK cells increased

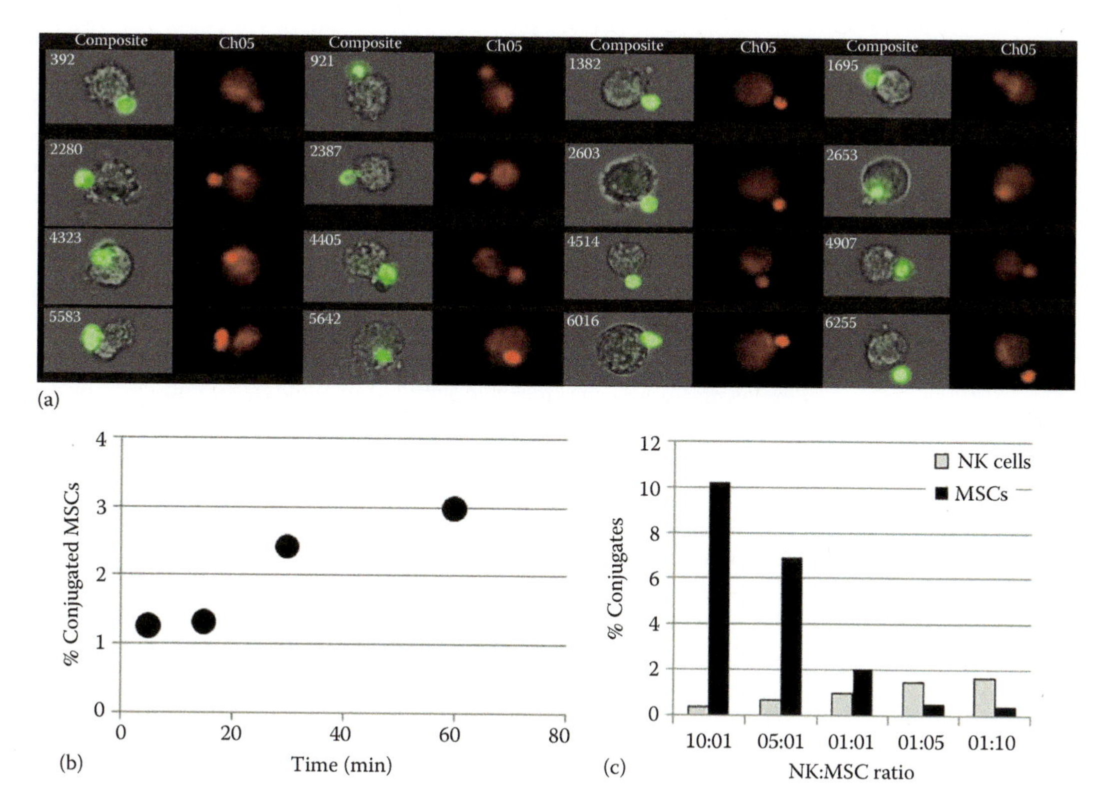

Figure 2.3 NK cells form conjugates with MSCs. Human NK cells were labeled with PKH67 (in green), incubated with human bone marrow MSCs, and fixed with 4% paraformaldehyde. Cells were then stained with DRAQ5 prior to analysis with ImageStream^X (Amnis). NK cells could be distinguished on the basis of labeling and MSCs on the basis of size. (a) Examples of conjugates formed between NK cells and MSCs and (b) kinetics of conjugation. Cells were fixed after the indicated time points at a 1:1 NK:MSC ratio. (c) Cells were incubated for 30 min at the indicated ratios.

when the number of MSC increased, whereas the number of conjugated MSC increased for a higher NK cell number. It would not be possible to confidently quantify such a small number of conjugates with other available techniques. Imaging flow cytometry will undoubtedly be a valuable tool to further develop into mechanistic studies on what determines the outcome of NK cell–MSC interactions.

After immune cells adhere, proteins may reorganize at the intercellular contact, forming an immune synapse. Advanced imaging technologies are essential to study this supramolecular structure. In the past, to quantify protein accumulation, reorganization, and colocalization at the synapse, images of a limited number of cells would be acquired and analyzed [74]. This time-consuming strategy only allowed analysis of a relatively small number or events, selected by the experimenter according to defined criteria. For example, in our own studies of the NK cell synapse, we imaged conjugates between NK cells and target cells transfected to express GFP-tagged MHC class I proteins and considered the protein to be clustered when the fluorescence within a part of the intercellular contact was at least twice the average fluorescence intensity at the cell membrane away from the contact [74]. Imaging flow cytometry brought the possibility to more easily and robustly quantify protein accumulation at the cell–cell interface and thus perform more powerful and significant analysis. Indeed, the kinetics of Lck and CD3ε accumulation at the synapse formed between T cells and antigen presenting cells (APC) has been analyzed [75]. Others have also used imaging flow cytometry to carefully compare the percentage of CD8+ T cells from either chronic lymphocytic leukemia patients or healthy controls accumulating CD8 at the interface with cytomegalovirus-peptide-loaded B cells and found no differences [76]. Most recently, methods have been described that quantify in detail actin and LFA-1 rearrangements at the immune synapse formed by human T cells from freshly obtained blood samples not subjected to any purification step other than red blood cell lysis that were incubated with Raji B cells as APC [77], or at the synapse formed by purified murine CD4+ T cells and DCs [78]. It was observed that there is an increase in the actin content of T cells conjugated to peptide-loaded APC [77].

It is only a question of time until more robust analyses on the molecular mechanisms occurring at the immune synapse are performed. The main limitations of imaging flow cytometry for these studies are the impossibility to obtain time-lapse movies, the relatively low resolution, and the impossibility to obtain 3D images and thus obtain *en face* views of the synapse, which is important to clearly see protein reorganization.

Future

The interest in imaging flow cytometry has been steadily increasing due to the easiness to perform quantitative image analysis in a large number of cells. It is likely that many more publications will report the use of this technology in the near future as the analyses that can be performed can still be combined in endless ways. It is now possible to analyze subtle differences in molecular events, which may have a significant impact. Therefore, imaging flow cytometry can also become a useful tool for diagnostics or for drug screening tests.

The recent developments in flow cytometry have been toward increasing the number of colors that can be measured without a need to compensate. At the same time, imaging has evolved toward super-resolution technologies. Thus, we might expect that imaging flow cytometry will also evolve to improve the capacity to image more colors, and with a better resolution. At the same time, it would be desirable to be able to obtain 3D images and to sort cells.

Acknowledgments

Data shown here are the outcome from work performed at the b.IMAGE—Bioimaging Center for Biomaterials and Regenerative Therapies, INEB. Cells were isolated from buffy coats and bone marrow samples kindly donated by Centro Hospitalar de São João, Porto, Portugal. This work had the financial support of FCT/MEC through National Funds and, when applicable, cofinanced by the FEDER through the PT2020 Partnership Agreement under the 4293 Unit I&D and in the framework of project EXPL/BIM-MED/0022/2013.

References

1. Gopakumar, G. et al., Framework for morphometric classification of cells in imaging flow cytometry. *J Microsc*, 2015. **261**: 307–319.
2. Zhu, H. and A. Ozcan, Opto-fluidics based microscopy and flow cytometry on a cell phone for blood analysis. *Methods Mol Biol*, 2015. **1256**: 171–190.
3. Kim, H. et al., Development of on-chip multi-imaging flow cytometry for identification of imaging biomarkers of clustered circulating tumor cells. *PLoS One*, 2014. **9**(8): p. e104372.
4. Brosnahan, M.L. et al., Complexities of bloom dynamics in the toxic dinoflagellate revealed through DNA measurements by imaging flow cytometry coupled with species-specific rRNA probes. *Deep Sea Res Part 2 Top Stud Oceanogr*, 2014. **103**: 185–198.
5. Wong, T.T. et al., Asymmetric-detection time-stretch optical microscopy (ATOM) for ultrafast high-contrast cellular imaging in flow. *Sci Rep*, 2014. **4**: 3656.
6. Regmi, R., K. Mohan, and P.P. Mondal, MRT letter: Light sheet based imaging flow cytometry on a microfluidic platform. *Microsc Res Tech*, 2013. **76**(11): 1101–1107.
7. Gorthi, S.S. and E. Schonbrun, Phase imaging flow cytometry using a focus-stack collecting microscope. *Opt Lett*, 2012. **37**(4): 707–709.
8. Schonbrun, E., S.S. Gorthi, and D. Schaak, Microfabricated multiple field of view imaging flow cytometry. *Lab Chip*, 2012. **12**(2): 268–273.
9. Barteneva, N.S., E. Fasler-Kan, and I.A. Vorobjev, Imaging flow cytometry: Coping with heterogeneity in biological systems. *J Histochem Cytochem*, 2012. **60**(10): 723–733.
10. Heisler, J. et al., Morphological effects of natural products on schizosaccharomyces pombe measured by imaging flow cytometry. *Nat Prod Bioprospect*, 2014. **4**(1): 27–35.
11. Patterson, J.O., M. Swaffer, and A. Filby, An imaging flow cytometry-based approach to analyse the fission yeast cell cycle in fixed cells. *Methods*, 2015. **82**: 74–84.
12. Pan, Y. and L. Kaatz, Use of image-based flow cytometry in bacterial viability analysis using fluorescent probes. *Curr Protoc Microbiol*, 2012. **Chapter 2**: Unit 2C 5.
13. Erdbrugger, U. et al., Imaging flow cytometry elucidates limitations of microparticle analysis by conventional flow cytometry. *Cytometry A*, 2014. **85**(9): 756–770.
14. Headland, S.E. et al., Cutting-edge analysis of extracellular microparticles using ImageStream(X) imaging flow cytometry. *Sci Rep*, 2014. **4**: 5237.

15. Samsel, L. and J.P. McCoy, Jr., Imaging flow cytometry for the study of erythroid cell biology and pathology. *J Immunol Methods*, 2015. **423**: 52–59.
16. McGrath, K.E., T.P. Bushnell, and J. Palis, Multispectral imaging of hematopoietic cells: Where flow meets morphology. *J Immunol Methods*, 2008. **336**(2): 91–97.
17. McGrath, K.E. et al., Enucleation of primitive erythroid cells generates a transient population of "pyrenocytes" in the mammalian fetus. *Blood*, 2008. **111**(4): 2409–2417.
18. Thom, C.S. et al., Trim58 degrades Dynein and regulates terminal erythropoiesis. *Dev Cell*, 2014. **30**(6): 688–700.
19. Konstantinidis, D.G. et al., Signaling and cytoskeletal requirements in erythroblast enucleation. *Blood*, 2012. **119**(25): 6118–6127.
20. Getman, M. et al., Extensively self-renewing erythroblasts derived from transgenic beta-yac mice is a novel model system for studying globin switching and erythroid maturation. *Exp Hematol*, 2014. **42**(7): 536–546.
21. Beers, E.J. et al., Imaging flow cytometry for automated detection of hypoxia-induced erythrocyte shape change in sickle cell disease. *Am J Hematol*, 2014. **89**(6): 598–603.
22. McGrath, K.E., Utilization of imaging flow cytometry to define intermediates of megakaryopoiesis in vivo and in vitro. *J Immunol Methods*, 2015. **423**: 45–51.
23. Niswander, L.M. et al., Improved quantitative analysis of primary bone marrow megakaryocytes utilizing imaging flow cytometry. *Cytometry A*, 2014. **85**(4): 302–312.
24. Megrelis, L. and J. Delon, Rapid and robust analysis of cellular and molecular polarization induced by chemokine signaling. *J Vis Exp*, 2014. (94).
25. Zhou, Z.H. et al., Oversulfated chondroitin sulfate binds to chemokines and inhibits stromal cell-derived factor-1 mediated signaling in activated T cells. *PLoS One*, 2014. **9**(4): e94402.
26. Molinos, M. et al., Improvement of bovine nucleus pulposus cells isolation leads to identification of three phenotypically distinct cell subpopulations. *Tissue Eng Part A*, 2015. **21**(15–16): 2216–2227.
27. Zuba-Surma, E.K. et al., Morphological characterization of very small embryonic-like stem cells (VSELs) by ImageStream system analysis. *J Cell Mol Med*, 2008. **12**(1): 292–303.
28. Kucia, M. et al., A population of very small embryonic-like (VSEL) CXCR4(+) SSEA-1(+)Oct-4+ stem cells identified in adult bone marrow. *Leukemia*, 2006. **20**(5): 857–869.
29. Miyanishi, M. et al., Do pluripotent stem cells exist in adult mice as very small embryonic stem cells? *Stem Cell Reports*, 2013. **1**(2): 198–208.
30. Heider, A. et al., Murine and human very small embryonic-like cells: A perspective. *Cytometry A*, 2013. **83**(1): 72–75.
31. Benbijja, M., A. Mellouk, and P. Bobe, Sensitivity of leukemic T-cell lines to arsenic trioxide cytotoxicity is dependent on the induction of phosphatase B220/CD45R expression at the cell surface. *Mol Cancer*, 2014. **13**: 251.
32. Arakawa, M. et al., A novel evaluation method of survival motor neuron protein as a biomarker of spinal muscular atrophy by imaging flow cytometry. *Biochem Biophys Res Commun*, 2014. **453**(3): 368–374.
33. Castaneda, J.T. et al., Differential expression of intracellular and extracellular CB(2) cannabinoid receptor protein by human peripheral blood leukocytes. *J Neuroimmune Pharmacol*, 2013. **8**(1): 323–332.
34. Hennies, C.M., M.A. Lehn, and E.M. Janssen, Quantitating MHC class II trafficking in primary dendritic cells using imaging flow cytometry. *J Immunol Methods*, 2015. **423**: 18–28.
35. Garcia-Vallejo, J.J. et al., The consequences of multiple simultaneous C-type lectin-ligand interactions: DCIR alters the endo-lysosomal routing of DC-SIGN. *Front Immunol*, 2015. **6**: 87.

36. Martorelli, D. et al., A natural HIV p17 protein variant up-regulates the LMP-1 EBV oncoprotein and promotes the growth of EBV-infected B-lymphocytes: Implications for EBV-driven lymphomagenesis in the HIV setting. *Int J Cancer*, 2015. **137**(6): 1374–1385.
37. Franzen, C.A. et al., Characterization of uptake and internalization of exosomes by bladder cancer cells. *Biomed Res Int*, 2014. **2014**: 619829.
38. Shimoni, O. et al., Shape-dependent cellular processing of polyelectrolyte capsules. *ACS Nano*, 2013. **7**(1): 522–530.
39. Phanse, Y. et al., Functionalization of polyanhydride microparticles with dimannose influences uptake by and intracellular fate within dendritic cells. *Acta Biomater*, 2013. **9**(11): 8902–8909.
40. D'Mello, V. et al., The urokinase plasminogen activator receptor promotes efferocytosis of apoptotic cells. *J Biol Chem*, 2009. **284**(25): 17030–17038.
41. Smirnov, A. et al., An improved method for differentiating cell-bound from internalized particles by imaging flow cytometry. *J Immunol Methods*, 2015. **423**: 60–69.
42. Phanse, Y. et al., Analyzing cellular internalization of nanoparticles and bacteria by multi-spectral imaging flow cytometry. *J Vis Exp*, 2012. **8**(64): e3884.
43. Lopes, C.D., M. Gomez-Lazaro, and A.P. Pego, Seeing is believing but quantifying is deciding. *Nanomedicine (Lond)*, 2015. **10**(15): 2307–2310.
44. Vranic, S. et al., Deciphering the mechanisms of cellular uptake of engineered nanoparticles by accurate evaluation of internalization using imaging flow cytometry. *Part Fibre Toxicol*, 2013. **10**: 2.
45. Johnston, A.P. et al., Targeting cancer cells: Controlling the binding and internalization of antibody-functionalized capsules. *ACS Nano*, 2012. **6**(8): 6667–6674.
46. Nogueira, E. et al., Enhancing methotrexate tolerance with folate tagged liposomes in arthritic mice. *J Biomed Nanotechnol*, 2015. **11**(12): 2243–2252.
47. Summers, H.D. et al., Statistical analysis of nanoparticle dosing in a dynamic cellular system. *Nat Nanotechnol*, 2011. **6**(3): 170–174.
48. Hazin, J. et al., A novel method for measuring cellular antibody uptake using imaging flow cytometry reveals distinct uptake rates for two different monoclonal antibodies targeting L1. *J Immunol Methods*, 2015. **423**: 70–77.
49. Skovbakke, S.L. et al., The proteolytically stable peptidomimetic Pam-(Lys-β NSpe)$_6$-NH$_2$ selectively inhibits human neutrophil activation via formyl peptide receptor 2. *Biochem Pharmacol*, 2015. **93**(2): 182–195.
50. Hem, C.D. et al., T cell specific adaptor protein (TSAd) promotes interaction of Nck with Lck and SLP-76 in T cells. *Cell Commun Signal*, 2015. **13**: 31.
51. Beum, P.V. et al., Quantitative analysis of protein co-localization on B cells opsonized with rituximab and complement using the ImageStream multispectral imaging flow cytometer. *J Immunol Methods*, 2006. **317**(1–2): 90–9.
52. Erie, A.J. et al., MHC class II upregulation and colocalization with Fas in experimental models of immune-mediated bone marrow failure. *Exp Hematol*, 2011. **39**(8): 837–849.
53. Dias, A.M. et al., Dysregulation of T cell receptor N-glycosylation: A molecular mechanism involved in ulcerative colitis. *Hum Mol Genet*, 2014. **23**(9): 2416–2427.
54. Dias, A.M. et al., Studying T cells N-glycosylation by imaging flow cytometry. *Imag Flow Cytom: Methods Protoc*, 2016. **1389**: 167–176.
55. Silva, A.M. et al., Resveratrol as a natural anti-tumor necrosis factor-alpha molecule: Implications to dendritic cells and their crosstalk with mesenchymal stromal cells. *PLoS One*, 2014. **9**(3): e91406.
56. Cerveira, J. et al., An imaging flow cytometry-based approach to measuring the spatiotemporal calcium mobilisation in activated T cells. *J Immunol Methods*, 2015. **423**: 120–130.
57. Caplan, A.I. and D. Correa, The MSC: An injury drugstore. *Cell Stem Cell*, 2011. **9**(1): 11–15.

58. Singer, N.G. and A.I. Caplan, Mesenchymal stem cells: Mechanisms of inflammation. *Annu Rev Pathol*, 2011. **6**: 457–478.
59. Vivier, E. et al., Functions of natural killer cells. *Nat Immunol*, 2008. **9**(5): 503–510.
60. Rasmusson, I. et al., Mesenchymal stem cells inhibit the formation of cytotoxic T lymphocytes, but not activated cytotoxic T lymphocytes or natural killer cells. *Transplantation*, 2003. **76**(8): 1208–1213.
61. Poggi, A. et al., Interaction between human NK cells and bone marrow stromal cells induces NK cell triggering: Role of NKp30 and NKG2D receptors. *J Immunol*, 2005. **175**(10): 6352–6360.
62. Sotiropoulou, P.A. et al., Interactions between human mesenchymal stem cells and natural killer cells. *Stem Cells*, 2006. **24**(1): 74–85.
63. Spaggiari, G.M. et al., Mesenchymal stem cells inhibit natural killer-cell proliferation, cytotoxicity, and cytokine production: Role of indoleamine 2,3-dioxygenase and prostaglandin E2. *Blood*, 2008. **111**(3): 1327–1333.
64. Spaggiari, G.M. et al., Mesenchymal stem cell-natural killer cell interactions: Evidence that activated NK cells are capable of killing MSCs, whereas MSCs can inhibit IL-2-induced NK-cell proliferation. *Blood*, 2006. **107**(4): 1484–1490.
65. Götherström, C. et al., Fetal and adult multipotent mesenchymal stromal cells are killed by different pathways. *Cytotherapy*, 2010. **13**(3): 269–278.
66. Jewett, A. et al., Strategies to rescue mesenchymal stem cells (MSCs) and dental pulp stem cells (DPSCs) from NK cell mediated cytotoxicity. *PLoS One*, 2010. **5**(3): e9874.
67. Tseng, H.C. et al., Increased lysis of stem cells but not their differentiated cells by natural killer cells; de-differentiation or reprogramming activates NK cells. *PLoS One*, 2010. **5**(7): e11590.
68. Aggarwal, S. and M.F. Pittenger, Human mesenchymal stem cells modulate allogeneic immune cell responses. *Blood*, 2005. **105**(4): 1815–1822.
69. Krampera, M. et al., Role for interferon-gamma in the immunomodulatory activity of human bone marrow mesenchymal stem cells. *Stem Cells*, 2006. **24**(2): 386–398.
70. Selmani, Z. et al., Human leukocyte antigen-G5 secretion by human mesenchymal stem cells is required to suppress T lymphocyte and natural killer function and to induce CD4+CD25highFOXP3+ regulatory T cells. *Stem Cells*, 2008. **26**(1): 212–222.
71. Pradier, A. et al., Human bone marrow stromal cells and skin fibroblasts inhibit natural killer cell proliferation and cytotoxic activity. *Cell Transplant*, 2010. **20**(5): 681–691.
72. Casado, J.G., R. Tarazona, and F.M. Sanchez-Margallo, NK and MSCs crosstalk: The sense of immunomodulation and their sensitivity. *Stem Cell Rev*, 2013. **9**(2): 184–189.
73. Almeida, C.R. et al., Enhanced mesenchymal stromal cell recruitment via natural killer cells by incorporation of inflammatory signals in biomaterials. *J R Soc Interface*, 2012. **9**: 261–271.
74. Almeida, C.R. and D.M. Davis, Segregation of HLA-C from ICAM-1 at NK cell immune synapses is controlled by its cell surface density. *J Immunol*, 2006. **177**(10): 6904–6910.
75. Ahmed, F. et al., Numbers matter: Quantitative and dynamic analysis of the formation of an immunological synapse using imaging flow cytometry. *J Immunol Methods*, 2009. **347**(1–2): 79–86.
76. te Raa, G.D. et al., CMV-specific CD8+ T-cell function is not impaired in chronic lymphocytic leukemia. *Blood*, 2014. **123**(5): 717–724.
77. Wabnitz, G.H. et al., InFlow microscopy of human leukocytes: A tool for quantitative analysis of actin rearrangements in the immune synapse. *J Immunol Methods*, 2015. **423**: 29–39.
78. Markey, K.A. et al., Imaging the immunological synapse between dendritic cells and T cells. *J Immunol Methods*, 2015. **423**: 40–44.

Chapter 3 Live-cell imaging
Seeing is believing

Patrícia Castro and
Patrícia Carneiro

Contents

Introduction

The past several decades have witnessed major breakthroughs in the understanding of life and the molecular underpinnings of biological systems. Emerging advances in live-cell imaging have bridged gaps between biochemical, genomic, and developmental biology approaches, providing critical insights into the molecular dynamics and structure inside living cells. Indeed, the widely used adage "A picture is worth a thousand words" sums up in a few words the importance of microscopy in the field of life sciences research. Developed in the 1600s, the microscope allowed mankind to define cells as basic units of structure and function, as well as to discover bacteria and mitochondria. It has since evolved giving rise to a plethora of imaging techniques, from brightfield and differential interference contrast to modern fluorescence up to the advent of super-resolution microscopy [1].

Super-resolution microscopy has arisen on the need to circumvent the diffraction limit of light, allowing resolution at the nanoscale. Super-resolution fluorescence techniques have evolved greatly and three main approaches are currently used to tackle the diffraction barrier of ~200 nm: (1) stimulated emission depletion (STED) microscopy, (2) structured illumination microscopy (SIM), and (3) single-molecule localization microscopy (SMLM) [2].

Approaches

Imaging of live cells and tissues nowadays spans all fields of life sciences research and it even extends to physical sciences. However, optimization of image-acquisition conditions is required to achieve high-quality images while preserving healthy cells, thus harnessing the full potential that live-cell microscopy offers. Further, when choosing a microscopy system for imaging live cells four things should be considered: (1) sensitivity of detection, (2) speed of acquisition, (3) resolution, and (4) specimen viability.

Live cells are very sensitive to light, which is toxic in high amounts yielding artifacts. Moreover, photodamage starts gradually before obvious signs occur [3,4]. When performing live-cell imaging, the experimental settings must be finely optimized to minimize heat and avoid excitation with short wavelengths. Further optimization involves applying dyes at the lowest concentration possible, acquiring images at the lowest possible resolution to image specific details, using high-sensitive detectors in order to optimize the photon yield, namely through number of collected signals versus *incoming* photons, and choosing efficiency fluorescence filters and a shutter system to minimize the exposure of the cells to harmful excitation light.

Light transmission microscopy

The main technique used in light microscopy for the past three centuries, light transmission brightfield microscopy, relies on the changes in light absorption. As light passes through the specimen, regions that differentially absorb the light generate variations in light intensity that when the rays are gathered and focused by the objective results in an image with contrast. Resolution in a brightfield system depends on both the objective and condenser numerical apertures, and an immersion medium is often required on both sides of the specimen for numerical aperture combinations exceeding a value of 1.0.

Unstained specimens, such as live cells, are relatively transparent and thus nearly invisible through brightfield microscopy. To surpass this lack of contrast, several optical contrast techniques have been developed (e.g., phase contrast, differential interference contrast, polarized light,

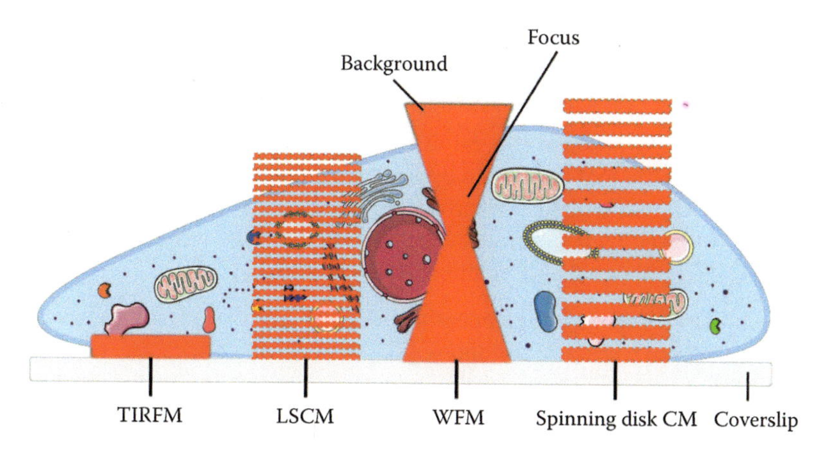

Figure 3.1 Live-cell fluorescence-imaging approaches. (Adapted from Education in Microscopy and Digital Imaging, Carl Zeiss http://zeiss-campus.magnet.fsu.edu/articles/livecellimaging/techniques.html.)

and darkfield microscopy) (Figure 3.1). These techniques rely on changes in intensity, phase, polarization, and direction of light when it interacts with the specimen.

Widefield fluorescence microscopy

Widefield fluorescence microscopy (WFM) relies on the epi illumination of the whole field of view, with excitation light and detection of emitted light from that area recorded at once on the chip of a high-sensitive camera (charge-coupled device [CCD], *electron-multiplying charge-coupled device* [EMCCD], or scientific complementary metaloxide semiconductor [sCMOS]). Widefield microscopy achieves its best results when imaging cells in cultures or thin samples, since the detection of out-of-focus light reduces the image contrast when imaging thick samples.

The application of genetically encoded fluorescent proteins has dramatically expanded the capabilities of live-cell imaging in fluorescence microscopy by enabling researchers to precisely target subcellular regions-of-interest (ROI) [5].

Laser scanning confocal microscopy

Laser scanning confocal microscopy (LSCM) offers several advantages over widefield fluorescence microscopy, namely controllable depth of field, the elimination of out-of-focus information, and the ability to collect serial optical sections from thick specimens [6]. To increase the specific signal-to-noise ratio for the features of interest, LSCM rejects the light from other focal planes, despite the fact that absolute signal values are way below those observed with widefield microscopy.

However, it is challenging to optimize LSCM settings when working with live tissue due to localized phototoxicity caused by the highly focused laser light [4].

Further, WFM systems can be faster and more sensitive than LSCM systems, as light that is rejected by the pinhole on a confocal microscope is collected on a widefield microscope. As such, for some applications, WFM followed by restorative deconvolution might be more appropriate than confocal laser-scanning microscopes (CLSM).

For in-depth information concerning CLSM for live-cell microscopy please refer to a specific handbook [7].

Spinning disk confocal-microscopy

The spinning disk confocal microscope is a multipoint LSCM, which captures dynamic events in a wide spectrum of timescales. The spinning disk spreads the illumination over a larger area, which in turn will be more efficient in avoiding phototoxicity comparing to LSCM. Moreover, data are collected with an EMCCD array from simultaneous locations using a disk that has either slits or thousands of pinholes, allowing light to selectively excite fluorescence in multiple regions of the sample and thus speeding image acquisition [8].

Multiphoton microscopy

Multiphoton fluorescence excitation is inherently confocal given that it is based on the simultaneous absorption of more than one near-infrared (NIR) photon of light by a single fluorophore molecule that only occurs in the highly photon-dense laser focal volume [9].

Multiphoton confocal laser-scanning microscopes (MP-CLSM) present unique advantages for intravital imaging in thick tissues, in live animals or in tissue-slice cultures [10,11].

MP-CLSM is more efficient in avoiding chromatic aberrations because, rather than focusing multiple lasers of different colors, a single laser excitation can be used for multiple fluorophores. To minimize photobleaching in the focal plane and subsequent phototoxicity, highly localized excitation energy is used, thus increasing sample viability and the duration of experiments in live tissues.

In addition, the application of NIR excitation wavelengths reduces the high degree of light scattering that is observed at shorter wavelengths, allowing deeper penetration into biological materials, and thus enabling imaging approaches on thick living tissue samples, such as brain slices and developing embryos. Strikingly, optimization of scattered emissions and the use of ultrashort-pulsed lasers emitting brief but high-energy pulses of NIR light have endowed multiphoton imaging in intact brain to be close to 1 mm depth [12,13].

Further, intravital imaging of genetically encoded fluorescence proteins combined with probe-free second-harmonic imaging that generates signals directly in ordered protein structures, such as collagen fibers, has provided ground-breaking insights in cancer research, such as concerning migration of tumor cells and the relationship between podosomes and invadopodia [14]. However, MP-CLSM is an expensive tool that requires specialized expertise, and for which many dyes have not yet been characterized.

Applications

Imaging gene expression in live cells

The advent of green fluorescent proteins (GFPs) and its derivatives made the use of fluorescence in live-cell imaging a routine technique.

Imaging tissues from mice following injection with tumor cell lines expressing GFP proved useful in detecting the location of tumors and its metastases [15]. GFP-transfected tumor cells are fundamental tools for real-time visualization of tumor growth and metastases, as well as to study anticancer drug effects in tumor inhibition, which can be imaged and quantified for rapid anticancer drug screening [16].

Although fluorescence optical imaging has obvious inherent disadvantages such as poor tissue penetration, there are always new imaging tools and instrumentation emerging, as well as new fluorescent proteins and dyes [17,18], awarding optical imaging an irreplaceable place in basic and preclinical research.

Fluorescence recovery after photobleaching—monitoring protein dynamics

In fluorescence recovery after photobleaching (FRAP), the protein of interest is attached to a fluorescent protein, then a ROI is exposed to high intensities of light destroying the fluorescence in that particular region and the recovery of signal is recorded along the time. The fluorescence recover into the bleached area will depend on the movement of the protein of interest from other cellular areas, providing researchers with an insight into the intracellular transport dynamics.

Using FRAP for visualizing protein movement can be challenging because photobleaching can lead to the creation of free radicals, which causes damage to cells. In order to overcome this problem, one should limit the photobleached area to a minimum.

FRAP techniques have proven very useful in evaluating vesicular trafficking of proteins in neurons, where Nakata et al. reported reduction of background signal from stationary proteins [19].

Förster resonance energy transfer—quantifying protein–protein interactions

Protein–protein interactions play a fundamental role in regulating cellular signal transduction pathways. Those interactions might be weak or transient making biochemical approaches very challenging. Förster resonance energy transfer (FRET) is useful for imaging and quantifying molecular activity, such as protein–protein interactions, protein–DNA interactions, and protein conformational changes. In order for FRET to occur the two fluorescent molecules have to be very close. The energy flows from one molecule (donor) to the other (acceptor) by nonradiative transfer.

FRET is usually performed using derivatives of GFP such as cyan fluorescent protein (CFP) and yellow fluorescent protein (YFP), each is attached to proteins of interest. Once the proteins in study are closer than 10 nm, the CFP will act as a donor and transfer the energy that is emitted in the form of light to the YFP, which acts as an acceptor. The measurements are made in terms of a shift from blue fluorescence emitted from the CFP to yellow fluorescence emitted from the YFP.

FRET has been used to establish the microdomains in the plasma membrane and in other membranes, although the plasma membrane is the most accessible and easy-to-observe in an intact cell [20].

Total internal reflection fluorescence microscopy—observing processes close to the cell membrane

To observe events located close to or in the plasma membrane of a cell, researchers can use total internal reflection fluorescence microscopy (TIRFM). TIRF microscopy uses an evanescent field for fluorochrome excitation, which only penetrates the cell 60–200 nm, providing a good Z-resolution without the interference of the fluorescence from molecules inside the cell.

TIRF has proved useful in the study of cell adhesion and migration [21,22], exocytosis and endocytosis [23,24]. TIRF presents a reduced phototoxicity because only those molecules that are close to the cover glass will be excited.

When using confluent cell monolayers, TIRF illumination can lead to artifacts, which can be overcome by the use of specific filters to avoid laser-light reflections and interference patterns.

Fluorescence lifetime imaging microscopy—spatial measurements in live cells

In fluorescence lifetime imaging microscopy (FLIM), fluorescence decay (lifetime) of a fluorophore is measured in the frequency domain by a modulated light source and detector; or in the time domain using time-correlated single-photon counting. FLIM is not influenced by photobleaching and concentration variation, because the fluorescence lifetime is determined for each pixel in a field of view. This method is very useful in the research of extracellular and intracellular environmental modifications as they lead to an alteration of the fluorescence lifetime.

FLIM is also used to track changes in local tissue microenvironment, such as those in pH and in O_2 and Ca^{2+} concentrations, since fluorescence lifetimes are affected by the transfer of energy from a fluorophore to its surroundings [25].

Future

Super-resolution microscopy

Super-resolution microscopy has provided important insights of biological structures and processes, by allowing resolution beyond conventional fluorescence microscopy. Indeed, through a focused laser beam *STED*, it is possible to overcome the diffraction limit by sharpening the point-spread function of the microscope [26].

STED can be successfully applied to assess dynamic processes; several live samples to study slow morphing; and movements of organelles such as reticulum endoplasmic or microtubules [27], subcellular organization [28], and synaptic structures [29] in living cells.

SIM is a widefield-based microscopy approach, which is used to enhance spatial resolution. SIM is compatible with standard fluorophores and multiplexing protocols [30].

Despite its many advantages that make it useful for live cell and 3D imaging, including rapid image acquisition at low doses of light, superresolution structured illumination microscopy (SR-SIM) provides only an approximately two-fold enhancement over diffraction-limited resolution [31].

Throughout the past years, the emergence of super-resolution approaches has allowed the investigation of subcellular organizations at the nanometer-scale resolutions. Despite that, some of these methods can now be performed in many research facilities; we hope that in the near future they will be routinely used in most settings.

References

1. Fujita, K., Follow-up review: Recent progress in the development of super-resolution optical microscopy. *Microscopy (Oxford)*, 2016. **65**(4): 275–281.
2. Turkowyd, B., D. Virant, and U. Endesfelder, From single molecules to life: Microscopy at the nanoscale. *Anal Bioanal Chem*, 2016. **408**(25): 6885–6911.
3. Hopt, A. and E. Neher, Highly nonlinear photodamage in two-photon fluorescence microscopy. *Biophys J*, 2001. **80**(4): 2029–2036.

4. Koester, H.J. et al., Ca2+ fluorescence imaging with pico- and femtosecond two-photon excitation: Signal and photodamage. *Biophys J*, 1999. **77**(4): 2226–2236.
5. Day, R.N. and M.W. Davidson, The fluorescent protein palette: Tools for cellular imaging. *Chem Soc Rev*, 2009. **38**(10): 2887–2921.
6. Shaw, P.J., Comparison of widefield/deconvolution and confocal microscopy for three-dimensional imaging, in *Handbook of Biological Confocal Microscopy*, J. Pawley (Ed.), New York: Springer. pp. 453–467, 2006.
7. Dailey, M.E., Manders, E., Soll, D.R. and M. Terasaki, Confocal microscopy of living cells, in *Handbook of Biological Confocal Microscopy*, J.B. Pawley (Ed.), New York: Springer. pp. 381–403, 2006.
8. Ichihara, A., Tanaami, T., Isozaki, K., Sugiyama, Y., Kosugi, Y., Mikuriya, K., Abe, M. and I. Uemura, High-speed confocal fluorescence microscopy using a Nipkow scanner with microlenses for 3D imaging of single fluorescent molecule in real time. *Bioimages*, 1996. **4**: 52–62.
9. Oheim, M. et al., Principles of two-photon excitation fluorescence microscopy and other nonlinear imaging approaches. *Adv Drug Deliv Rev*, 2006. **58**(7): 788–808.
10. Piston, D.W., The coming of age of two-photon excitation imaging for intravital microscopy. *Adv Drug Deliv Rev*, 2006. **58**(7): 770–772.
11. Rocheleau, J.V. and D.W. Piston, Two-photon excitation microscopy for the study of living cells and tissues. *Curr Protoc Cell Biol*, 2003. Chapter 4: Unit 4 11.
12. Helmchen, F. and W. Denk, Deep tissue two-photon microscopy. *Nat Methods*, 2005. **2**(12): 932–940.
13. Wilt, B.A. et al., Advances in light microscopy for neuroscience. *Annu Rev Neurosci*, 2009. **32**: 435–506.
14. Yamaguchi, H., J. Wyckoff, and J. Condeelis, Cell migration in tumors. *Curr Opin Cell Biol*, 2005. **17**(5): 559–564.
15. Hoffman, R.M., Visualization of GFP-expressing tumors and metastasis in vivo. *Biotechniques*, 2001. **30**(5): 1016–1022, 1024–1026.
16. Luker, G.D. et al., In vitro and in vivo characterization of a dual-function green fluorescent protein—HSV1-thymidine kinase reporter gene driven by the human elongation factor 1 alpha promoter. *Mol Imaging*, 2002. **1**(2): 65–73.
17. Shaner, N.C. et al., Improved monomeric red, orange and yellow fluorescent proteins derived from Discosoma sp. red fluorescent protein. *Nat Biotechnol*, 2004. **22**(12): 1567–1572.
18. Shu, X. et al., Mammalian expression of infrared fluorescent proteins engineered from a bacterial phytochrome. *Science*, 2009. **324**(5928): 804–807.
19. Nakata, T., S. Terada, and N. Hirokawa, Visualization of the dynamics of synaptic vesicle and plasma membrane proteins in living axons. *J Cell Biol*, 1998. **140**(3): 659–674.
20. Kusumi, A., I. Koyama-Honda, and K. Suzuki, Molecular dynamics and interactions for creation of stimulation-induced stabilized rafts from small unstable steady-state rafts. *Traffic*, 2004. **5**(4): 213–230.
21. Adams, M.C. et al., Signal analysis of total internal reflection fluorescent speckle microscopy (TIR-FSM) and wide-field epi-fluorescence FSM of the actin cytoskeleton and focal adhesions in living cells. *J Microsc*, 2004. **216**(2): 138–152.
22. Choi, C.K. et al., Actin and alpha-actinin orchestrate the assembly and maturation of nascent adhesions in a myosin II motor-independent manner. *Nat Cell Biol*, 2008. **10**(9): 1039–1050.
23. Nagamatsu, S. and M. Ohara-Imaizumi, Imaging exocytosis of single insulin secretory granules with TIRF microscopy. *Methods Mol Biol*, 2008. **440**: 259–268.

24. Schneckenburger, H., Total internal reflection fluorescence microscopy: Technical innovations and novel applications. *Curr Opin Biotechnol*, 2005. **16**(1): 13–18.
25. Chen, H.M. et al., Time-resolved autofluorescence spectroscopy for classifying normal and premalignant oral tissues. *Lasers Surg Med*, 2005. **37**(1): 37–45.
26. Hell, S.W. and J. Wichmann, Breaking the diffraction resolution limit by stimulated emission: Stimulated-emission-depletion fluorescence microscopy. *Opt Lett*, 1994. **19**(11): 780–782.
27. Hein, B., K.I. Willig, and S.W. Hell, Stimulated emission depletion (STED) nanoscopy of a fluorescent protein-labeled organelle inside a living cell. *Proc Natl Acad Sci USA*, 2008. **105**(38): 14271–14276.
28. Westphal, V. et al., Video-rate far-field optical nanoscopy dissects synaptic vesicle movement. *Science*, 2008. **320**(5873): 246–249.
29. Nagerl, U.V. et al., Live-cell imaging of dendritic spines by STED microscopy. *Proc Natl Acad Sci USA*, 2008. **105**(48): 18982–18987.
30. Godin, A.G., B. Lounis, and L. Cognet, Super-resolution microscopy approaches for live cell imaging. *Biophys J*, 2014. **107**(8): 1777–17784.
31. Gustafsson, M.G., Surpassing the lateral resolution limit by a factor of two using structured illumination microscopy. *J Microsc*, 2000. **198**(2): 82–87.

Chapter 4 Atomic force microscopy

A tool for imaging, manipulating biomolecules, and quantifying single-molecule interactions in biological systems

Catarina S. Lopes,
Filomena A. Carvalho,
and Nuno C. Santos

Contents

Introduction

Atomic force microscopy (AFM) is a technique that can be used for high-resolution real-time studies of properties of molecules at the nanoscale. It has been applied to structural and functional studies of dynamic processes in biological systems.[1–4] AFM has contributed significantly to the nanotechnology field, with developments in nanodiagnostic and therapeutics, contributing to the improvement of the state of the art in health care,[5] pathology,[6,7] identification of biomolecules[8] and cellular micro-environments, physiology and functions of living cells and organisms,[9] understanding of the structure–function relationship of DNA,[10] cell and molecular recognition, signaling, adhesion, and fusion.[11] AFM is a tool that allows imaging and force probing biological samples with nanometer (nm)[3] topographical and piconewton (pN) force resolution.[12] This microscope central element is a small cantilever, which distance from the sample is controlled in the z-axis by a piezoelectric crystal. A sharp tip is usually mounted at the end of the microfabricated flexible cantilever (typically 20–200 μm long), used as a kind of spring to measure the force between the tip and the sample surface.[2,13] AFM is a surface probe technique, which uses the tip of this soft cantilever to image surfaces, or to measure or apply forces while interacting with them.[14] The cantilever behaves similarly to a Hookean spring when the AFM is used for biological applications; it has spring constants on the order of 0.1 N/m.[15] On bringing the tip in close proximity to the sample surface, attractive and repulsive forces cause the deflection of the cantilever.[3,16] These deflections cause changes in the position of the reflection of the light beam pointed to the cantilever's top surface. This information may be translated into applied force through Hooke's Law: $F = -K\Delta x$. The cantilever deflection *versus* scanner displacement curve data can be converted into a force–distance curve[17]. Consequently, these means the force (F) needed to extend the cantilever with spring constant (K) depends in a linear way on the cantilever deflection (Δx).[16,17] This deflection is detected and processed as a function of the position on the (x, y) plane.[3,5,17–22] The bending and torsion of the cantilever are measured based on the light beam reflection point of incidence on a four-quadrant photodetector.[16,23–27] An optical lever is created by this light beam, from a laser or light-emitting diode, reflected by the cantilever onto the position-sensitive photodiode.[13] The tip movement relative to the sample can follow different operation modes, including contact mode (in which the tip is in continuous contact with the sample's surface), noncontact mode (the cantilever vibrates and variations from its resonance frequency are used to generate images), and intermittent contact (the cantilever moves rapidly, with a large oscillation, between repulsive and attractive forces).[3,16] When the tip contacts the sample, the deflection of the cantilever provides information on the mechanical properties of the sample.[27,28] AFM has other applications, based on force spectroscopy, where attractive and repulsive forces are measured (Figure 4.1).[24]

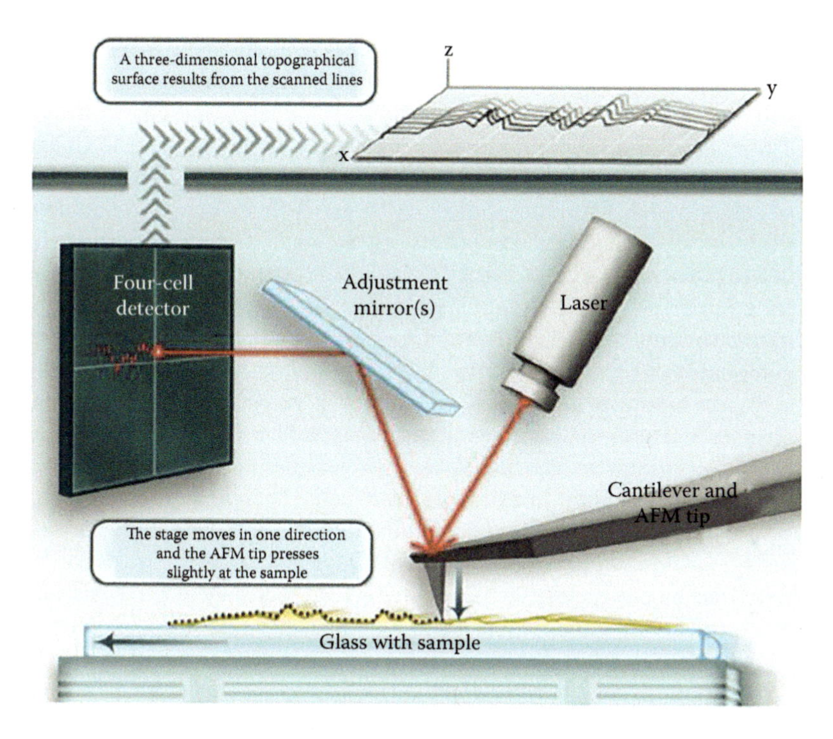

Figure 4.1 Schematic illustration of the imaging mode using a sample scanner AFM: A laser beam is reflected from the cantilever, the interaction forces of the tip with the piezo-positioned samples are deflected and the adjustment mirror directs the signaling for a segmented photodetector. A pseudo three-dimensional topographical image of the sample surface results from the scanned lines. (From Schön, P., *Methods*, 103, 25–33, 2016.)

Applications

AFM can be applied in various modalities to study different biological systems. In addition to imaging, one of the most promising areas of applications is the quantification of the forces resulting in the interaction between the tip (or something attached to it) and a sample, taking advantage of the pN sensitivity of the equipment. These approaches are generally termed *force spectroscopy*. By measuring the variations of the force exerted on the sample, AFM enables the detection of specific interaction forces at the single-molecule level. The possibility of modifying the surface and manipulating individual molecules made AFM an ideal tool for biological and biomedical applications.[5]

Imaging

AFM has been applied on the characterization of nanometric features at the surfaces of the samples and also on the mapping of the spatial distribution of their physicochemical properties. These equipments have the capacity to visualize nanotopographical features and correlate them to surface-charge density and potential relationships between topography and local characteristics of biomaterials (e.g., charge and conductivity). The contact and oscillation modes are most commonly used to image reconstituted membrane proteins and native membranes. AFM imaging of cells generates structural information about them, enabling the identification of functional components within such structures, despite the heterogeneity of the cell surface in terms of protein composition and distribution.[3,15,16,29,30] AFM imaging of living mammalian cells remains limited to resolutions in the 50–100 nm range, meaning that the individual components of the cell-surface machinery cannot be observed. Living cells are highly dynamic, and continuously respond to environmental changes.[30] These methodologies can be applied to blood cells, such as human erythrocytes and platelets from patients or healthy donors.[2,5,8] They are also used for imaging of cancer cells, including lymphoma Raji cells[31] and human lung adenocarcinoma cells[32] and, recently, neuronal cells from mice with chronic epilepsy.[33] Other typical applications can be found for biomolecules, such as proteins (e.g., fibrinogen[13]), DNA,[10] and RNA (including its assemblies and aggregates[12]), and microorganisms, such as viruses, their membrane and capsid,[34] bacteria[35,36] and mycobacteria.[30] Membrane channels, such as potassium channels and microfilaments (polymerization and depolymerization),[29] phospholipid bilayers, and their perturbation by antibiotics,[37] viral fusion inhibitors,[38,39] singlet oxygen production[34] or antibodies,[40] and antimicrobial peptides[41–43] were also studied by these approaches (Figure 4.2).

Atomic force microscopy-based force spectroscopy

AFM-based force spectroscopy allows the measurement of inter- and intramolecular interaction forces required to separate the tip from the sample. It is seldom used to quantify the interaction between the tip and a specific spot of the sample.[5,13,17]

The interaction force depends on the nature of the sample, the probe tip, and the distance between them. The force–distance curve depends on these characteristics and of the medium composition, and there are obvious differences between curves obtained in air and in liquid medium. In force spectroscopy measurements, the cantilever moves in the vertical direction (z-axis) toward the surface and then in the opposite direction. During this procedure, the cantilever deflection as a function of the vertical displacement of the piezoscanner can be recorded. The result is a cantilever-deflection *versus* scanner-displacement curve, which can be converted into a force–distance curve after applying the Hooke's law of elasticity.[17] To calculate the quantitative parameters, an accurate calibration of the spring constant of the cantilevers used is necessary.[13]

This technique is a highly sensitive, rapid, and low operation cost nanotool for the diagnostic and unbiased functional evaluation of the

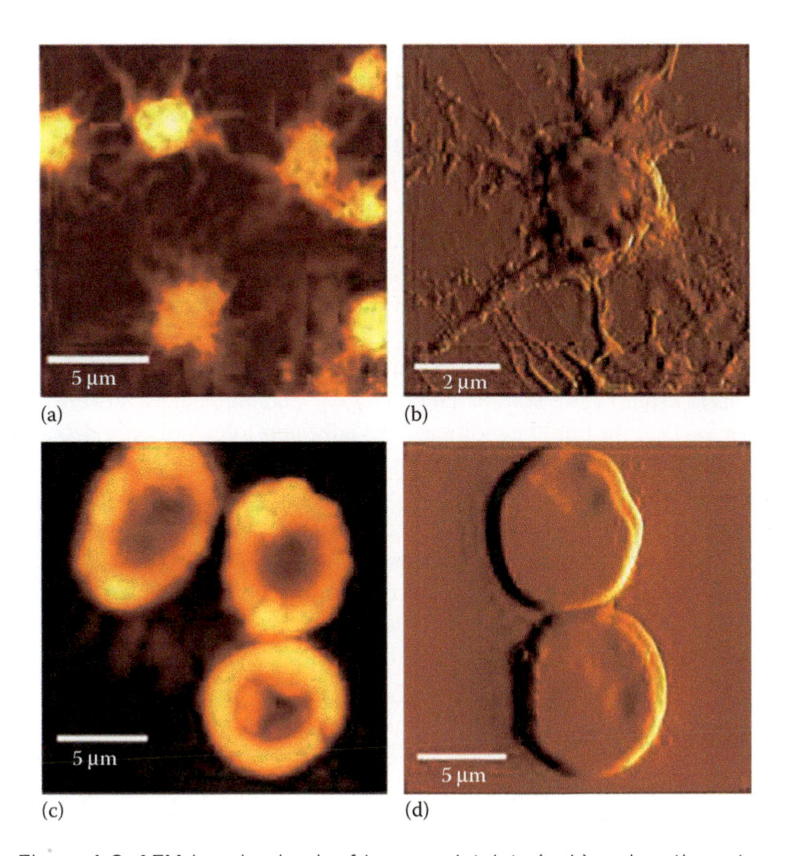

(a) (b) (c) (d)

Figure 4.2 AFM imaging in air of human platelets (a, b) and erythrocytes (c, d) from healthy donors. (a) and (c) are height images, whereas (b) and (d) are error signal images. (Adapted from Carvalho, F.A. et al., *ACS Nano,* 4, 4609–4620, 2010.)

severity of hematological diseases arising from genetic mutations. AFM-based force spectroscopy can be used to measure the binding force between fibrinogen and cell receptors, a strategy successfully used on the identification of the fibrinogen receptor on human erythrocytes.[8] Different types of forces can be studied using AFM-based force spectroscopy, both attractive and repulsive.[24]

The cycle begins with the tip away from the cell surface, and the forces measured by the cantilever deflection can change as it is moved toward or away from the sample on its neutral position, at 0 pN of force, from which it starts moving down toward the surface, reaching the contact point. On the approach curve, the van der Waals interactions are the main type of force present on the approaching of two hard surfaces in the absence of long-range interactions. This force is characterized by a small deflection of the cantilever, at the approach curve, before the contact point. The jump to contact causes instability in the position of the cantilever because it occurs when the gradient of force between the tip and the sample exceeds the stiffness of the cantilever. If the approach curve has a smooth and exponentially increasing repulsive force, it is expectable that electrostatic or polymer-brush forces are present. Viscoelastic properties of some biological systems have been determined using AFM force curves. A waiting time should be kept before starting the retraction curve. Afterward, the tip and cantilever begin the upward movements away from the sample, in the opposite direction, reaching the contact/adhesion point. In a retraction curve where adhesion occurs, the force depends on the sample and appears as a deflection of the cantilever below the zero-deflection line. The central basis of adhesion forces is the development of a capillary bridge between the tip and the sample. This capillary force depends on whether the measurements are made in air or in liquid. In air, samples usually have several nanometers of water molecules adsorbed to their surface. In liquid conditions, the adhesion force not only depends on the interaction energies between the tip and the sample, but also on the solution used. If no binding occurs between the molecules attached to the tip and the cell surface, the tip continues its upward movements and reaches back the neutral position at a defined z-distance. If a bond is formed between the tip and the sample, as the cantilever moves upward, it bends down to negative values in force. Cantilever stiffness depends on the shape and on the material properties of the cantilever.[16,24]

AFM-based force spectroscopy can be used for diverse applications.

Single-molecule interactions

Single-molecule force spectroscopy (SMFS) is an extremely powerful tool for detecting and localizing single-molecule recognition events, and for exploring the energy landscape of molecular interactions. Different molecules, or their domains, can be attached to the tip, and each part can break contact separately or altogether. After attaching a molecule to the tip and/or to the substrate surface, the unfolding, stretching, or adhesion of single molecules can be studied. The tips are commonly functionalized with one or a small amount of probe molecules. These molecules can recognize a specific target molecule on the sample surface. When analyzing the stretching of biomolecules, and in order to measure specific and strong interactions between tip and sample, it is necessary to specifically attach to the tip the biomolecule under study. The attachment should be firm enough to avoid reallocations, but it should maintain some autonomy of the molecule to change its conformation during or before the interaction. In force measurements, both the tip on the cantilever and the surface can be chemically modified to form specific and strong bonds.[24] AFM has been used as a nanodiagnostic tool for patient cells, such as the interaction between fibrinogen and erythrocytes in chronic heart failure patients, which affects blood microcirculatory flow conditions.[44] With AFM-based force spectroscopy, fibrinogen–erythrocyte interactions were assessed at the single-molecule level using the adhesion profiles obtained on each force curve to evaluate the increased cardiovascular risk.[5,8,44,45]

Another study by this methodology was the assessment of the cranberry juice constituents that most strongly influence *Escherichia coli* adhesion forces and adhesive properties of an antibiotic-resistant clinical bacterial strain.[46] Studies of folding and unfolding kinetics of proteins on the membrane of the same bacteria were also successfully conducted.[24,36] It was possible to compare the barriers observed for unfolding from the N-terminus with those from the C-terminus. The barrier positions and heights found in bacteriorhodopsin when probed from both sides were located not only in or at the ends of stiff α-helical rods, but also in loops that are not well resolved by other structural biology methods.[28] This approach was also used for lipid bilayers of dioleoylphosphatidylcholine (DOPC), for therapeutic drugs or protein molecules that target specific receptors.[11] It was also used for understanding the molecular determinants and identifying the ligand for the dengue virus capsid protein on intracellular lipid droplets and plasma lipoproteins (Figure 4.3).[17,47–50]

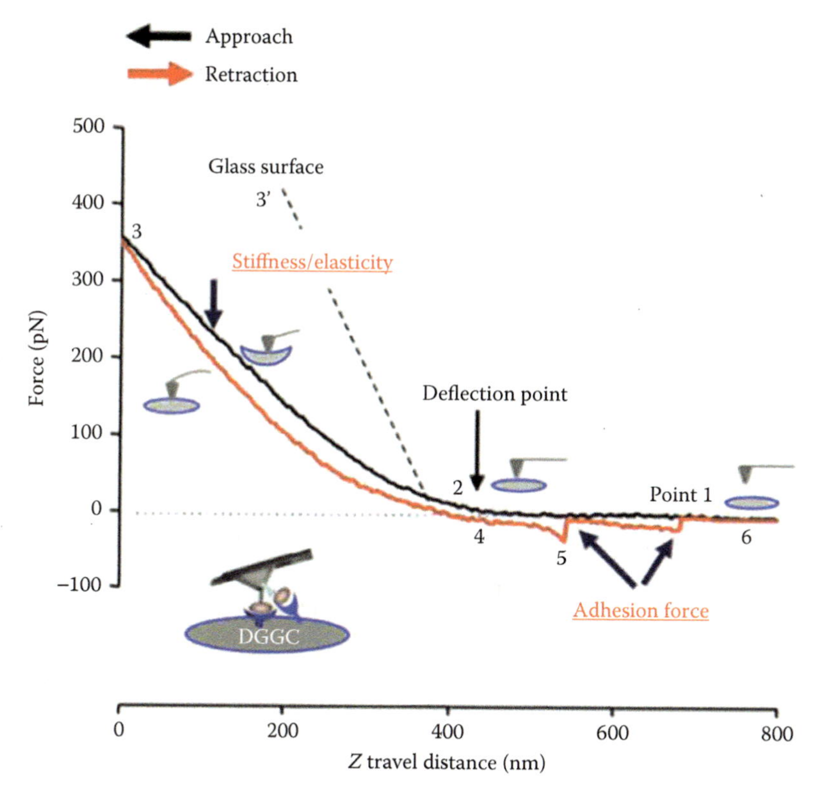

Figure 4.3 Example of the force curves data from fibronectin (FN)-coated AFM probe on dentate gyrus granule cells (DGGC) approach/attaching (black trace) and retract/withdrawal (red trace) from single DGGC. The stages of attaching and withdrawal are shown on points 1–6. On the approach curve, it is possible to measure the stiffness or/and elasticity from the sample. On the retraction curve, it is possible to measure the adhesion force or interaction between cell and molecule under study. (Adapted from Wu, X. et al., *Front. Aging Neurosci.*, 8, 1–12, 2016.)

Indentation

Nanoindentation experiments, for cell elasticity assessment, can be carried out on live cells. AFM-based detection of stiffness is highly dependent on the appropriate use of theoretical models. Cell elasticity can be measured by nanoindentation with the most common AFM cantilevers, via the Hertzian theory. This theory of elastic contact is the most widely used approach to estimate the elastic properties of cells from force indentation curves, using the depth of indentation to assess elasticity in terms of the Young's elastic modulus. AFM can measure the apparent stiffness of mammalian cells, which ranges between 1 to 10 kPa, and cells with cell wall (~100 to 1000 kPa).[9] The stiffness measurements depend on cell type, on probe geometry, rate of force application, and force magnitude.[51] Specifically, this method allows characterizing the elasticity of biological structures, comparing different types of cells or even organelles, matching experimental conditions concerning the indenters' shape, or the thickness of the sample.[17]

Indentation was used for the tensile testing and bulge testing to determine the elastic modulus of the cornea and other eye components.[52] Erythrocytes from patients with type 2 diabetes demonstrated significant aggregation of surface proteins, increased tip–cell adhesion, as well as increased stiffness in comparison with healthy cells.[51] Erythrocytes from chronic heart failure patients are also stiffer than those from healthy donors.[44] This method can be applied in human lung adenocarcinoma cells in different medium conditions for anticancer effect studies.[32]

Single-cell adhesion studies

AFM-based single-cell force spectroscopy (SCFS) is used to quantify the contribution of cell-adhesion molecules to the binding of cells to specific substrates at both the cell and single-molecule level.[53] In AFM-based SCFS, a single cell is attached to a cantilever, commonly facilitated by an adhesive coating (e.g., concanavalin A, poly-L-lysine, or CellTak). The attached cell is lowered (in the approach) onto a substrate, which can be a protein-coated surface, another cell or a biomaterial, until a set force is reached. Then, the cell is kept stationary for a set time to allow the formation of adhesive interactions. During the subsequent retraction of the of the cantilever, the force acting on the cell and the distance between cell and substrate are recorded in a force–distance curve. The force range that can be detected with AFM-based SCFS is typically from ~10 pN up to ~100 nN; thereby, SCFS allows both the overall cell adhesion and the contribution of single adhesion receptors to be quantified.[53] During initial cantilever retraction, the upward force acting on the cell increases until the force needed to initiate cell deadhesion is reached and unbinding events occur. At the cellular level, single cells can be ingeniously fixed onto the cantilever, becoming the probe for the dynamic quantification of cell–substrate interactions. Parameters such as maximum detachment force and work necessary for the entire cell detachment, as well as the number of detachment events of single cellular tethers have already been successfully quantified. The maximum force is termed as the adhesion force and is a measure of how strong the cell adhered to the substrate. Single unbinding events can be characterized from individual rupture (jumps) or tether events depicted on the force–distance curve. The analysis of these unbinding events may be used to characterize the strength of single bonds and cell membrane properties.[54,55]

Cell adhesion is a fundamental aspect on both health and disease. In the past decade, single-cell adhesion studies have contributed to the understanding of adhesion proteins and their regulation. AFM-based SCFS has been used to quantify adhesion of numerous cell types to a diverse set of substrates, including extracellular matrix (ECM) proteins, biomaterials, and cell–cell adhesion proteins.[54] SCFS has been applied as a tool to quantify cell adhesion between two cells to identify key proteins regulating the differential adhesive behavior of zebrafish mesendodermal progenitor cells to fibronectin, thus providing an insight into the germ layer formation and separation between gastrulation zebrafish cells.[53,56] It has also contributed for the understanding of differential cell adhesion and cell-cortex tension in germ layer organization in Chinese hamster ovary (CHO) cells with integrin activators.[53] AFM allows recording of the actual adhesion force between a bacterium (*Pseudomonas aeruginosa*) and *Candida albicans.* Bacterial adhesion to hyphae was always accompanied by strong adhesion forces, but did not occur on yeast cells (Figure 4.4).[57]

Combining with fluorescence studies

AFM has been used simultaneously with optical-fluorescence imaging in cell biology. These techniques are usually combined by successive applications and led to the implementation of new modalities in AFM to allow easy image superposition. AFM indentation can be performed while recording a fluorescence signal. AFM combined with fluorescence microscopy has helped to determine the forces that the immune synapse exert between immune cells, such as photostimulated T-cells, gaining insight on their mechanical properties, when activating a small GTPase, Rac, in real time.[14] Approaches combining AFM with fluorescence microscopy were also used to characterize peptide-planar–lipid bilayer interactions[39] and to study hepatocytes infected by the malaria parasite.[58]

Scanning probe lithography

AFM lithography is a method in which it is possible to draw a pattern on a solid surface, by scratching, oxidation, or reduction, using the AFM tip. In this method, selected molecules are adsorbed by capillarity onto the

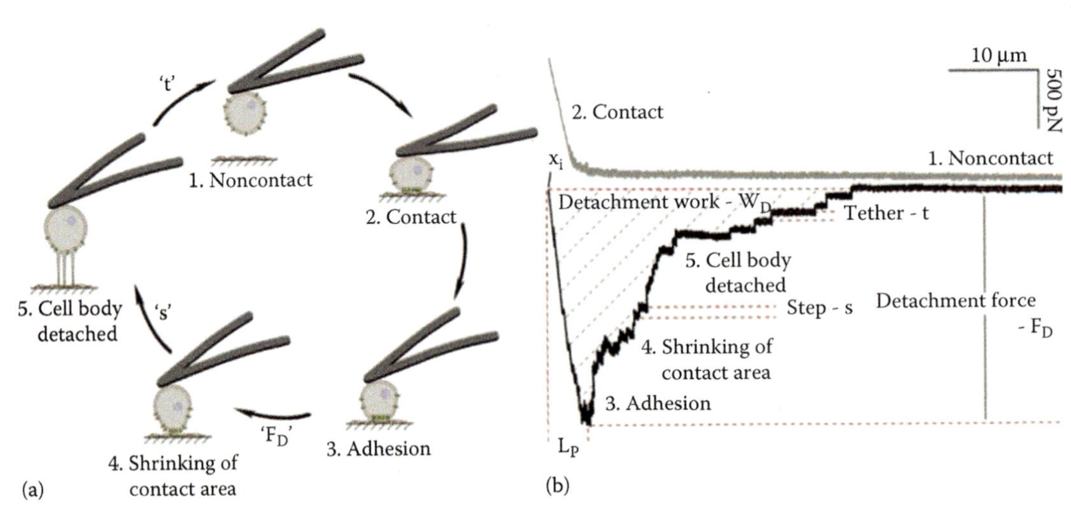

Figure 4.4 Schematic illustration of a single-cell force spectroscopy experiment and of the adhesion events detected (a, b). In (b), the force–distance curve shows steps 1–5, corresponding to those outlined in (a). The AFM cantilever catches a single cell (1. noncontact [a, b]) and approaches to a substrate (1–2, a, b). In contact (2, gray line in b), cell adhesion molecules diffuse into the contact zone. The adhesive force between the cell and substrate increases. After a predefined contact time, the cell is retracted (black line in b) and the cantilever bends due to the adhesive force between cell and substrate (3). Once the force of the cantilever exceeds that of the interactions between cell and substrate, the cell starts to detach (3–4). The force at this point corresponds to the maximum detachment force (F_D). On further retraction of the cantilever, the contact area between the cell and substrate shrinks (4). The cell sequentially detaches from the substrate (5) and this force is generally followed by step-like events that correspond to the unbinding single-cell adhesion molecules from the substrate (step and tether events). The cell and substrate are completely separated again (1). (Adapted from Friedrichs, J. et al., *Methods,* 60, 169–178, 2013.)

AFM tip, which is successively used in contact mode to transfer predetermined molecular patterns to the surface, with a resolution of 30–100 nm.[3] This technique has been optimized on atomically flat and contaminant-free surfaces (e.g., gold and silicon). It can now count on a dramatically improved patterning efficiency and flexibility in depositing different ink molecules.[3] In contact mode lithography on Si surfaces, an area is oxidized to SiO_2 when a positive voltage is applied to the tip, and an oxidized pattern can be drawn.[17] This approach was applied to microarrays composed by DNA-directed protein immobilization, with a precise pattern, used to investigate the recruitment of transmembrane receptors in living cells.[3] Different AFM nanolithography modes, such as dip-pen nanolithography, electrochemical AFM nanolithography, thermal AFM nanolithography, nanografting, nanoshaving, and tapping-mode AFM nanolithography of alkanethiols can be done.[17] On a related matter, the AFM tip has also been successfully used to reconstitute nanometer-scale defined areas of supported phospholipid bilayers.[17,59] However, its translation to medically relevant materials and applications may still be far, mainly because of the nonatomic planarity of most biomaterials and the practical need to pattern areas whose dimensions lay beyond the range of this technique. Nevertheless, the capacity of creating specific molecular patterns will undoubtedly benefit more fundamental studies in biomaterials surface science, focused, for example, on furthering the investigation of cellular response to specific chemical cues (e.g., chemotaxis) and/or on the validation of functional molecular nanostructures.[3]

Limitations

With the advance of this technique, AFM has been an ideal tool to study biological interactions, but every technique has its limitations.

Despite the development of high-speed atomic force microscopes, the time resolution of AFM applications is a limitation in several biological questions, as numerous biological interactions occur faster than the time required by the AFM to probe, in approximately 0.001–1 s.[60]

The shape of the AFM tip can also have a drastic effect on the images that are acquired. In a real imaging situation, sample compression and deformation also need to be taken into consideration. The part of the tip that interacts with the sample is often a critical source of artifacts in AFM. Double tips may also occur on worn tips that have been damaged during scanning. When an interaction force is measured by the cantilever, the probe always also exerts some force on the sample. This can cause problems of distortion or damage to the sample, which may be deformed under this force, particularly since many biological samples are soft and delicate, and require particularly careful AFM imaging.[16]

Another limitation of the AFM application is a maximum scan speed in imaging rate. On increasing the imaging rate for capturing multiple frames per second, a major error associated to tip–sample interaction forces can be caused. Major limitations for fast scanning concern the mechanical properties of the cantilever. Fast and accurate regulation of the cantilever deflection is required to maintain a constant interaction between the tip and the sample. Additionally, this cantilever deflection regulation is needed to prevent force-induced conformational changes or damages of the sample. In liquid, to achieve a fast imaging for monitoring the dynamics of biomacromolecules, high-resonance frequencies in stiff cantilevers are necessary.[28] When the substrates are soft, the determination of the tip area contacting the substrate and spring-constant calculation are difficult.[13] Furthermore, the AFM tip can be easily contaminated by the adsorption of molecules coming from the sample surface.[60]

A problem with oscillation mode in liquid is the low quality factor of a cantilever. It is still debated whether the image quality is significantly improved compared to conventional feedback by tuning other scan parameters, or by scanning with small amplitudes and stiff cantilevers. The low quality factor in liquid also gives rise to unclean excitation spectra when exciting the cantilever via its chip or the complete cantilever holder. As a result of the liquid coupling, system resonances of the AFM are superimposed to the cantilever spectrum.[28]

In cell adhesion studies, the limitations are in conducting *n* approach and retraction cycles between cells, which need to be repeated a sufficient number of times to obtain reproducible results, due to the complexity of living cells.[6] Cells may also be in different states and thus show distinct adhesive properties that may be difficult to compare.[53]

Future

AFM can be an important tool for nanotechnology and nanomedicine in biological systems. The microscope can be used alone or combined with other techniques (for example, optical microscopies) that provide complementary information, creating an even more powerful tool. The combination of nanophysics with cell biology establishes a mechanical assay that relates and is used for qualitatively and quantitatively characterizing specific interactions of surfaces from living cells, which can open an enormous variety of applications.

The evolution of AFM applications has been very important to the advance of science. This potential is based on the versatility of AFM force detection and mapping methods, as well as its *toolbox*, which allows functionalizing the AFM tip to tackle biological and biomedical questions that are difficult or impossible to address by other methods. AFM can be used to image living cells in aqueous solutions with nanometer resolution, so that the dynamic changes of cellular ultramicrostructures on single cells in response to a given molecule introduced on the volume of liquid surrounding the sample can be monitored by AFM. It can simultaneously obtain multiple physicochemical parameters, such as morphology, elasticity, adhesion, deformation, and energy dissipation of biological systems in a relatively short time (approximately several minutes).

AFM imaging will continue to be the simplest AFM methodology to use, as it is quite more straightforward to the user, while enabling a structural and morphological evaluation of the sample. Requiring additional training and a thorough optimization for each new type of sample, AFM-based force spectroscopy has become an important tool on the study of biological systems. Single-molecule interaction studies have allowed, through the combination of nanophysics with cells, the establishment of a mechanical assay that relates qualitatively cooperative molecular processes during contact formation, or even quantitatively the expression of genes, to the function of its product in cell adhesion.[53,54,61] This type of force spectroscopy performed on live cells is directly applicable to a variety of different cell-adhesion systems.[9,61] An extensive field of application for this cell-based molecular assay is expectable, for instance, when investigating mutated cell-adhesion proteins or the coupling of cell-adhesion molecules to the cytoskeleton and also when evaluating adhesion-blocking drugs.[31,61]

Still regarding cell adhesion, the additional molecular insight offered by single-cell techniques, together with the continuing development and improvement of quantitative single-cell methods, is providing stimulating original insight into the dynamic regulation of cell–cell adhesion during the formation of multicellular organisms.[62] Since the dysregulation of adhesion receptors and their signaling function also can lead to an extensive variety of pathological defects, these tools can also improve our understanding of how changes in cell adhesion can contribute to different diseases, namely oncologic processes.[9,32] Furthermore, the extended effective pulling range of the instrument leads to the ability to detect the (un)binding forces associated to interactions going from short-range, unspecific interactions, up to specific tether-associated interactions. Most importantly, this method can be operated with an experimental microenvironment as close to truly physiologic conditions in organisms as possible. Most of the studies can be conducted at controlled conditions of temperature, pH, ionic strength, and other physiological parameters.

In future studies, it is important to understand how cells are stimulated, how these cells interact between them, and how their properties can change under various conditions. AFM is expected to be used for these purposes in more and more routine studies, namely due to its special adequacy to understand the biochemical and biophysical mechanisms underlying different biological processes.

Acknowledgments

This work was supported by Fundação para a Ciência e a Tecnologia—Ministério da Ciência, Tecnologia e Ensino Superior (FCT–MCTES, Portugal) grants PTDC/BBB-BMD/6307/2014 and PTDC/BBB-BQB/3494/2014.

References

1. Kim, H., Arakawa, H., Osada, T. and Ikai, A. Quantification of cell adhesion force with AFM: Distribution of vitronectin receptors on a living MC3T3-E1 cell. *Ultramicroscopy* **97**, 359–363 (2003).
2. Karagkiozaki, V., Logothetidis, S., Laskarakis, A., Giannoglou, G. and Lousinian, S. AFM study of the thrombogenicity of carbon-based coatings for cardiovascular applications. *Mater. Sci. Eng. B Solid-State Mater. Adv. Technol.* **152**, 16–21 (2008).
3. Variola, F. Atomic force microscopy in biomaterials surface science. *Phys. Chem. Chem. Phys.* **17**, 2950–2959 (2015).
4. Santos, N. C. and Castanho, M. A. An overview of the biophysical applications of atomic force microscopy. *Biophys. Chem.* **107**, 133–149 (2004).
5. Carvalho, F. A., Freitas, T. and Santos, N. C. Taking nanomedicine teaching into practice with atomic force microscopy and force spectroscopy. *Adv. Physiol. Educ.* **39**, 360–366 (2015).

6. Simon, A. and Durrieu, M.-C. Strategies and results of atomic force microscopy in the study of cellular adhesion. *Micron* **37**, 1–13 (2006).
7. Dong, C., Hu, X. and Dinu, C. Z. Current status and perspectives in atomic force microscopy-based identification of cellular transformation. *Int. J. Nanomedicine* **11**, 2107–2118 (2016).
8. Carvalho, F. A. et al. Atomic force microscopy-based molecular recognition of a fibrinogen receptor on human erythrocytes. *ACS Nano* **4**, 4609–4620 (2010).
9. Dufrêne, Y. F. and Pelling, A. E. Force nanoscopy of cell mechanics and cell adhesion. *Nanoscale* **5**, 4094–4104 (2013).
10. Lyubchenko, Y. L. and Shlyakhtenko, L. S. Imaging of DNA and protein-DNA complexes with atomic force microscopy. *Crit. Rev. Eukaryot. Gene Expr.* **26**, 63–96 (2016).
11. Attwood, S. J., Choi, Y. and Leonenko, Z. Preparation of DOPC and DPPC supported planar lipid bilayers for atomic force microscopy and atomic force spectroscopy. *Int. J. Mol. Sci.* **14**, 3514–3539 (2013).
12. Schön, P. Imaging and force probing RNA by atomic force microscopy. *Methods* **103**, 25–33 (2016).
13. Averett, L. E. and Schoenfisch, M. H. Atomic force microscope studies of fibrinogen adsorption. *Analyst* **135**, 1201–1209 (2010).
14. Cazaux, S. et al. Synchronizing atomic force microscopy force mode and fluorescence microscopy in real time for immune cell stimulation and activation studies. *Ultramicroscopy* **160**, 168–181 (2016).
15. Fotiadis, D. Atomic force microscopy for the study of membrane proteins. *Curr. Opin. Biotechnol.* **23**, 510–515 (2012).
16. JPK Instruments AG. *NanoWizard® AFM Handbook*. pp. 1–55 (2012).
17. Carvalho, F. A. and Santos, N. C. Atomic force microscopy-based force spectroscopy–Biological and biomedical applications. *IUBMB Life* **64**, 465–472 (2012).
18. Rappaz, B. et al. Comparative study of human erythrocytes by digital holographic microscopy, confocal microscopy, and impedance volume analyzer. *Cytometry. A* **73**, 895–903 (2008).
19. Wang, Y., Wang, H., Bi, S. and Guo, B. Nano-Wilhelmy investigation of dynamic wetting properties of AFM tips through tip-nanobubble interaction. *Sci. Rep.* **6**, 1–14 (2016).
20. Schmitt, L., Ludwig, M., Gaub, H. E. and Tampé, R. A metal-chelating microscopy tip as a new toolbox for single-molecule experiments by atomic force microscopy. *Biophys. J.* **78**, 3275–3285 (2000).
21. De Oliveira, R. R. L. Albuquerque, D. A. C., Cruz, T. G. S, Yamaji, F.M. and Leite, F. L. Measurement of the nanoscale roughness by atomic force microscopy: Basic principles and applications. *At. Force Microsc.–Imaging, Meas. Manip. Surfaces At. Scale* 147–174 (2012). doi:10.5772/37583
22. Zeidan, A. and Yelin, D. Reflectance confocal microscopy of red blood cells: Simulation and experiment. *Biomed. Opt. Express* **6**, 4335–4343 (2015).
23. Torre, B., Braga, P.C. and Ricci, D. How the Atomic Force Microscope Works? - Atomic Force Microscopy in Biomedical Research–Methods and Protocols, *Humana Press–Springer Science* **736**, 3–18 (2011). doi: 10.1007/978-1-62703-239-1_1
24. Carvalho, F. A., Martins, I. C. and Santos, N. C. Atomic force microscopy and force spectroscopy on the assessment of protein folding and functionality. *Arch. Biochem. Biophys.* **531**, 116–127 (2013).
25. Behary, N. and Perwuelz, A. Atomic force microscopy–for investigating surface treatment of textile fibers. *Atomic Force Microscopy–Imaging, Measuring and Manipulating Surfaces at the Atomic Scale* 231–256 (2012). doi: 10.5772/35656

26. Morris, V. J. *Atomic force microscopy (AFM) and related tools for the imaging of foods and beverages on the nanoscale. Nanotechnology in the Food, Beverage and Nutraceutical Industries* (Woodhead Publishing Limited, 2012). doi:10.1533/9780857095657.1.99
27. Ramachandran, S., Teran Arce, F. and Lal, R. Potential role of atomic force microscopy in systems biology. *Wiley Interdiscip. Rev. Syst. Biol. Med.* **3**, 702–716 (2011).
28. Frederix, P. L. T. M., Bosshart, P. D. and Engel, A. Atomic force microscopy of biological membranes. *Biophys. J.* **96**, 329–338 (2009).
29. Jung, S.-H., Park, D., Park, J. H., Kim, Y.-M. and Ha, K.-S. Molecular imaging of membrane proteins and microfilaments using atomic force microscopy. *Exp. Mol. Med.* **42**, 597–605 (2010).
30. Müller, D. J. and Dufrêne, Y. F. Atomic force microscopy: A nanoscopic window on the cell surface. *Trends Cell Biol.* **21**, 461–469 (2011).
31. Li, M., Liu, L., Xi, N. and Wang, Y. Nanoscale monitoring of drug actions on cell membrane using atomic force microscopy. *Acta Pharmacol. Sin.* **36**, 1–14 (2015). doi:10.1038/aps.2015.28.
32. Bernardes, N. et al. Modulation of membrane properties of lung cancer cells by azurin enhances the sensitivity to EGFR-targeted therapy and decreased β1 integrin-mediated adhesion. *Cell Cycle* **15**, 1415–1424 (2016).
33. Wu, X., Muthuchamy, M. and Reddy, D. S. Atomic force microscopy protocol for measurement of membrane plasticity and extracellular interactions in single neurons in epilepsy. *Front. Aging Neurosci.* **8**, 1–12 (2016).
34. Hollmann, A. et al. Effects of singlet oxygen generated by a broad-spectrum viral fusion inhibitor on membrane nanoarchitecture. *Nanomed.* **11**, 1163–1167 (2015).
35. Fang, H. H., Chan, K. Y. and Xu, L. C. Quantification of bacterial adhesion forces using atomic force microscopy (AFM). *J. Microbiol. Methods* **40**, 89–97 (2000).
36. Domingues, M. M. et al. Antimicrobial protein rBPI21-induced surface changes on gram-negative and gram-positive bacteria. *Nanomedicine* **10**, 543–551 (2014).
37. Santos, N. C., Ter-Ovanesyan, E., Zasadzinski, J. A., Prieto, M. and Castanho, M. A. R. B. Filipin-induced lesions in planar phospholipid bilayers imaged by atomic force microscopy. *Biophys. J.* **75**, 1869–1873 (1998).
38. Franquelim, H. G., Veiga, A. S., Weissmüller, G., Santos, N. C. and Castanho, M. A. Unravelling the molecular basis of the selectivity of the HIV-1 fusion inhibitor sifuvirtide towards phosphatidylcholine-rich rigid membranes. *Biochim. Biophys. Acta* **1798**, 1234–1243 (2010).
39. Franquelim, H. G., Gaspar, D., Veiga, A. S., Santos, N. C. and Castanho, M. A. Decoding distinct membrane interactions of HIV-1 fusion inhibitors using a combined atomic force and fluorescence microscopy approach. *Biochim. Biophys. Acta* **1828**, 1777–1785 (2013).
40. Franquelim, H. G. et al. Anti-HIV-1 antibodies 2F5 and 4E10 interact differently with lipids to bind their epitopes. *AIDS* **25**, 419–428 (2011).
41. Migliolo, L. et al. Structural and functional evaluation of the palindromic alanine-rich antimicrobial peptide Pa-MAP2. *Biochim. Biophys. Acta* **1858**, 1488–1498 (2016).
42. Bravo-Ferrada, B. M. et al. Study of surface damage on cell envelope assessed by AFM and flow cytometry of Lactobacillus plantarum exposed to ethanol and dehydration. *J. Appl. Microbiol.* **118**, 1409–1417 (2015).
43. Cardoso, M. H. et al. A polyalanine peptide derived from polar fish with anti-infectious activities. *Sci. Rep.* **6**, 21385 (2016).
44. Guedes, A. F. et al. Atomic force microscopy as a tool to evaluate the risk of cardiovascular diseases in patients. *Nat. Nanotechnol.* **11**, 687–692 (2016).

45. Carvalho, F. A., de Oliveira, S., Freitas, T., Gonçalves, S. and Santos, N. C. Variations on fibrinogen-erythrocyte interactions during cell aging. *PLoS One* **6**, e18167 (2011).
46. Gupta, P., Song, B., Neto, C. and Camesano, T. A. Atomic force microscopy-guided fractionation reveals the influence of cranberry phytochemicals on adhesion of Escherichia coli. *Food Funct.* **7**, 2655–2666 (2016).
47. Carvalho, F. A. et al. Dengue virus capsid protein binding to hepatic lipid droplets (LD) is potassium ion dependent and is mediated by LD surface proteins. *J. Virol.* **86**, 2096–2108 (2012).
48. Faustino, A. F. et al. Understanding dengue virus capsid protein interaction with key biological targets. *Sci. Rep.* **5**, 10592 (2015).
49. Faustino, A. F. et al. Understanding dengue virus capsid protein disordered N-terminus and pep14-23-based inhibition. *ACS Chem. Biol.* **10**, 517–526 (2015).
50. Faustino, A. F. et al. Dengue virus capsid protein interacts specifically with very low-density lipoproteins. *Nanomedicine* **10**, 247–255 (2014).
51. Haase, K. and Pelling, A. E. Investigating cell mechanics with atomic force microscopy. *J. R. Soc. Interface* **12**, 1–16 (2015).
52. Last, J. A., Russell, P., Nealey, P. F. and Murphy, C. J. The applications of atomic force microscopy to vision science. *Investig. Ophthalmol. Vis. Sci.* **51**, 6083–6094 (2010).
53. Friedrichs, J. et al. A practical guide to quantify cell adhesion using single-cell force spectroscopy. *Methods* **60**, 169–178 (2013).
54. Yu, M., Strohmeyer, N., Wang, J., Müller, D. J. and Helenius, J. Increasing throughput of AFM-based single cell adhesion measurements through multi-substrate surfaces. *Beilstein J. Nanotechnol.* **6**, 157–166 (2015).
55. Bowman, K., Saffell, J. Measuring the cell-cell adhesion force exerted by a cell adhesion molecule. *JPK Instruments AG* - Application Note, 1–4 (2012).
56. Puech, P.-H. et al. Measuring cell adhesion forces of primary gastrulating cells from zebrafish using atomic force microscopy. *J. Cell Sci.* **118**, 4199–4206 (2005).
57. Ovchinnikova, E. S., Krom, B. P., Busscher, H. J. and van der Mei, H. C. Evaluation of adhesion forces of Staphylococcus aureus along the length of Candida albicans hyphae. *BMC Microbiol.* **12**, 281 (2012).
58. Eaton, P., Zuzarte-Luis, V., Mota, M. M., Santos, N. C. and Prudêncio, M. Infection by Plasmodium changes shape and stiffness of hepatic cells. *Nanomedicine* **8**, 17–19 (2012).
59. Santos, N. C., Ter-Ovanesyan, E., Zasadzinski, J. A and Castanho, M. A. Reconstitution of phospholipid bilayer by an atomic force microscope tip. *Biophys. J.* **75**, 2119–2120 (1998).
60. Muller, D. J., Helenius, J., Alsteens, D. and Dufrene, Y. F. Force probing surfaces of living cells to molecular resolution. *Nat Chem Biol* **5**, 383–390 (2009).
61. Benoit, M. and Gaub, H. E. Measuring cell adhesion forces with the atomic force microscope at the molecular level. *Cells Tissues Organs* **172**, 174–189 (2002).
62. Kashef, J. and Franz, C. M. Quantitative methods for analyzing cell–cell adhesion in development. *Dev. Biol.* **401**, 165–174 (2015).

Chapter 5 Confocal Raman microscopy *Imaging the chemistry*

M. Gomez-Lazaro, A. Freitas, and C.C. Ribeiro

Contents

Introduction

Confocal Raman microscopy is an optical spectroscopic technique, which allows for the noninvasive measurement of chemical compounds in samples, without any previous labeling. The Raman spectrum provides a molecular fingerprint of a sample, because the specific vibrational modes of the molecules are influenced by their structure, environment, bond order, substituents, geometry, and hydrogen bonds. The technique allows both qualitative and quantitative analysis, since the intensity of the bands in a Raman spectrum is proportional to the concentration of the correspondent molecules. Another advantageous aspect of Raman microscopy is the possibility of selectively exciting a needed portion of a molecule by changing the excitation wavelength using near-infrared (NIR), visible, or ultraviolet light.

This technique is based on the so-called Raman effect, first published in 1928 by Sir Chandrasekhara Venkata Raman [1], which relies on the inelastic light scattering at the chemical bonds of a molecule, giving information of its vibrational and rotational states. In short, when a photon interacts with a molecule it induces its excitation, leading to light scattering. The scattered light might have the same or different wavelength as

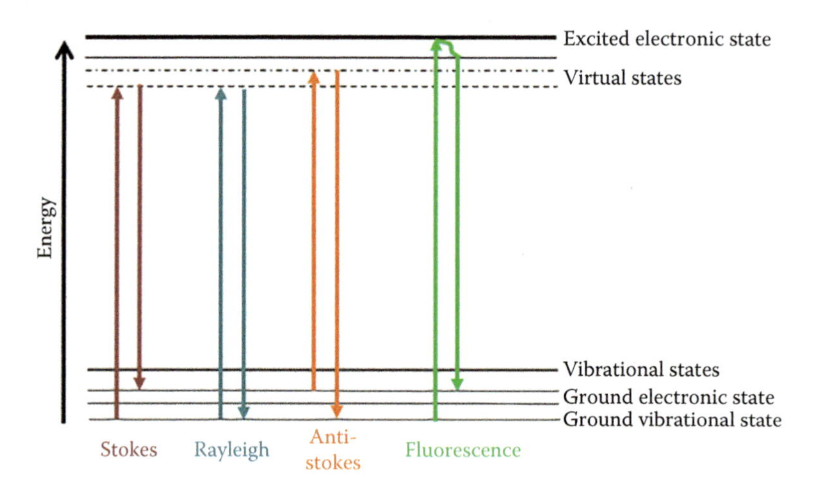

Figure 5.1 Jablonski diagram comparing Rayleigh scattering, Raman scattering, and fluorescence. Raman and Rayleigh scattering occur when a photon excites a molecule to a virtual state (which has less energy than a real electronic transition). The molecule then emits a photon, which can relax back from the virtual state. This scattering can be elastic (Rayleigh scattering) or inelastic (Raman scattering). During Rayleigh scattering, the incident light does not lose or gain energy (maintaining the same wavelength). On the other hand, Raman scattering is inelastic, leading to loss or gain of energy during scattering—increasing (Stokes) or decreasing (anti-Stokes)—its wavelength. Fluorescence, on the other hand, occurs when the molecule is excited to another electronic state.

the incident light (Figure 5.1). The Raman shift (Ω) is expressed in cm^{-1} and it is related to the initial (λ_0) and final wavelengths (λ):

$$\Omega = \left(\frac{1}{\lambda_0} - \frac{1}{\lambda} \right)$$

The light scattered can be either (1) elastic, where the Rayleigh scattered light oscillates at the same frequency as the incident light or (2) inelastic, the Raman scatter. The latter can be further differentiated in Stokes and anti-Stokes Raman scattering (Figure 5.1). The Raman bands commonly used for chemical analysis are the Stokes-shifted bands that are more intense [2], although both anti-Stokes and Stokes Raman bands can give valuable chemical information.

The Raman effect is very weak considering that less than one in a million excitation photons give rise to a single Raman photon, contrary to what happens in fluorescence that is much more likely to occur [3]. In this regard, background fluorescence is often present in the Raman spectrum of biological samples and higher laser powers are required in these conditions together with longer integration times, making necessary the use of postprocessing algorithms.

Relevant for the acquisition of chemical information in biological samples is the combination of the spectral and the spatial information within the sample. In order to combine these two types of information, a Raman spectrometer can be coupled to a microscope, allowing high-magnification visualization of a sample area. Then it is possible to obtain a Raman spectrum from a micrometer laser spot area using the objective of the microscope as the light-collecting device. By working with high numerical aperture (NA) objectives, we can enhance the efficiency of the acquisition of the Raman signal through the increase of the photon density per area, but also the collection efficiency (Table 5.1). However, depending on the laser wavelength, intensity, and acquisition time, the utilization of an objective may lead to sample heating due to focusing of the laser beam. Therefore, care must be taken not to destroy the sample. By using a confocal Raman microscope, additional advantages arise such as improvement in the lateral resolution and increased sensitivity. Lateral resolution can be narrowed down to the light diffraction limit. For a confocal setup, the lateral spatial resolution can be calculated as follows, relying on the laser wavelength (λ) and the numerical aperture (NA) of the objective:

$$\delta_{\text{lat}} = \frac{0.62\lambda}{\text{NA}}$$

Although the axial resolution can be obtained from the following formula, where λ is the laser wavelength, NA is the numerical aperture, and n is the refractive index of the immersion media, since the immersion media will also contribute to the overall spectrum:

$$\delta_{\text{ax}} = \frac{2\lambda n}{(\text{NA})^2}$$

With the use of a confocal microscope, we are also able to suppress the contribution of the out-of-focus planes to the spectrum by using a point illumination of the focal plane and a pinhole (Figure 5.2). Nonetheless, this will also represent a decrease in the Raman signal due to the smaller sample volume analyzed. As an advantage, we can bleach a fluorophore in the focal plane, therefore decreasing the contribution of the fluorescence background to the Raman spectra. Another benefit of the confocal Raman microscope setup is that the sample can be scanned point-by-point and line-by-line. In this case, every pixel of the image corresponds to a complete Raman spectrum, which in turn can give information of the spatial distribution of the different chemical components, as cells and tissues are not homogeneous samples at the microscopic level. By collecting the Raman spectra from different points in a sample we can perform *Raman mapping* to create a distribution map based on the chemical information obtained in the spectra, combining the lateral information on the subcellular level, acquiring then a *chemical image* of the sample [4]. In addition, 3D information of the chemical composition of the sample

Table 5.1 Parameters that Influence the Quality of a Raman Spectrum

Parameter	Definition	How to Improve
Focal length of the spectrometer	Distance between the entrance slit of the spectrograph and the collimated mirror.	Longer lengths result in higher resolution.
Diffraction grating	The part of the spectrometer that disperses the Raman signal into the CCD by reflecting the different wavelengths with different angles. Different gratings are appropriated for specific wavelength ranges because they display a reflection maximum in certain spectral ranges. It can be optimized by the user.	The higher groove density of the grating (number of grooves/mm), the higher will be the spectral resolution because present higher dispersion, distributing the signal to a large number of pixels in the CCD. Typical gratings used for Raman vary from 300gr/m (low resolution) through to 1800gr/mm (high resolution). However, when using higher grove density gratings it is necessary to use longer acquisition times because they give less intense spectra. A 1800 gr/mm will give a scanning range from 0 to 950 nm, whereas the 600 and the 300 gr/mm will give ranges between 0–2800 and 0–5700 nm, respectively.
Pixel size on the CCD camera	The CCD camera converts the information of the arriving photons into electrons, a signal that will be then used by the acquisition software.	Smaller pixel size increases the resolution. Important feature of this device is the quantum efficiency or the percentage of detected photons.
Entrance spectrometer confocal hole (Pinhole)	Can be optimized by the user.	Smaller pinhole will increase the lateral resolution. However, it will decrease the intensity of the Raman signal. By decreasing the sample volume we can reduce the fluorescence contribution.
Acquisition time	The sensitivity determines the total time employed to acquire a spectrum with a given signal-to-noise ratio.	The longer the acquisition time the better signal-to-noise ratio.
Excitation wavelength	Wavelength of the laser used to excite the molecules. The Raman scattered signal is proportional to the excitation wavelength. The excitation wavelength also affects the spatial resolution. Can be optimized by the user.	It can be chosen to decrease fluorescence of the sample by using a wavelength at which the sample shows no fluorescence. Lower excitation wavelength give rise to higher Raman signal. Higher spatial resolution can be achieved by using shorter excitation wavelengths. In addition, shorter wavelengths are more energetic and can damage the sample more than longer wavelengths.

(*Continued*)

Table 5.1 (*Continued*) Parameters that Influence the Quality of a Raman Spectrum

Parameter	Definition	How to Improve
Numerical aperture (NA) of the objective	Corresponds to the light gathering capacity of an objective. It is defined by the refractive index of the medium and the half angle of the maximum cone of light that can enter or exit the lens.	Although higher NA objectives (i.e., 100X) will collect a bigger Raman signal, the depth of focus will be higher with a macro objective (i.e., 10X) collecting the signal from a bigger volume with the later. The NA of the objective also influences the spatial resolution of the Raman spectra.
Lateral resolution	It is proportional to $\lambda/(2NA)$ in nm.	The smaller the excitation wavelength (λ), the smaller the lateral resolution in nm.
Integration time	Total time employed measuring each point spectrum.	Higher integration times.
Notch filter	Filter that attenuates light within a specific wavelength range while transmitting specific wavelengths with minimal intensity loss. They are used to block the excitation light of the laser and are placed in the detection channel of the setup.	To obtain good signal-to-noise ratios.
Density filter	To decrease the intensity of the laser power used for excitation.	
Magnification of the sample	To estimate the magnification of the sample, the objective magnification has to be multiplied by a factor that depends on the tube length of the microscope objective and the focal length of the lens that form the image (i.e., for a 100X objective with an objective microscope tube length of 180 mm and the focal length of 250 mm, the magnification will be calculated by $100 \times (250/180)$.	

can also be obtained by acquiring z-stacks [5]. Figure 5.3 [6] shows an example of a Raman map of an air-dried human sperm, obtained with a step of 100 nm, using a LabRAM Aramis Raman system in confocal mode with the He–Ne laser (632.8 nm). By assigning a specific color to each of the chemical signatures, three different regions of the sperm can

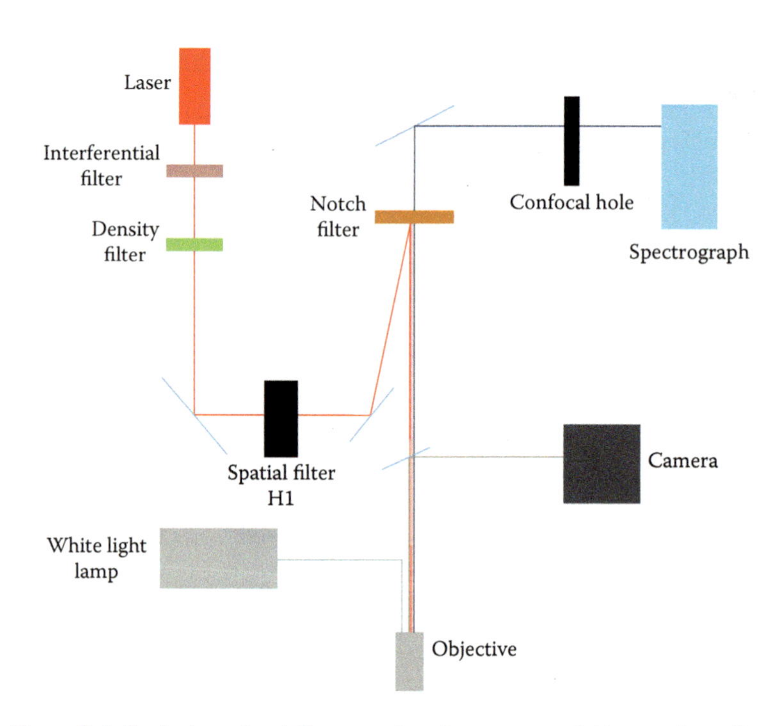

Figure 5.2 Typical confocal Raman setup. In a representative configuration, the laser beam is directed towards an interferential pass-band filter, specific for every exciting laser line, to filter out the plasma line of the laser. The intensity of the laser can be controlled by the use of different density filters, after which the laser is focused onto a pinhole H1, and the beam directed toward the sample. The Raman signal will be then collected by the microscope objective. The reflected light (Rayleigh scattered) will be reflected by a notch filter, whereas the Raman signal is transmitted with minimal loss through this filter towards the confocal hole into the spectrograph. Every laser matches to a specific notch filter. The Raman light will enter the spectrograph, which in turn is able to spatially separate the different wavelengths and detect them on a multichannel detector (charge-coupled device, CCD). Important characteristics of the spectrograph are its sensitivity and spectral resolution. By modifying several features of the spectrograph, we can improve the quality of our spectra (see Table 5.1). By using a white lamp, it is possible to observe the sample on a TV camera.

be identified corresponding to the proximal head (containing DNA), the distal head (Acrosome), and the tail of the sperm (Flagellum). A sharp peak at 1000 cm^{-1} assigned to phenylalanine is seen only in the spectra from the tail and this, together with the sharp 1447 cm^{-1} methylene deformation peak indicates the presence of protein. The other two spectra contain a prominent 785 cm^{-1} peak associated with thymine, cytosine, and the DNA backbone but only the distal head region contained a peak at 1092 cm^{-1}, indicative of the PO_4 backbone. The high-spatial resolution of the Raman map allows the detection of small irregularities in the sperm head such as vacuoles (yellow circles).

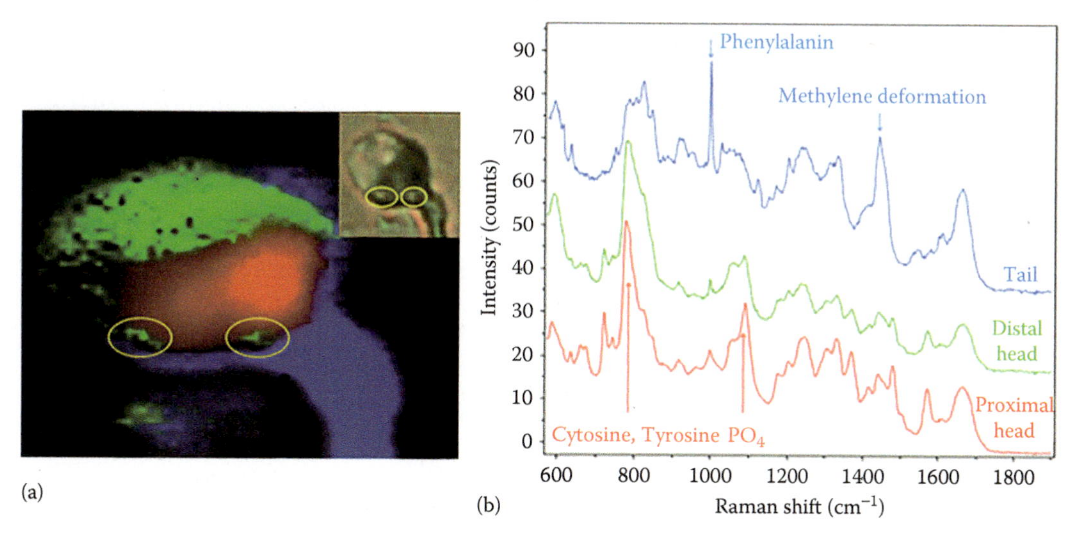

Figure 5.3 Raman map (a) and the characteristic Raman spectra of native sperm (the video image of the sperm in the inset) (b). Three distinct regions in the sperm are observed: proximal head, distal head and tail, as well as small details as vacuoles. (Courtesy of Mallidas, C. and J. Wistuba, *Raman analysis of sperm nuclear DNA integrity.* Horiba Scientific Application Note, 2013.)

Experimental procedures

Sample processing: Needs and troubles

One of the major advantages claimed for the use of Raman spectroscopy is the inexistent or minimal sample processing or staining procedures needed. Yet, in the life-sciences field, sample processing could be impossible to avoid, and these procedures might interfere with the typical Raman spectrum of a sample. Examples of procedures and stainings commonly used in the life sciences and its effects on the spectrum are listed in Table 5.2. For example, when working with biological material, some procedures are frequently used in order to preserve the samples; for a comprehensive review of the most common sample processing methods and their effect on the Raman spectrum, see Reference 7. The tissues are often embedded in specific media for section cutting, as well as fixatives. When this happens, dewaxing steps are undergone before Raman analysis. But again all these reagents will have an influence on the spectra [8].

On the other hand, for specific analysis, certain procedures might be advantageous in search of particular unique peaks, which is especially useful because with Raman spectroscopy we can define delimited regions in macromolecular complexes [15]. For example, Raman spectroscopy has been successfully used to assess protein posttranslational modifications (PTMs) such as carbonylation. By conjugating a sample with dinitrophenyl hydrazine (DNPH), it is even possible to quantify the

Table 5.2 Effects of Sample Processing and Preservation Procedures on the Raman Spectrum of Biological Samples

Procedure	Effect	Reference
Paraffin embedding	Tissue: The use of organic solvents might induce degradation or even loss of some cellular components. Bands affected: 1063, 1130, 1296, and 1436 cm^{-1}.	[7]
Flash freezing	Tissue: No need for the use of organic solvents. Difficult to obtain, deteriorates rapidly. Overall deterioration of the spectra (reduction in the intensity at 1002 cm^{-1} (C–C aromatic ring stretching), 1447 cm^{-1} (CH_2 bending mode of proteins and lipids), and 1637 cm^{-1} (Amide I band), and an additional contribution at 1493 cm^{-1}. Wills et al. found no differences in comparison with fresh tissue.	[7,9,10]
Formalin fixation	Tissue: Decrease in overall Raman intensities. Formalin peaks can be detected at 907, 1041, and 1492 cm^{-1}. Reduction in the intensity of the amide I band (1637 cm^{-1}). Preserves Raman signal of the lipids. Cells: Signal loss 1600–1500 and 1448–1127 cm^{-1}. Overall spectrum decreased intensity.	[7,9,11–13]
Drying	Tissue: Disruption of protein vibrational modes. Decreased intensity of the 1645 and the 930 bands cm^{-1}. Cells: Signal loss 1253–1127 cm^{-1}. Decreased intensity at 1612 and 1180 cm^{-1}. Desiccation produces a slightly less intense spectral profile compared to air-drying.	[7,9,12–14]
Dewaxed and formalin fixed paraffin (FFPP)	Tissue: Soaking in xylene before wax embedding adds several strong peaks at 620, 1002, 1032, 1203, and 1601 cm^{-1}, and reduction in the 1585 cm^{-1} band. Wax is incompletely removed since its contribution can be seen at 1063, 1130, 1296, and 1436 cm^{-1}.	[9]
Ethanol fixation	Tissue: Induces alterations in the spectra. Addition of a peak at 1071 cm^{-1}. Decreased intensity of the 1645 cm^{-1} band.	[14]
Glycerol fixation	Tissue: Induces alterations in the spectra. Decreased intensity of the 1645 cm^{-1} band. Strong peak at 1450 cm^{-1}.	[14]
Aralite and Eponate embedding	Tissue: Strong peak at 1450 cm^{-1}. Small peak at 960 cm^{-1}.	[14]
Technovit embedding	Tissue: Strong peak at 1450 cm^{-1}. Moderate peak at 960 cm^{-1}. Peak at 1667 cm^{-1}.	[14]
Glycol methacrylate, Polymethyl methacrylate, and LR white embedding	Tissue: Strong peak at 1450 cm^{-1}. Strong peak at 960 cm^{-1}. Peak at 1667 cm^{-1}.	[14]

number of carbonyls in a protein sample [16]. Not only other PTMs, but also other modifications such as protein fibrillation and aggregation can be detected with Raman spectroscopy [17–19].

Spectrum processing and data analysis

The data acquired in a confocal Raman microscope contains different information such as spatial coordinates (*x*, *y*, and *z*), spectral position (in wavenumber [cm^{-1}]); relative wavenumber (rel. cm^{-1}) or wavelength (nm), intensity (counts per second [cps]), and time. An example of a Raman spectrum obtained from the mice striatum can be seen in Figure 5.4. Biological samples frequently display high background signal and poor signal-to-noise ratio due to the presence of autofluorescent components, which as stated before can be reduced by using longer wavelengths as excitation source. A classical preprocessing of the spectra involves background subtraction, smoothing, and cosmic ray removal (Figure 5.4). The so-called cosmic rays represent high-energy particles that might impact the detector during the spectrum acquisition, inducing the appearance of spikes in the spectrum. The fact that they are easy to perceive due to its narrow and sharp shape and high intensity, allows for manual removal from the spectrum, thus avoiding false identifications. Nevertheless, we can easily avoid them by acquiring the spectrum more than one time, since it is highly unlikely that two cosmic rays affect the detector in the same position.

Smoothing represents a method to reduce the noise from the spectrum to increase the signal-to-noise ratio; however, it might decrease the

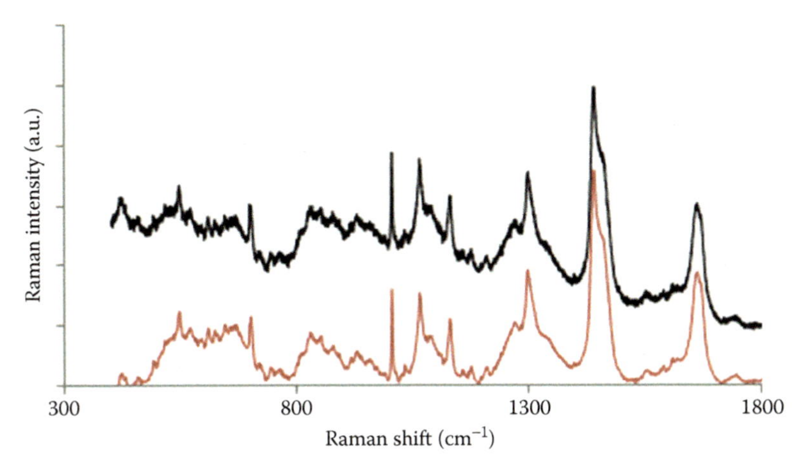

Figure 5.4 Example of a Raman spectrum obtained from decellularized mouse brain. The Raman spectrum is displayed as Raman shift units (ν) that arise from the conversion of the wavelength values of the Raman spectrum ν (cm^{-1}) = 1/λ (nm). It is shown as a raw Raman spectrum acquired from a single point in the decellularized brain slice (black) and the same spectrum after background subtraction and smoothing is shown in red.

Table 5.3 Parameters that Can be Evaluated from a Raman Spectrum and the Respective Information Obtained

Parameter	Information
Peak intensity	Quantification of a specific compound.
Peak shift	Identification of stress and strain states.
Peak width	Evaluation of the degree of crystallinity.
Polarization state	Crystal symmetry and orientation.

resolution of the spectrum. These algorithms are based on the fact that, contrary to noise, the spectral data between adjacent points in the spectra changes progressively. In this direction, the noise value is exchanged for the value calculated by the algorithm, taking into account the values of the neighboring spectral points [20]. In the Raman spectrum, the background signals can be the resultant from the charge-coupled device (CCD) camera itself, the autofluorescence present in the sample and the contribution of the devices used for sample immobilization. In the latter, it is important to use low Raman active materials for sample immobilization such as quartz or calcium fluoride microscope slides. Examples of parameters that can be extracted from a Raman spectrum are listed in Table 5.3.

When we acquire a dataset of Raman spectra, additional processing is required to generate the chemical image (Figure 5.3); examples of these analyses are the univariate and multivariate image generation. In the case of the univariate analysis, every pixel of the image corresponds to one spectrum, whereas in the multivariate analysis the entire hyperspectral dataset is used for the determination of every pixel.

Applications

Ever since Raman and Krishnan first reported the Raman effect in 1928 [21], it has gained popularity in both fundamental and clinical research, due to the many advantages associated to this technique. Absence of sample preparation for Raman measurements (Experimental procedures), for instance, is one of the many assets that have been successfully applied *in situ*, *in vitro*, and *in vivo* in the life sciences context. Some examples are listed below.

In the life-sciences field, the use of cell cultures as models of different human pathologies is widely extended, due to their easy handling and lower costs, but most importantly, aiding in the reduction of the number of animals for further physiological studies. Confocal Raman microscopy is now applied to cell-culture studies due to its capacity to reveal molecular-specific information within a sample, adding spatial resolution to the acquired data. One major advantage is that this technique can give valuable information from single cells [22], and it is even possible to determine differences in cell-differentiation states [23]. Given that these cell cultures are also commonly used as an assessment tool for drug efficacy testing, it is possible to apply a confocal Raman microscope setup, in order to monitor anticancer drug efficacy in dried and living cells [24,25]. The authors were able to detect changes in the cell nucleus due to drug treatment in dry samples. In this scenario, they showed differences in specific Raman peaks within the spectrum whose intensity was decreased after the treatment, mainly at 785 cm^{-1} (OPO stretch DNA), 1086 cm^{-1} (PO_2 stretch DNA/RNA), and 1575 cm^{-1} (guanine, adenine) corresponding to nuclear degradation and decreased DNA/RNA content [24,25]. These results were then confirmed in living cells subjected to the same treatment [24].

Although traditionally 2D cell cultures have been used to gain an insight into cellular physiology and behavior prior to *in vivo* trials, 3D cell cultures are now accepted as better predictors [26]. By using 3D cultures that recapitulate the heterogeneity of avascular tumors (organoids), it was shown that it is possible to define differentiated spatial regions within the organoids with Raman spectroscopy [27]. The necrotic core displayed higher autofluorescence, whereas higher protein and peptides bands were localized away from this core. Ahlf and coworkers also took advantage of confocal Raman microscope and were able to define three distinct regions within the organoids: (1) a proliferative outer region, (2) a necrotic core, and (3) a transition intermediate zone. In the field of regenerative therapies, confocal Raman microscopy has also been successfully used in the study of the extracellular matrix deposition in bioengineered scaffolds developed by Kunstar and coworkers [28]. The Raman bands at 937 and 1062 cm^{-1}, which correspond to collagen and sulphated glycosaminoglycans were used to monitor the extracellular matrix formed by the cells growing in these scaffolds.

Not only the study of mammalian cells can benefit with this technology; in fact, it has also been shown that it is possible to distinguish different bacteria by their specific Raman spectrum [29]. In addition, it can aid in the study of biofilm formation [30].

In addition to the qualitative and quantitative evaluation of the molecular compositions of solid, liquid, and gaseous samples, Raman spectroscopy currently encompasses several *in vivo* research and diagnostic applications ranging from (1) neurodegenerative diseases; (2) skin studies; (3) cancer detection and severity assessment (breast, bladder, cervical, and colorectal); (4) allograft rejection (cardiac and renal); and (5) immunology [31]. It has also been employed to track nanomaterials moving through the circulation and inhaled nanoparticles in the respiratory tract, as reviewed in Tu et al. [32].

Raman spectroscopy also constituted a successful working tool in the characterization of a rat model of spinal cord injury (SCI). Saxena and coworkers showed that there are specific differences within the Raman spectrum at the injury site, in comparison to the uninjured spinal cord. Mainly, they found a decrease in the peaks at 700 and 1450 cm^{-1} pointing out to axon demyelination and cell death. In addition being indicators of the formation of the glial scar, they found a large increase in the chondroitin sulphate proteoglycan peak (CSPG) below 600 cm^{-1} and the peak area within the region from 950 to 1000 cm^{-1} [33].

Some *straightforward* clinical applications are emerging for clinical dermatology, due to easy skin accessibility and the unnecessary use of invasive probes. For example, Raman spectroscopy can be used for diagnosis of skin pathologies to aid in the decision-making for the best treatment in melasma patients [34]. With Raman spectroscopy, it is possible to recognize nonrespondent patients in which their skin Raman spectrum shows alterations in the peaks associated with melanin (1350 and 1580 cm^{-1}), and with protein and lipids (1450–1544 cm^{-1}).

The greatest efforts to apply Raman spectroscopy are performed within the cancer research field, since thousands of publications have appeared showing the use of this technology in oncology and in the search of biomarkers. In the earliest 1990s, Alfano and coworkers first employed Raman spectroscopy to assess the differences between cancerous and noncancerous samples from human breast tissue [35]. In the same direction, Brozek–Pluskaa [36] team has postulated that Raman spectra and mapping images are sensitive indicators of the distribution of lipids, proteins, and carotenoids in breast tissue for its use as putative Raman biomarkers. Advances in cancer diagnosis, using Raman spectroscopy, have been accomplished by Huang group where in the endoscopy they employed a fiber-optic probe for the *in vivo* diagnosis of esophageal squamous cell carcinoma and gastric dysplasia [37,38]. Under the scope of human brain cancer and the development of new tools for improving surgery procedures in oncological patients, Jermyn et al. developed a contact Raman probe that enabled the local detection and differentiation of the neoplasic *versus* normal cells, with a sensitivity of 93% and a specificity of 91% [39], which could represent a significantly important tool for surgical resection and decision-making in real time.

A further promising application of Raman spectroscopy in the clinics is the detection of cardiac allograft rejection. By detecting the Raman signal arising from serotonin within the endomyocardial biopsies from the heart-transplanted patients, during regular surveillance procedures, it is possible to monitor this rejection biomarker [40].

Another illustration for the potential use of Raman spectroscopy into the clinics is exemplified by a work published in the *Journal of the Royal Society of Chemistry*, where the authors showed the promising use of this technology in the diagnosis of malaria infection severity. The malaria pigment (hemozoin), a substance difficult to degrade, accumulates within macrophages during malaria infection. Through Raman spectroscopy it has been shown that it is possible to identify this substance without any previous sample processing (1638 cm^{-1} peak), and the measurement of its amount, within the cell, can be used as a predictor for disease severity [41].

In addition, Raman can be used as an adjunct to computed tomography, ultrasound, and magnetic resonance imaging in addition to cystoscopy, urine cytology, or bladder biopsy—providing more accurate differential diagnoses [28,42].

Limitations

Biological macromolecules display low Raman scattering cross sections together with high levels of autofluorescence, leading to a spectrum with low signal-to-noise ratio [43]. This limitation can be reduced when using wavelengths within the 700 to 900 nm range, since biological samples display lower autofluorescence under these conditions. However, these particular wavelengths might lead to a spectrum with a poor signal, when utilized as excitation sources. To circumvent this situation, several Raman enhancement mechanisms have arisen, which are mainly focused on two features: (1) polarizability and electric field by resonance Raman scattering (RR) and (2) surface-enhanced Raman scattering (SERS).

In RR, the wavelength needed for the electronic transition of a molecule is used as the excitation wavelength. However, with this method we increase not only the inelastic scattered light (by up to eight orders of magnitude) improving the signal-to-noise ratio and allowing for the detection of low concentrated molecules (10^{-3}–10^{-5} M), but we can also enhance fluorescence and luminescence. To exclude fluorescence ultrafast, shutters must be employed [44]. Although in nonresonant Raman we can obtain information on the ground-state structures of a molecule, with resonant Raman the geometry of the excited state can be described. The resulting Raman spectrum is less complex, as only the modes that are coupled to the specific excitation wavelength are enhanced.

SERS is based on the property of noble metal nanoparticles that present strong surface curvature and roughness, enhancing its Raman scattering by up to 15 orders of magnitude (which might allow for single-molecule detection) [45]. With this technique, it has been described the detection of ultratrace levels of analytes down to single molecules [46]. The technology is widely used in combination with antibodies for the detection of specific biomolecules in tissues [47]. The demonstration that Raman signals from pyridine were significantly enhanced when adsorbed onto a roughened Ag electrode was an important turning point in the employment of SERS, as a technique for single-molecule detection [42] and overall research for biomedical application. In that regard, it has been utilized in cancer diagnostics, identification of new biomarkers, single-nucleotide polymorphisms, and circulating tumor cells [48] as well as nucleic acid, ribonucleic acid, proteins, and lipids [48]. Moreover, SERS may be a central imaging-based immunoassay technique, as Lee and colleagues [49] demonstrated, by manufacturing gold-patterned microarray chip and hollow gold nanospheres (HGNs). They were able to target and image specific cancer markers in the surface of live cells and validate that this technique could be also be employed for reproducible immunoanalysis in sandwich ELISA microarray.

Coherent anti-Stokes Raman spectroscopy (CARS) is more sensitive than spontaneous Raman scattering making it more convenient for *in vivo* Raman imaging [50]. This technique involves the simultaneous use of two or three laser lines focused on the same spot at the sample. When the difference of the laser beam frequencies matches one of the Raman active molecules, a stronger anti-Stokes signal is generated [51]. As an example, this technology has been employed for the analysis of lipid droplet composition in a mouse model of nonalcoholic steatohepatitis in nonfixed samples [52]. The intensity ratio of the unsaturated peak CH_2 (1444 cm^{-1}) to the unsaturated C=C (1660 cm^{-1}) and the presence of a peak at 1730 cm^{-1}, revealed that the majority of the fatty acids in the lipid droplets are saturated in an esterified form.

Stimulated Raman scattering (SRS) microscopy is also a technique that uses two laser beams focused at the sample, increasing the efficiency of Raman spectroscopy [53,54]. Briefly, when the difference between both frequencies matches a specific vibrational frequency, the spontaneous Raman signal is amplified. An advantage of this technology is also that the contribution of tissue autofluorescence might be diminished. For a comprehensive review on biomedical applications, Reference 53.

Another drawback of this technology is the limit in resolution that is inherent to the use of a confocal microscope. In this regard, different technological combinations expand our possibilities regarding not only increase in resolution but also in signal strength. Among them are (1) tip-enhanced Raman scattering (TERS), (2) surface-enhanced resonance Raman scattering (SERRS), (3) Raman + scanning near-field optical microscopy (SNOM), and (4) the Raman-scanning electron microscope (SEM) combination.

TERS is the combination of SERS with atomic force microscopy (AFM) or scanning probe microscopy (SPM). With this technique, the chemical sensitivity or Raman is coupled to the spatial and topological resolution of the AFM, decreasing the lateral resolution below the diffraction limit [45]. In brief, the AFM tip must be metal coated and to enhance the Raman spectra, the laser is focused onto the tip apex. The characteristics of the tip apex will determine the final lateral resolution.

SERRS combines the resonance Raman with the SERS. With this technology, it is possible to acquire stronger signals than the ones obtained by SERS and multiplexing has also been addressed [55]. In the bioengineering field, Harmsen and colleagues described the use of gold nanostars for quantitative detection of four different particles for *in vitro* and *ex vivo* studies. In cancer imaging, SERRS technology has been also successfully applied; for a short review in this field see Reference 56.

With the combination of Raman and SNOM it is possible to overcome the light diffraction limit encountered when working with confocal Raman microscopy where we can obtain high-resolution images down to 60 nm [57].

Finally, the Raman–SEM combination is a technique that allows for the chemical detection of structures in the nanometer range [58]. As an example under the biomedicine scope, van Apeldoorn and coworkers demonstrated that the extracellular matrix of the early *in vitro* bone resembled that of the mature bone.

Another limitation for the application of this technology in the life-sciences field is the lack of specific databases and libraries with information on biomolecular spectra and peak assignment contrary to what happens within chemistry and pharmacology fields.

Future

Raman spectroscopy has drum up the interest in the biomedical and life-sciences fields recently due to its great informative potential regarding chemical composition. In the past few years there have been great efforts to improve the lateral resolution of a confocal Raman microscope by combination with other technologies. Interestingly, in addition to sample characterization, it is also a possible quantification for comparison purposes, even until the ultratrace level. However, there are some drawbacks when using this technology within the life-science field. The application of Raman spectroscopy within the chemistry field is not new and there are several libraries with Raman information for different reference spectra (i.e., small inorganic compounds, minerals, etc.), which is quite convenient, for example, in the pharmaceutics field, where it is relatively simple to analyze the spectra of a given tablet formulation. However, biological samples are extremely complex and such reference libraries are scarce, which makes peak assignment quite a complicated task. In this regard, there is an urgent need for the development of libraries from biological compounds spectrum, together with databases with peak assignments. In this direction, Movasaghi and coworkers published a list of peak assignment on tissues and cells based on published work [59]; yet, further and up-to-date information are needed not only regarding characterization under healthy conditions but also in the disease context.

This technology lacks the availability of more user-friendly equipment together with automated analysis software. Probably the biggest milestone for this technology might be the development of a high-throughput platform, which will favor an exponential growth in its application in both life-sciences and biomedical fields, namely in drug discovery, diagnostics, and the clinics.

Acknowledgments

This work was financed by FEDER—Fundo Europeu de Desenvolvimento Regional funds through the COMPETE 2020—Operacional Programme for Competitiveness and Internationalisation (POCI), Portugal 2020, and by Portuguese funds through FCT—Fundação para a Ciência e a Tecnologia/Ministério da Ciência, Tecnologia e Inovação in the framework of the project *Institute for Research and Innovation in Health Sciences* (POCI-01-0145-FEDER-007274). Ana Freitas acknowledges FCT for her PhD scholarship (SFRH/BD/111423/2015). The authors thank Dr. Con Mallidis and Dr. Joachim Wistuba from the Center for Reproductive Medicine and Andrology, from the University Clinic of Munster in Germany, the grant of use of data on sperm mapping.

References

1. Raman, C.V. and K.J. Krishnan, A new type of secondary radiation. *Nature*, 1928. **121**: 2.
2. Larkin, P., Chapter 2–Basic principles, in *Infrared and Raman Spectroscopy*, P. Larkin, (Ed.) 2011, Oxford: Elsevier. pp. 7–25.
3. Pettinger, B., in *Adsorption at Electrode Surface*, J. Lipkowski and. P.N. Ross, (Eds.) 1992. pp. 285.
4. Schaeberle, M.D. et al., Peer reviewed: Raman chemical imaging spectroscopy. *Anal Chem*, 1999. **71**(5): 175A–181A.
5. Majzner, K. et al., 3D confocal Raman imaging of endothelial cells and vascular wall: Perspectives in analytical spectroscopy of biomedical research. *Analyst*, 2013. **138**(2): 603–610.
6. Mallidas, C. and J. Wistuba, *Raman analysis of sperm nuclear DNA integrity*. Horiba Scientific Application Note, 2013.
7. Lyng, F., E. Gazi, and P. Gardner, Preparation of tissues and cells for infrared and Raman spectroscopy and imaging in *Biomedical Applications of Synchrotron Infrared Microspectroscopy*, D. Moss, (Ed.) 2011. pp. 147–185.
8. Meade, A.D. et al., Studies of chemical fixation effects in human cell lines using Raman microspectroscopy. *Anal Bioanal Chem*, 2010. **396**(5): 1781–1791.
9. Faolain, E.O. et al., Raman spectroscopic evaluation of efficacy of current paraffin wax section dewaxing agents. *J Histochem Cytochem*, 2005. **53**(1): 121–129.

10. Wills, H. et al., Raman spectroscopy detects and distinguishes neuroblastoma and related tissues in fresh and (banked) frozen specimens. *J Pediatr Surg*, 2009. **44**(2): 386–391.
11. Huang, Z. et al., Effect of formalin fixation on the near-infrared Raman spectroscopy of normal and cancerous human bronchial tissues. *Int J Oncol*, 2003. **23**(3): 649–655.
12. Shim, M.G. and B.C. Wilson, The effects of ex vivo handling procedures on the near-infrared Raman spectra of normal mammalian tissues. *Photochem Photobiol*, 1996. **63**(5): 662–671.
13. Mariani, M.M. et al., Impact of fixation on in vitro cell culture lines monitored with Raman spectroscopy. *Analyst*, 2009. **134**(6): 1154–1161.
14. Yeni, Y.N. et al., Effect of fixation and embedding on Raman spectroscopic analysis of bone tissue. *Calcif Tissue Int*, 2006. **78**(6): 363–371.
15. Carey, P.R., Raman spectroscopy, the sleeping giant in structural biology, awakes. *J Biol Chem*, 1999. **274**(38): 26625–26628.
16. Zhang, D. et al., Ratiometric Raman spectroscopy for quantification of protein oxidative damage. *Anal Biochem*, 2009. **391**(2): 121–126.
17. Gryniewicz, C.M. and J.F. Kauffman, Multivariate calibration of covalent aggregate fraction to the raman spectrum of regular human insulin. *J Pharm Sci*, 2008. **97**(9): 3727–3734.
18. Ortiz, C. et al., Analysis of insulin amyloid fibrils by Raman spectroscopy. *Biophys Chem*, 2007. **128**(2–3): 150–155.
19. Xie, Y. et al., The Raman detection of peptide tyrosine phosphorylation. *Anal Biochem*, 2004. **332**(1): 116–121.
20. Dieing, T. and W. Ibach, Software requirements and data analysis in confocal Raman microscopy. *Confocal Raman Microscopy*, ed. Springer series in Optical Sciences. Vol. 158. 2010. Berlin Heidelberg: Springer.
21. Osterberg, Clinical and investigative applications of Raman spectroscopy in urology and andrology. *Transl Androl Urol*, 2013. **3**(1): 84–88.
22. Neugebauer, U. et al., Identification and differentiation of single cells from peripheral blood by Raman spectroscopic imaging. *J Biophotonics*, 2010. **3**(8–9): 579–587.
23. Ichimura, T. et al., Visualizing cell state transition using Raman spectroscopy. *PLoS One*, 2014. **9**(1): e84478.
24. Brautigam, K. et al., Raman spectroscopic imaging for the real-time detection of chemical changes associated with docetaxel exposure. *Chemphyschem*, 2013. **14**(3): 550–553.
25. Krafft, C. et al., Mapping of single cells by near infrared Raman microspectroscopy. *Vib Spectrosc*, 2003. **32**(1): 75–83.
26. Edmondson, R. et al., Three-dimensional cell culture systems and their applications in drug discovery and cell-based biosensors. *Assay Drug Dev Technol*, 2014. **12**(4): 207–218.
27. Ahlf, D.R. et al., Correlated mass spectrometry imaging and confocal Raman microscopy for studies of three-dimensional cell culture sections. *Analyst*, 2014. **139**(18): 4578–4585.
28. Kunstar, A. et al., Label-free Raman monitoring of extracellular matrix formation in three-dimensional polymeric scaffolds. *J R Soc Interface*, 2013. **10**(86): 20130464.
29. Beier, B.D., R.G. Quivey, and A.J. Berger, Identification of different bacterial species in biofilms using confocal Raman microscopy. *J Biomed Opt*, 2010. **15**(6): 066001.
30. Lanni, E.J. et al., Correlated imaging with C60-SIMS and confocal Raman microscopy: Visualization of cell-scale molecular distributions in bacterial biofilms. *Anal Chem*, 2014. **86**(21): 10885–10891.
31. Choo-Smith, L.P. et al., Medical applications of Raman spectroscopy: From proof of principle to clinical implementation. *Biopolymers*, 2002. **67**(1): 1–9.

32. Tu, Q. and C. Chang, Diagnostic applications of Raman spectroscopy. *Nanomedicine*, 2012. **8**(5): 545–558.
33. Saxena, T. et al., Raman spectroscopic investigation of spinal cord injury in a rat model. *J Biomed Opt*, 2011. **16**(2): 027003.
34. Moncada, B. et al., Raman spectroscopy analysis of the skin of patients with melasma before standard treatment with topical corticosteroids, retinoic acid, and hydroquinone mixture. *Skin Res Technol*, 2016. **22**(2): 170–173.
35. Alfano, R.R. et al., Human breast tissues studied by IR Fourier transform Raman spectroscopy. *Lasers Life Sci*, 1991. **4**(1): 6.
36. Brozek-Pluska, B. et al., Raman spectroscopy and imaging: Applications in human breast cancer diagnosis. *Analyst*, 2012. **137**(16): 3773–3780.
37. Wang, J. et al., Simultaneous fingerprint and high-wavenumber fiber-optic Raman spectroscopy improves in vivo diagnosis of esophageal squamous cell carcinoma at endoscopy. *Sci Rep*, 2015. **5**: 12957.
38. Wang, J. et al., Comparative study of the endoscope-based bevelled and volume fiber-optic Raman probes for optical diagnosis of gastric dysplasia in vivo at endoscopy. *Anal Bioanal Chem*, 2015. **407**(27): 8303–8310.
39. Jermyn, M. et al., Intraoperative brain cancer detection with Raman spectroscopy in humans. *Sci Transl Med*, 2015. **7**(274): 274ra19.
40. Chung, Y.G. et al., Raman spectroscopy detects cardiac allograft rejection with molecular specificity. *Clin Transl Sci*, 2009. **2**(3): 206–210.
41. Hobro, A.J. et al., Label-free Raman imaging of the macrophage response to the malaria pigment hemozoin. *Analyst*, 2015. **140**(7): 2350–2359.
42. Osterberg, E.C. et al., Clinical and investigative applications of Raman spectroscopy in urology and andrology. *Transl Androl Urol*, 2013. **3**(1): 84–88.
43. Monici, M., Cell and tissue autofluorescence research and diagnostic applications. *Biotechnol Annu Rev*, 2005. **11**: 227–256.
44. Smith, Z.J. et al., Rejection of fluorescence background in resonance and spontaneous Raman microspectroscopy. *J Vis Exp*, 2011. (51): 2592.
45. Kneipp, K. et al., Single molecule detection using surface-enhanced Raman scattering (SERS). *Phys Rev Lett*, 1997. **78**(9): 1667–1670.
46. Culha, M. et al., Surface-enhanced Raman scattering as an emerging characterization and detection technique. *J Nanotechnol*, 2012. **2012**: 15.
47. Lutz, B. et al., Raman nanoparticle probes for antibody-based protein detection in tissues. *J Histochem Cytochem*, 2008. **56**(4): 371–379.
48. Ito, H. et al., Use of surface-enhanced Raman scattering for detection of cancer-related serum-constituents in gastrointestinal cancer patients. *Nanomedicine*, 2014. **10**(3): 599–608.
49. Lee, M. et al., Highly reproducible immunoassay of cancer markers on a gold-patterned microarray chip using surface-enhanced Raman scattering imaging. *Biosens Bioelectron*, 2011. **26**: 2135–2141.
50. Evans, C.L. et al., Chemical imaging of tissue in vivo with video-rate coherent anti-Stokes Raman scattering microscopy. *Proc Natl Acad Sci USA*, 2005. **102**(46): 16807–16812.
51. Begin, S. et al., In vivo optical monitoring of tissue pathologies and diseases with vibrational contrast. *J Biophotonics*, 2009. **2**(11): 632–642.
52. Adkins, Y. et al., A novel mouse model of nonalcoholic steatohepatitis with significant insulin resistance. *Lab Invest*, 2013. **93**(12): 1313–1322.
53. Freudiger, C.W. et al., Label-free biomedical imaging with high sensitivity by stimulated Raman scattering microscopy. *Science*, 2008. **322**(5909): 1857–1861.
54. Yakovlev, V.V. et al., Stimulated Raman scattering: Old physics, new applications. *J Mod Opt*, 2009. **56**(18–19): 1970–1973.
55. Harmsen, S. et al., Surface-enhanced resonance Raman scattering nanostars for high-precision cancer imaging. *Sci Transl Med*, 2015. **7**(271): 271ra7.

56. Feuillie, C., Raman spectroscopy in biomedicine: New advances in SERRS cancer imaging. *Ann Transl Med*, 2015. **3**(22): 347.
57. Kaupp, G., Scanning near-field optical microscopy on rough surfaces: Applications in chemistry, biology, and medicine. *Int J Photoenergy*, 2006. **2006**: 1–22.
58. van Apeldoorn, A.A. et al., Parallel high-resolution confocal Raman SEM analysis of inorganic and organic bone matrix constituents. *J R Soc Interface*, 2005. **2**(2): 39–45.
59. Movasaghi, Z., S. Rehman, and I.U. Rehman, Raman spectroscopy of biological tissues. *Applied Spectroscopy Reviews*, 2007. **42**(5): 493–541.

Chapter 6 Tracking cancer

In vivo *imaging techniques*

Patrícia Carvalho,
André Roma,
and Sérgia Velho

Contents

Introduction

For decades, the oncology research has been focusing on the search for specific targets not only for therapeutic but also for diagnosis and imaging purposes. Consequently, there is an obvious need to fill the gap between *in vitro* models and their possible clinical significance. In this line, small animal models have acquired a decisive role in cancer research being helpful from basic research to drug development and testing and also disease follow up [1,2]. There are several available mouse models to study cancer: ectopic xenografts, orthotopic models, germ-line transgenic and conditional transgenic models (also known as genetically engineered mouse models [GEMMs]), primary human tumor grafts, and various carcinogen-induced models, which are used to study specific tumors and answer different questions [2]. Thus, the existence of accurate and sensitive imaging techniques for these living models becomes essential and also advantageous as they represent noninvasive techniques that permit the assessment of tumor development and whole-body disease burden evaluation without the need of necropsy. This allows the implementation of longitudinal studies, leading to a reduction in the number of animals required for an experiment, and providing increased statistical power, as each animal functions as its own control. In addition, the use of *in vivo* imaging complies with the principle of the 3 *R's*—replacement, reduction, and refinement—within the context of animal welfare, as these are less invasive techniques and allow a reduction in the number of sacrificed animals [1,3].

Mouse imaging techniques have emerged by *scaling-down* clinically used devices, including computed tomography (CT), magnetic resonance imaging (MRI), positron emission tomography (PET), single photon emission computed tomography (SPECT), and ultrasound (US), whereas others, such as optical imaging with fluorescent and bioluminescent markers, were *scaled-up* from *in vitro* microscopic cellular imaging [3,4]. Each technique has different inherent characteristics and serves different purposes (Table 6.1). As so, a careful selection of both the most appropriate animal model and imaging systems, together with an optimal study design, are crucial decisions that will determine the translational impact of the research [1].

In vivo imaging has been witnessing great technological advances in the past decades, enabling the extension of their applications' list to the point that some of the techniques have sufficient in-depth to study various cancer hallmarks [5].

In this chapter, we will explore the main techniques used for cancer imaging in mouse models, addressing their major applications, strengths, and drawbacks and also some promising future perspectives.

Applications and limitations

Imaging systems can be categorized considering different characteristics, such as the energy used to derive visual information (X-rays, positrons, photons, and sound waves), the spatial resolution attained (macro-, meso-, and microscopic), or the type of information obtained (anatomic, physiological, and molecular/cellular). CT, MRI, and US are macroscopic imaging systems that can provide anatomic and physiological information and are widespread used techniques in clinical and preclinical settings. Molecular imaging systems, such as PET and SPECT, are also used in both clinical and experimental contexts, whereas bioluminescence imaging and fluorescence-based techniques are restricted to experimental use [6]. Each imaging modality has its own characteristics that should be cautiously taken into account during study design. However, there are some questions that are transversal to all techniques, for instance, the need to have anesthetized animals,

Table 6.1 Summary of the Main Strengths, Limitations, and Applications of Mouse Imaging Techniques

Imaging Technique	Detection Method	Strengths	Limitations	Main Applications
Computed Tomography	Ionizing radiation (X-rays)	• High resolution in tissues with inherent contrast • Availability of contrast agents	• Radiation dose • Poor soft-tissue contrast	• Tumor/metastasis detection (contrast agents are needed for tissues with no natural radiographic contrast) • Tumor angiogenesis characterization
Magnetic Resonance Imaging	Electromagnetism	• High resolution • Good tissue contrast • Availability of different variants and contrast agents	• High space and hardware requirements	• Tumor/metastasis detection
Ultrasound	Sound waves	• High spatial and temporal resolution • Good soft-tissue contrast • Availability of molecularly targeted contrast agents	• Directed imaging (impossibility of whole-body imaging)	• Tumor/metastasis detection and measurement • Molecular characterization of angiogenesis (Molecular US)
Positron Emission Tomography/Single Photon Emission Computed Tomography	Ionizing radiation (γ-rays)	• Highly sensitive and specific • Molecular targets	• Poor spatial resolution/ no anatomical context (frequently need to be combined with other techniques) • Radioactive probes	• Tumor/metastasis detection • Tumor metabolism assessment • Therapy response evaluation with molecular targets
Bioluminescence	Bioluminescent light	• High sensitive • Low background • Molecular targets	• Low spatial resolution • No clinical application	• Tumor and metastasis detection • Molecular evaluation of events, such as apoptosis and receptor activity
Fluorescence Imaging	Fluorescent light	• Nonradioactive probes • Permit to follow biological processes on its natural environment (in the case of IVM) • Molecular targets	• High scattering of photons • Tissue autofluorescence • Invasive procedures are needed (in the case of IVM)	• Tumor and metastasis detection • Therapy response evaluation (through the molecular evaluation of processess such as apoptosis) • Tumor microenvironment and cell–cell interactions study

which require close monitoring, during the time of data acquisition in order to reduce motion artifacts and obtain optimal image quality, and a careful design of the three essential steps of experimental workflow: planning, imaging, and analysis [4].

Computed tomography

CT is a three-dimensional (3D) X-ray imaging method that consists of obtaining X-ray projection images at many rotation angles of the body and then applying a tomographic reconstruction algorithm to generate a stack of thin tomographic images of contiguous transaxial slices through the body. There are three levels of microscopic CT based on their spatial resolutions: mini-CT (200–50 μm), micro-CT (50–1 μm), and nano-CT (1–0.1 μm) [7]. Micro-CT is one of the most widely used imaging techniques, due to its relatively low cost, high resolution, and scanning efficiency [8,9]. In small animal models, it is successfully used for the detection and tracking of tumors, being particularly useful for lung and bone tumors and metastasis detection, as these tissues display a natural radiographic contrast [10,11]. Furthermore, micro-CT applications extend beyond tumor detection, for instance, it can be used to quantitatively study the kinetics of tumor development and response to therapy and measuring the total tumor burden, as done in nonsmall cell lung carcinoma GEEMs [12,13]. Moreover, the use of contrast agents, for example, iodine-containing nanoparticles and gold nanoparticles, adds some value to the applications' list as it allows imaging of soft tissues, such as the liver, that has no inherent contrast [9,14]. The use of contrast agents is also intimately related to tumor vasculature characterization, allowing the quantification of the blood flow, blood volume, microvessel density, and vascular permeability [15–18]. The poor soft tissue contrast and the radiation dose associated with X-ray imaging constitute the major limitations of this technique. In addition, working with small-sized animals imposes some limitations related to equipment and system resolution constraints: to obtain a higher resolution, a significant time and radiation dose are needed. Still, all of these can be overpassed, resorting, for example, to several image reconstruction strategies and contrast agents, enabling the acquisition of high-quality images in anesthetized rats and mice with short X-ray exposure intervals (typically <10 min) [8,19].

Magnetic resonance imaging

MRI is a digital technique that permits the acquisition of 3D images with excellent tissue contrast and spatial resolution, allowing the visualization of the whole animal or its organs in any plane [20]. In comparison to X-rays that rely on tissue density differences to obtain contrast, MRI provides excellent soft tissue contrast, displaying its physical characteristics without the need for ionizing radiation. In this technique, the source of the signal comes from the polarization of protons (e.g., ^{1}H) in tissues when subjected to a strong magnetic field, with the subsequent emission of weak radio waves that can be detected by a MR scanner during the relaxation phase. In order to obtain the same level of detail as in humans, rodents MRI devices use higher field strengths, that result in higher spatial resolution, signal-to-noise parameters, but on the other hand introduce some imaging artifacts and represent a significant hardware and space requirement [4]. Challenges related to small animal MRI instruments scaling will further need physics and engineering support [3,4].

MRI techniques can be divided in essentially two groups: (1) those that use extrinsic contrast agents (e.g., dynamic–contrast–enhanced [DCE]-MRI and MRI with targeted agents) and (2) those that do not (e.g., arterial spin labeling [ASL] and blood oxygenation level-dependent [BOLD] imaging) [21]. This imaging modality is used in tumor detection, for example, of brain tumors in early stages [22] and, most commonly DCE-MRI and ASL, are particularly used for tumor angiogenesis imaging. DCE-MRI (mainly with intravenous injection of gadolinium-based contrast) provide information about microvasculature characteristics, including vascular permeability/surface area product and extravascular volume by measuring the leakage rate of low molecular weight MR contrast agents into tumors [23]. For instance, it was used to functionally and structurally characterize and distinguish angiogenesis induced by the overexpression of different vascular endothelial growth factor (VEGF) isoforms in murine xenograft models [24]. Furthermore, DCE-MRI is used to evaluate and monitor the efficacy of antiangiogenic therapies in various rodent tumor models, such as subcutaneous lung cancer model [25] and orthotopic thyroid cancer model [26]. Moreover, DCE-MRI has increasingly been recognized as a good method to provide biomarkers to assess the radioresponsiveness and metastatic potential of tumors [27–30]. ASL is used to noninvasively quantify the perfusion (delivery of oxygen and nutrients to tissue by means of blood flow) by magnetically labeling the arterial water in the incoming blood as a freely diffusible endogenous tracer [31]. This technique has been successfully used to measure tumor volume, blood volume, and tumor blood flow in rat glioma models [32,33] and also to evaluate the response to antiangiogenic therapy in other models [34].

Ultrasound

US imaging is achieved by exposing tissues to high-frequency US waves (20–60 MHz in animals; 2–10 MHz in humans) when placing a transducer (which contains crystals that vibrate when exposed to small electrical currents and produce sound waves) on the skin and detect the US reflections from the internal organs. This technique provides a dynamic real-time image from which structural and functional information can be obtained, and offers good soft tissue contrast with good spatial and temporal resolution, without the use of radiation or contrast agents. Other advantages of this technique rely on its versatility, portability, safety, and cost-effectiveness [35–37]. However, as it is a directed imaging technique, it limits the field of view and cannot be used for whole-animal imaging [4]. Despite being a 2D technique, the 3D reconstruction of US images significantly expands its value beyond the tumor detection, since it turns possible to measure tumor volume and monitor tumor progression. With this intent, US imaging has been applied in various mouse models, such as orthotopic mouse models of pancreatic cancer [37], genetically engineered [38] and orthotopic [39] prostate tumor mouse models, genetically engineered mouse [36] and orthotopic mouse and rat [40] mammary tumor models, and orthotopic mouse models of bladder tumors [41]. Moreover, US imaging has been used to distinguish and characterize mammary fibroadenomas and carcinomas [42], track the growth of liver metastases, and evaluate potential chemotherapeutics [43], and it is the method of choice for guiding the injection/delivery of therapies [44–46].

Recent advances have made possible to upgrade US imaging into a molecular, multimodal, and theranostic technique that is moving forward toward clinical translation. Molecular US imaging or targeted contrast-enhanced ultrasound has emerged by the introduction of novel molecularly-targeted contrast agents, most commonly microbubbles (1 and 4 μm sized gas bubbles that are composed of an insoluble gas stabilized by a shell) that remain in the vascular compartment and can be functionalized to recognize specific molecular markers. The use of microbubbles offer the opportunity to molecularly characterize angiogenesis and evaluate therapy effects and besides that, it opens a window to theranostic delivery of targeted therapies [47–50]. Some examples of the applications of this technology are: (1) human kinase insert domain receptor (KDR)-targeted microbubbles were used to monitor antiangiogenic therapy in a colon cancer xenograft mice model [51] and (2) microbubbles were used to assess the expression dynamics of the angiogenic markers $\alpha V\beta 3$ integrin, endoglin, and vascular endothelial growth factor receptor type 2 (VEGFR2) in tumor vascular endothelial cells of subcutaneous breast, ovarian, and pancreatic cancer xenografted mice [52]. Most importantly, a molecular ultrasound contrast agent has already entered in clinical trials to visualize expression levels of VEGFR2 in prostate cancer patients [50].

Positron emission tomography and single photon emission computed tomography

PET and SPECT are nuclear imaging modalities that rely on the administration of radiolabeled tracers to acquire information on physiological and biological processes. These techniques allow the detection of disease-related biochemical and physiological abnormalities before the appearance of anatomical changes that are visualized by other imaging modalities such as CT and MRI. PET uses radioisotopes that decay via emission of positrons, whereas SPECT uses radioisotopes that decay by electron capture and/or gamma emission [53]. These are highly sensitive (can detect picomolar concentrations of probe) and specific (the more specific is the probe the less normal anatomy is visualized) imaging modalities, although displaying poor spatial resolution. Owing to this limitation, PET and SPECT are frequently used in combination with CT to provide anatomic context. PET–CT, SPECT–CT, and even PET–SPECT–CT hybrid instruments are already available [4].

Radioligands used in PET and SPECT imaging are designed with the same rationale: target and take advantage of specific aspects of tumor biology, including molecular biomarkers, such as overexpression of growth factor receptors and protein kinases or biological events like angiogenesis, apoptosis, hypoxia, and tumor proliferation. These two imaging techniques can have outstanding value in many aspects of the drug development process, providing useful information that can improve the efficiency and cost-effectiveness of clinical trials [53]. PET with the radiotracer 2-deoxy-2-(^{18}F) fluoro-D-glucose (^{18}F-FDG), a glucose analog, are the most commonly used in human and mice cancer imaging. This method exploits the increased expression of glucose transporters and the activation of the glycolytic pathway in cancer cells to preferentially metabolize and retain ^{18}F-FDG. Besides being used to tumor detection purposes, PET–FDG has been used in various cancer models, for instance, in lung cancer [54], glioblastoma, and lymphoma [55] to assess tumor metabolism and evaluate the effects of anticancer drugs, demonstrating its value in pharmacodynamics studies. Recently, in a prostate xenograft model, caged FDG glycosylamines were used to explore the characteristic acidic pH of tumor microenvironments to develop a prodrug strategy that targets this characteristic, with potential clinical translation [56]. Several other probes, designed to target different cancer cell markers [57–60], that can even be functionalized with antibodies [61] or antibody mimetics [62], have been developed and are available for cancer and metastasis imaging in various models. SPECT has, in the same line of PET, been successfully used in tumor and metastasis detection and therapy response evaluation, using probes that target, for example, human epidermal growth factor receptor 1, 2, and 3 [63–65], urokinase-type plasminogen activator receptor [66], carcinoembryonic antigen [67] and $\alpha_5\beta_1$ integrin [68]. Most importantly, SPECT imaging of a radiolabeled probed targeting claudin-4 (a protein that is overexpressed in several premalignant precursor lesions) allowed the detection of precancerous aplastic lesions in mouse models of breast cancer, establishing a new early detection tool [69].

Bioluminescence

Bioluminescence imaging makes use of the light-emitting reaction that occurs between luciferase (*luc*) and its substrate, luciferin, in the presence of oxygen, ATP, and Mg^{2+}. *Luc* from the firefly *Photinus pyralis* is used in most cases, and its wavelength of emitted light ranges between 510 and 650 nm, peaking at 562 nm. Such range of wavelength is one of the main advantages of the use of *luc*, as near-red light is less prone to scattering and absorption by tissues [70]. This system is widely used in cancer research, with many studies being carried out in mice and rats. *Luc* vectors can be inserted into cells *in vivo*, using viral or nonviral vectors, or *in vitro* before implantation into the mice. Luciferin is administrated intraperitoneally or intravenously, being the latest the most efficient administration route, with a 4 to 10-fold higher light emission for the same administrated amount [71]. Light emission reaches its peak 5 to 20 min after luciferin injection, and is emitted for more than 1 h [72].

The emitted light is recorded inside a dark box, using a charge-coupled device (CCD) camera. Such cameras can be cooled to −120°C, not only to reduce thermal noise, but also to achieve higher sensitivity in the red range of the light, leading to the detecting of very low levels of emitted light [73]. Due to such sensitivity, luciferase-expressing tumors can be detected by this technique long before being palpable [72]. Signals at a depth up to 2 cm can be detected, although due to light scattering, this results in a low spatial resolution [73,74]. However, as luciferase is not endogenously expressed in mammalian cells, background signals are extremely low. Nevertheless, bioluminescence is strongly affected by tissues, in particular hemoglobin, which absorbs light in the green–blue regions [75]. Due to the high level of tumor vascularization and hypoxia with the increasing size of tumors, quantification of the tumor burden, and the possible analysis of therapy success, becomes more difficult [70].

The simplicity and versatility of this technique made it useful in different purposes. Both *Renilla* and *Photinus* luciferases were used to study tropism toward wounding and tumor microenvironment (TME) by mesenchymal stem cells, to track metastasis formation or to evaluate therapeutic response [76–78]. However, more complex systems can be created. To track metastasis formation by prostate cancer, Adams et al. [79] constructed a *luc* gene under transcriptional control of a prostate-specific promoter, and administrated it using a viral vector. Metastasis in lung and spine could be observed by bioluminescence in those places. In order to evaluate caspase-mediated apoptosis, Laxman et al. [80] made a fusion protein using *luc*, estrogen receptor regulatory domain and a target domain for caspases (DEVD sequence), resulting in the silencing of *luc* reporter. When caspase-mediated apoptosis was activated, the DEVD sequence was cleaved, allowing *luc* to catalyze the emission of light. Others have created transgenic mouse models that ubiquitously express *luc* associated with the activity of receptors, such as estrogen receptor, or mutated oncogenes, such as K-Ras and p53, for the study of cancer development and progression [81,82].

Planar imaging

Planar imaging techniques take advantage of the use of planar waves to illuminate a tissue, recording the emitted fluorescence. The approaches to collect the fluorescence emitted can be classified as epi-illumination (also referred as fluorescence-reflectance imaging [FRI]) or transillumination.

Epi-illumination or Fluorescence-reflectance imaging: FRI is a technique used to evaluate surface and subsurface fluorescence in the animals' body. Shine is directed toward side of the tissue and the emitted light from that same side is collected using a highly sensitive CCD camera.

Although FRI is a very popular technique due to its simplicity of use and relative low cost, it comes with significant drawbacks, due to the diffusive nature of photons and different biological characteristics, such as fluorochrome amount, depth, and degree of vascularization of the tumor. Owing to such limitations, this technique has restricted applications [6]. It can only be used in near-surface tumors, either in subcutaneous or in skin tumors, and in surgically exposed tumors, to evaluate size and location [6]. This can be attained by the use of probes such as cathepsin-sensitive fluorochromes or fluorochromes-conjugated antibodies to target specific proteins [83].

Transillumination: Transillumination is an alternative technique to epi-illumination and has been used since 1929 as a human breast evaluation technique called diaphanography. This method places the light source and detector in opposite sides of the tissue, shining light through it, and detecting the light attenuation or the fluorescence emitted.

Although transillumination has similar nonlinear dependencies as epi-illumination, it has the advantage of sampling the entire volume of interest. In addition, recent studies suggest that the normalization of measurements at the emission wavelength by measurements at the excitation wavelength improve quantification and contrast.

This technique has been applied to evaluate pharmacokinetics of photosensitizers [84] or as a complement for fluorescence molecular tomography (FMT) to evaluate the effects of chemotherapy in glioma [85].

Fluorescence molecular tomography: The pivotal principle behind tomographic techniques is shining light to a tissue from different projections and using the collected light to reconstruct, mathematically, the tissue in analysis. Specifically, in the case of FMT, the 3D reconstruction is based on the internal distribution of fluorochromes. Once again, due to the use of near-infrared (NIR) or visible radiation, there is high scattering of photons. Although, the tomographic nature of such technique can correct the errors that occur by the use of planar imaging techniques. Therefore, FMT is an important technique, able to identify specific zones where a biochemical process or disease is active due to the accumulation of a targeting probe. Reports of the use of FMT to evaluate proteases activity and integrin quantity, vascular volume, and specific tumor-related proteins are known [86–89].

On the technological point of view, the evolution from contact and fiber-based systems to noncontact technologies or flying spot illumination has recently been made. This made possible the rise of further advances, such as rapid whole-body imaging, complete projection tomography, or the combination with other imaging techniques [85,90].

Due to capacity of acquiring images from the whole body, this technique is applied on the evaluation of whole tumor characteristics, such as apoptosis as a result of chemotherapy administration [91] or evaluation of different aspects of the tumor simultaneously, such as tumor angiogenesis and expression of different receptors [92].

Intravital microscopy

Since its development, microscopy helped to elucidate our understanding of the biological processes. However, such system does not allow to fully understand the complexity of such mechanisms, as it is mostly performed *ex vivo* or *in vitro*, not fully reproducing what is observed *in vivo*. The development of intravital microscopy (IVM) can overcome such setbacks, allowing to follow biological processes on its natural environment [93].

In its primordial times, IVM involved bright-field illumination, which allowed the understanding of blood microcirculation, but with the advances on the use and detection of fluorescence, it made easy to study the localization, tracking, adhesion, and interaction of multiple cell types in their natural 3D environment [94].

Different microscopic technologies can be used. Confocal microscopy is cheap and simple to use, but due to limited tissue penetration, photobleaching, and photodamage it can cause problems when sampling *in vivo*. Multiphoton microscopy allows the acquisition of images in deeper planes by applying two low-energy photons to excite the fluorophore, reducing photobleaching and autofluorescence, making it preferable to use in *in vivo* studies [94].

Different approaches to access the tissue can be considered, such as implantation of windows, surgical procedures to make organs more accessible, or directly imaging of the tissue. Dorsal skin window chamber can be implanted in mice and used to image the tissue for extended periods. Cranial imaging windows can be used for evaluating brain tumor progression or metastization [95]. Other chambers such as abdominal [96,97] or mammary windows [98] are also frequently used in cancer research to evaluate metastization or extracellular matrix (ECM) integrity, for example. By adopting such approaches, different multifactorial studies can be applied to study processes such as signaling, metabolism or angiogenesis or even metastasis and migration, cell-cell interations, and therapy response.

Although a very useful technique, there are still some biological and technical limitations of this system. The current observations do not always evaluate the entire complexity of processes that occur *in vivo*. Thus, assessing a higher number of factors could help enlighten the knowledge about tissue organization and cell–cell communication. Another biological limitation is due to invasive procedures frequently required to access the tissues, some inflammatory and immune reactions are triggered [93]. When evaluating biological processes, such as administered drugs or TME responses, these reactions might be accounted for. From a technological point of view, the available intravital imaging technologies do not allow acquisition of high quality, and simultaneously dynamic, images of all tissues, and most of those that could be imaged are deep in the animal's body. Therefore, efforts to improve photon penetration and reducing the tissue motion are needed.

Multimodal imaging

Imaging techniques can be divided into those that provide structural information (CT and MRI) and those that provide functional or molecular information (PET, SPECT, and optical modalities). The use of different imaging techniques simultaneously can be complementary and advantageous, as combining one technique from each of the above groups can offer structural and functional knowledge at the same time, in the same individual [99].

The combination of imaging modalities is mostly used as testing platforms for therapeutic or diagnostic developments. The use of optical imaging and MRI has been reported in the study of glioma and pancreatic cancer response to chemotherapy [85,100], in the characterization of viable and necrotic tumors [101], and in the development of cell tracers and drug delivery vehicles [102–104]. Merging of optical imaging and CT is also referred as being used for diagnostic and imaging studies [105,106]. Combinations of CT with PET or SPECT are commonly used to evaluate radioactive molecules distribution [107–110], the effects of drugs in tumor therapy [111–113], as well as to track tumor associated cells [114]. Moreover, MRI, owing to its capacity to image the anatomy of organs, is also used in combination of PET or SPECT for similar purposes as SPECT/CT or PET/CT dual imaging [115–117].

Future perspectives

As elucidated throughout this chapter, small animal imaging techniques have witnessed remarkable advances in the past decades, from the development of new animal models to the application of new molecular targets and the multiple combinations of existing imaging techniques. However, there are still many difficulties in the whole-body imaging, mostly due to the inability to image deep into the tissues, but also due to the incapacity to image multiple and different cellular processes, important for tumor development and progression.

The targeting of specific molecules and processes has improved over the years, but there is still space for further developments in order to understand how the tumor development and progression is influenced by different and multiple factors that occur *in vivo*. Over the years, TME has demonstrated to play a decisive role in the tumorigenesis process. The imaging of such complex extracellular processes has gained some relevance and, indeed, some aspects of the TME context have been explored, even when applied to humans [118]. In mouse models, as mentioned throughout this chapter, some aspects of the TME have been approached, for example, the pH [56]. Many other strategies are emerging to study tumor–stroma interactions, namely transgenic nude mice with color-coded cancer and stromal cells, which are demonstrated to be a powerful method to study cellular interactions during tumor progression [119]. Ideally, with the development of these strategies, the knowledge could be further enhanced by the simultaneous imaging of the tumor location and the expression of specific molecules, cells, and/or biological processes, creating a global perspective of what happens in *in vivo* tumorigenesis.

Another major setback is the inability to image deep into the tissues with high reliability of localization and size of the signal. When considering fluorescence-based techniques, the use of near-red probes and the reduced scattering of these photons, improved significantly the quality of the signal obtained. However, due to the high scattering of the photons shining to and from the tissue, the depth that can be imaged is still limited. The use of techniques such as PET, CT, or MRI can overcome such depth restrictions, with the inability to be used in multichannel imaging. Further development of those imaging systems, as well as the labeling compounds must be attained to increase the versatility of such techniques.

However, in recent years, setbacks imposed by different imaging systems (e.g., diffuse nature of the photons in the tissues, the penetration depth, and localization of the signal) were improved by the combination of multiple techniques. No single imaging modality is perfect, so the synergies obtained by such combination of different imaging techniques, reporters, and labeling options boosted the visualization and the amount of information obtained about tumorigenesis *in vivo*. However, the implementation of this multimodal imaging is still in the beginning. Therefore, it is limited to small animals, and further developments are needed to image at the human scale. In addition, the combination of more than two imaging techniques, which have been developed and improved in more recent years, can further advance the imaging, quality, and quantity of information that can be obtained from a single imaging experiment. Moreover, besides being useful to detect and monitor tumor development, the potential combination of imaging to monitor tumor features associated to therapeutic responses, in particular, when we refer to nanoparticles, may allow *in vivo* imaging techniques to become *theranostic*.

In conclusion, the continuous development of *in vivo* imaging techniques in the experimental/preclinical context will continue to expand their applications. In the meantime, those advances will hopefully be translated to the clinic, allowing the implementation of the *bench to bedside* concept, with possible improvements of the diagnosis, disease evaluation, and therapy in cancer patients.

Acknowledgments

This work was financed by FEDER—Fundo Europeu de Desenvolvimento Regional funds through the COMPETE 2020—Operacional Programme for Competitiveness and Internationalisation (POCI), Portugal 2020, NORTE-01-0145-FEDER-000029, supported by Norte Portugal Regional Programme (NORTE 2020), under the PORTUGAL 2020 Partnership Agreement, through the European Regional Development Fund (ERDF), and by funds from the Investigator FCT programme IF/00136/2013/CP1184/CT0003.

References

1. de Jong, M., J. Essers, and W.M. van Weerden, Imaging preclinical tumour models: Improving translational power. *Nat Rev Cancer*, 2014. **14**(7): 481–493.
2. Ruggeri, B.A., F. Camp, and S. Miknyoczki, Animal models of disease: Pre-clinical animal models of cancer and their applications and utility in drug discovery. *Biochem Pharmacol*, 2014. **87**(1): 150–161.
3. Koba, W. et al., Imaging devices for use in small animals. *Semin Nucl Med*, 2011. **41**(3): 151–165.
4. Wang, Y. et al., Noninvasive imaging of tumor burden and molecular pathways in mouse models of cancer. *Cold Spring Harb Protoc*, 2015. **2015**(2): 135–144.
5. Ellenbroek, S.I. and J. van Rheenen, Imaging hallmarks of cancer in living mice. *Nat Rev Cancer*, 2014. **14**(6): 406–418.
6. Condeelis, J. and R. Weissleder, In vivo imaging in cancer. *Cold Spring Harb Perspect Biol*, 2010. **2**(12): a003848.
7. Ritman, E.L., Current status of developments and applications of micro-CT. *Annu Rev Biomed Eng*, 2011. **13**: 531–552.
8. Clark, D.P. and C.T. Badea, Micro-CT of rodents: State-of-the-art and future perspectives. *Phys Med*, 2014. **30**(6): 619–634.
9. Ashton, J.R., J.L. West, and C.T. Badea, In vivo small animal micro-CT using nanoparticle contrast agents. *Front Pharmacol*, 2015. **6**: p. 256.
10. De Clerck, N.M. et al., High-resolution X-ray microtomography for the detection of lung tumors in living mice. *Neoplasia*, 2004. **6**(4): 374–379.
11. Li, X.F. et al., Visualization of experimental lung and bone metastases in live nude mice by X-ray micro-computed tomography. *Technol Cancer Res Treat*, 2006. **5**(2): 147–155.

12. Haines, B.B. et al., A quantitative volumetric micro-computed tomography method to analyze lung tumors in genetically engineered mouse models. *Neoplasia*, 2009. **11**(1): 39–47.
13. Barck, K.H. et al., Quantification of tumor burden in a genetically engineered mouse model of lung cancer by micro-cT and automated analysis. *Transl Oncol*, 2015. **8**(2): 126–135.
14. Kim, H.W. et al., Micro-CT imaging with a hepatocyte-selective contrast agent for detecting liver metastasis in living mice. *Acad Radiol*, 2008. **15**(10): 1282–1290.
15. Tai, J.H. et al., Assessment of acute antivascular effects of vandetanib with high-resolution dynamic contrast-enhanced computed tomographic imaging in a human colon tumor xenograft model in the nude rat. *Neoplasia*, 2010. **12**(9): 697–707.
16. Eisa, F. et al., Dynamic contrast-enhanced micro-CT on mice with mammary carcinoma for the assessment of antiangiogenic therapy response. *Eur Radiol*, 2012. **22**(4): 900–907.
17. Clark, D.P. et al., In vivo characterization of tumor vasculature using iodine and gold nanoparticles and dual energy micro-CT. *Phys Med Biol*, 2013. **58**(6): 1683–1704.
18. Ehling, J. et al., Micro-CT imaging of tumor angiogenesis: Quantitative measures describing micromorphology and vascularization. *Am J Pathol*, 2014. **184**(2): 431–441.
19. Holdsworth, D.W. and M.M. Thornton, Micro-CT in small animal and specimen imaging. *Trends in Technology*, 2002. **20**(8): S34–S39.
20. Driehuys, B. et al., Small animal imaging with magnetic resonance microscopy. *ILAR J*, 2008. **49**(1): 35–53.
21. Barrett, T. et al., MRI of tumor angiogenesis. *J Magn Reson Imaging*, 2007. **26**(2): 235–249.
22. Suero-Abreu, G.A. et al., In vivo Mn-enhanced MRI for early tumor detection and growth rate analysis in a mouse medulloblastoma model. *Neoplasia*, 2014. **16**(12): 993–1006.
23. Ocak, I. et al., The biologic basis of in vivo angiogenesis imaging. *Front Biosci*, 2007. **12**: 3601–3616.
24. Yuan, A. et al., Functional and structural characteristics of tumor angiogenesis in lung cancers overexpressing different VEGF isoforms assessed by DCE- and SSCE-MRI. *PLoS One*, 2011. **6**(1): e16062.
25. Loveless, M.E. et al., Comparisons of the efficacy of a Jakl/2 inhibitor (AZD1480) with a VEGF signaling inhibitor (cediranib) and sham treatments in mouse tumors using DCE-MRI, DW-MRI, and histology. *Neoplasia*, 2012. **14**(1): 54–64.
26. Gule, M.K. et al., Targeted therapy of VEGFR2 and EGFR significantly inhibits growth of anaplastic thyroid cancer in an orthotopic murine model. *Clin Cancer Res*, 2011. **17**(8): 2281–2291.
27. de Lussanet, Q.G. et al., Dynamic contrast-enhanced magnetic resonance imaging of radiation therapy-induced microcirculation changes in rectal cancer. *Int J Radiat Oncol Biol Phys*, 2005. **63**(5): 1309–1315.
28. Ceelen, W. et al., Noninvasive monitoring of radiotherapy-induced microvascular changes using dynamic contrast enhanced magnetic resonance imaging (DCE-MRI) in a colorectal tumor model. *Int J Radiat Oncol Biol Phys*, 2006. **64**(4): 1188–1196.
29. Øvrebø, K.M. et al., Assessment of tumor radioresponsiveness and metastatic potential by dynamic contrast-enhanced magnetic resonance imaging. *Int J Radiat Oncol Biol Phys*, 2011. **81**(1): 255–261.
30. Øvrebø, K.M. et al., Dynamic contrast-enhanced magnetic resonance imaging of the metastatic potential of tumors: A preclinical study of cervical carcinoma and melanoma xenografts. *Acta Oncol*, 2013. **52**(3): 604–611.

31. Jahng, G.H. et al., Perfusion magnetic resonance imaging: A comprehensive update on principles and techniques. *Korean J Radiol*, 2014. **15**(5): 554–577.
32. Silva, A.C., S.G. Kim, and M. Garwood, Imaging blood flow in brain tumors using arterial spin labeling. *Magn Reson Med*, 2000. **44**(2): 169–173.
33. Moffat, B.A. et al., Inhibition of vascular endothelial growth factor (VEGF)-A causes a paradoxical increase in tumor blood flow and up-regulation of VEGF-D. *Clin Cancer Res*, 2006. **12**(5): 1525–1532.
34. Rajendran, R. et al., Early detection of antiangiogenic treatment responses in a mouse xenograft tumor model using quantitative perfusion MRI. *Cancer Med*, 2014. **3**(1): 47–60.
35. Coatney, R.W., Ultrasound imaging: Principles and applications in rodent research. *ILAR J*, 2001. **42**(3): 233–247.
36. Tilli, M.T. et al., Comparison of mouse mammary gland imaging techniques and applications: Reflectance confocal microscopy, GFP imaging, and ultrasound. *BMC Cancer*, 2008. **8**: 21.
37. Snyder, C.S. et al., Complementarity of ultrasound and fluorescence imaging in an orthotopic mouse model of pancreatic cancer. *BMC Cancer*, 2009. **9**: 106.
38. Wirtzfeld, L.A. et al., A new three-dimensional ultrasound microimaging technology for preclinical studies using a transgenic prostate cancer mouse model. *Cancer Res*, 2005. **65**(14): 6337–6345.
39. Singh, S. et al., Quantitative volumetric imaging of normal, neoplastic and hyperplastic mouse prostate using ultrasound. *BMC Urol*, 2015. **15**: 97.
40. Wirtzfeld, L.A. et al., Quantitative ultrasound comparison of MAT and 4T1 mammary tumors in mice and rats across multiple imaging systems. *J Ultrasound Med*, 2015. **34**(8): 1373–1383.
41. Rooks, V. et al., Sonographic evaluation of orthotopic bladder tumors in mice treated with TNP-470, an angiogenic inhibitor. *Acad Radiol*, 2001. **8**(2): 121–127.
42. Oelze, M.L. et al., Differentiation and characterization of rat mammary fibroadenomas and 4T1 mouse carcinomas using quantitative ultrasound imaging. *IEEE Trans Med Imaging*, 2004. **23**(6): 764–771.
43. Graham, K.C. et al., Three-dimensional high-frequency ultrasound imaging for longitudinal evaluation of liver metastases in preclinical models. *Cancer Res*, 2005. **65**(12): 5231–5237.
44. Le Pivert, P. et al., Ultrasound guided combined cryoablation and microencapsulated 5-Fluorouracil inhibits growth of human prostate tumors in xenogenic mouse model assessed by luminescence imaging. *Technol Cancer Res Treat*, 2004. **3**(2): 135–142.
45. Wang, T.Y. et al., Ultrasound-guided delivery of microRNA loaded nanoparticles into cancer. *J Control Release*, 2015. **203**: 99–108.
46. Geers, B. et al., Targeted liposome-loaded microbubbles for cell-specific ultrasound-triggered drug delivery. *Small*, 2013. **9**(23): 4027–4035.
47. Deshpande, N., A. Needles, and J.K. Willmann, Molecular ultrasound imaging: Current status and future directions. *Clin Radiol*, 2010. **65**(7): 567–581.
48. Kaneko, O.F. and J.K. Willmann, Ultrasound for molecular imaging and therapy in cancer. *Quant Imaging Med Surg*, 2012. **2**(2): 87–97.
49. Kiessling, F. et al., Recent advances in molecular, multimodal and theranostic ultrasound imaging. *Adv Drug Deliv Rev*, 2014. **72**: 15–27.
50. Abou-Elkacem, L., S.V. Bachawal, and J.K. Willmann, Ultrasound molecular imaging: Moving toward clinical translation. *Eur J Radiol*, 2015. **84**(9): 1685–1693.
51. Pysz, M.A. et al., Antiangiogenic cancer therapy: Monitoring with molecular US and a clinically translatable contrast agent (BR55). *Radiology*, 2010. **256**(2): 519–527.
52. Deshpande, N. et al., Tumor angiogenic marker expression levels during tumor growth: Longitudinal assessment with molecularly targeted microbubbles and US imaging. *Radiology*, 2011. **258**(3): 804–811.

53. Van Dort, M.E., A. Rehemtulla, and B.D. Ross, PET and SPECT imaging of tumor biology: New approaches towards oncology drug discovery and development. *Curr Comput Aided Drug Des*, 2008. **4**(1): 46–53.
54. Wang, Y. and A.L. Kung, 18F-FDG-PET/CT imaging of drug-induced metabolic changes in genetically engineered mouse lung cancer models. *Cold Spring Harb Protoc*, 2015. **2015**(2): 176–179.
55. Pollok, K.E. et al., In vivo measurements of tumor metabolism and growth after administration of enzastaurin using small animal FDG positron emission tomography. *J Oncol*, 2009. **2009**: 596560.
56. Flavell, R.R. et al., Caged [(18)F]FDG glycosylamines for imaging acidic tumor microenvironments using positron emission tomography. *Bioconjug Chem*, 2016. **27**(1): 170–178.
57. Nedrow, J.R. et al., Targeting PSMA with a Cu-64 labeled phosphoramidate inhibitor for PET/CT imaging of variant PSMA-expressing xenografts in mouse models of prostate cancer. *Mol Imaging Biol*, 2015. **18**(3): 402–410.
58. Carney, B. et al., Non-invasive PET imaging of PARP1 expression in glioblastoma models. *Mol Imaging Biol*, 2015. **18**(3): 386–392.
59. Bao, X. et al., The preclinical study of predicting radiosensitivity in human nasopharyngeal carcinoma xenografts by 18F-ML-10 animal-PET/CT imaging. *Oncotarget*, 2016. **7**(15): 20743.
60. Lau, J. et al., PET imaging of carbonic anhydrase IX expression of HT-29 tumor xenograft mice with (68)Ga-labeled benzenesulfonamides. *Mol Pharm*, 2016. **13**(3): 1137–1146.
61. England, C.G. et al., ImmunoPET imaging of insulin-like growth factor 1 receptor in a subcutaneous mouse model of pancreatic cancer. *Mol Pharm*, 2016. **13**(6): 1958–1966.
62. Garousi, J. et al., PET imaging of epidermal growth factor receptor expression in tumours using 89Zr-labelled ZEGFR: 2377 affibody molecules. *Int J Oncol*, 2016. **48**(4): 1325–1332.
63. McLarty, K. et al., Micro-SPECT/CT with 111In-DTPA-pertuzumab sensitively detects trastuzumab-mediated HER2 downregulation and tumor response in athymic mice bearing MDA-MB-361 human breast cancer xenografts. *J Nucl Med*, 2009. **50**(8): 1340–1348.
64. Razumienko, E.J., D.A. Scollard, and R.M. Reilly, Small-animal SPECT/CT of HER2 and HER3 expression in tumor xenografts in athymic mice using trastuzumab Fab-heregulin bispecific radioimmunoconjugates. *J Nucl Med*, 2012. **53**(12): 1943–1950.
65. Krüwel, T. et al., In vivo detection of small tumour lesions by multi-pinhole SPECT applying a ^{99m}Tc-labelled nanobody targeting the epidermal growth factor receptor. *Sci Rep*, 2016. **6**: 21834.
66. Zhang, X. et al., Imaging of human pancreatic cancer xenografts by single-photon emission computed tomography with ^{99m}Tc-Hynic-PEG-AE105. *Oncol Lett*, 2015. **10**(4): 2253–2258.
67. Rajkumar, V. et al., Texture analysis of (125)I-A5B7 anti-CEA antibody SPECT differentiates metastatic colorectal cancer model phenotypes and anti-vascular therapy response. *Br J Cancer*, 2015. **112**(12): 1882–1887.
68. Zhao, H. et al., ^{99m}Tc-HisoDGR as a potential SPECT probe for orthotopic glioma detection via targeting of integrin $\alpha_5\beta_1$. *Bioconjug Chem*, 2016. **27**(5): 1259–1266.
69. Mosley, M. et al., Claudin-4 SPECT imaging allows detection of aplastic lesions in a mouse model of breast cancer. *J Nucl Med*, 2015. **56**(5): 745–751.
70. Söling, A. and N.G. Rainov, Bioluminescence imaging in vivo–application to cancer research. *Expert Opin Biol Ther*, 2003. **3**(7): 1163–1172.

71. Wang, W. and W.S. El-Deiry, Bioluminescent molecular imaging of endogenous and exogenous p53-mediated transcription in vitro and in vivo using an HCT116 human colon carcinoma xenograft model. *Cancer Biol Ther*, 2003. **2**(2): 196–202.
72. Honigman, A. et al., Imaging transgene expression in live animals. *Mol Ther*, 2001. **4**(3): 239–249.
73. Massoud, T.F. and S.S. Gambhir, Molecular imaging in living subjects: Seeing fundamental biological processes in a new light. *Genes Dev*, 2003. **17**(5): 545–580.
74. Contag, P.R., Whole-animal cellular and molecular imaging to accelerate drug development. *Drug Discov Today*, 2002. **7**(10): 555–562.
75. Colin, M. et al., Haemoglobin interferes with the ex vivo luciferase luminescence assay: Consequence for detection of luciferase reporter gene expression in vivo. *Gene Ther*, 2000. **7**(15): 1333–1336.
76. Kidd, S. et al., Direct evidence of mesenchymal stem cell tropism for tumor and wounding microenvironments using in vivo bioluminescent imaging. *Stem Cells*, 2009. **27**(10): 2614–2623.
77. Thalheimer, A. et al., Noninvasive visualization of tumor growth in a human colorectal liver metastases xenograft model using bioluminescence in vivo imaging. *J Surg Res*, 2013. **185**(1): 143–151.
78. Köberle, M. et al., Monitoring disease progression and therapeutic response in a disseminated tumor model for non-hodgkin lymphoma by bioluminescence imaging. *Mol Imaging*, 2015. **14**: 400–413.
79. Adams, J.Y. et al., Visualization of advanced human prostate cancer lesions in living mice by a targeted gene transfer vector and optical imaging. *Nat Med*, 2002. **8**(8): 891–897.
80. Laxman, B. et al., Noninvasive real-time imaging of apoptosis. *Proc Natl Acad Sci USA*, 2002. **99**(26): 16551–16555.
81. Vantaggiato, C. et al., Bioluminescence imaging of estrogen receptor activity during breast cancer progression. *Am J Nucl Med Mol Imaging*, 2016. **6**(1): 32–41.
82. Ju, H.L. et al., Transgenic mouse model expressing P53(R172H), luciferase, EGFP, and KRAS(G12D) in a single open reading frame for live imaging of tumor. *Sci Rep*, 2015. **5**: 8053.
83. Weissleder, R. et al., In vivo imaging of tumors with protease-activated near-infrared fluorescent probes. *Nat Biotechnol*, 1999. **17**(4): 375–378.
84. Shirmanova, M. et al., In vivo study of photosensitizer pharmacokinetics by fluorescence transillumination imaging. *J Biomed Opt*, 2010. **15**(4): 048004.
85. McCann, C.M. et al., Combined magnetic resonance and fluorescence imaging of the living mouse brain reveals glioma response to chemotherapy. *Neuroimage*, 2009. **45**(2): 360–369.
86. Ntziachristos, V. et al., Fluorescence molecular tomography resolves protease activity in vivo. *Nat Med*, 2002. **8**(7): 757–760.
87. Kossodo, S. et al., Dual in vivo quantification of integrin-targeted and protease-activated agents in cancer using fluorescence molecular tomography (FMT). *Mol Imaging Biol*, 2010. **12**(5): 488–499.
88. Montet, X. et al., Tomographic fluorescence imaging of tumor vascular volume in mice. *Radiology*, 2007. **242**(3): 751–758.
89. Mazzocco, C. et al., In vivo imaging of prostate cancer using an anti-PSMA scFv fragment as a probe. *Sci Rep*, 2016. **6**: 23314.
90. Yang, M. et al., Whole-body optical imaging of green fluorescent protein-expressing tumors and metastases. *Proc Natl Acad Sci USA*, 2000. **97**(3): 1206–1211.
91. Ntziachristos, V. et al., Visualization of antitumor treatment by means of fluorescence molecular tomography with an annexin V-Cy5.5 conjugate. *Proc Natl Acad Sci USA*, 2004. **101**(33): 12294–12299.

92. Montet, X. et al., Tomographic fluorescence mapping of tumor targets. *Cancer Res*, 2005. **65**(14): 6330–6336.
93. Gavins, F.N. and B.E. Chatterjee, Intravital microscopy for the study of mouse microcirculation in anti-inflammatory drug research: Focus on the mesentery and cremaster preparations. *J Pharmacol Toxicol Methods*, 2004. **49**(1): 1–14.
94. Pittet, M.J. and R. Weissleder, Intravital imaging. *Cell*, 2011. **147**(5): 983–991.
95. Kienast, Y. et al., Real-time imaging reveals the single steps of brain metastasis formation. *Nat Med*, 2010. **16**(1): 116–122.
96. Ritsma, L. et al., Intravital microscopy through an abdominal imaging window reveals a pre-micrometastasis stage during liver metastasis. *Sci Transl Med*, 2012. **4**(158): 158ra145.
97. Tsuzuki, Y. et al., Pancreas microenvironment promotes VEGF expression and tumor growth: Novel window models for pancreatic tumor angiogenesis and microcirculation. *Lab Invest*, 2001. **81**(10): 1439–1451.
98. Kedrin, D. et al., Intravital imaging of metastatic behavior through a mammary imaging window. *Nat Methods*, 2008. **5**(12): 1019–1021.
99. Cherry, S.R., Multimodality in vivo imaging systems: Twice the power or double the trouble? *Annu Rev Biomed Eng*, 2006. **8**: 35–62.
100. Wang, P. et al., Predictive imaging of chemotherapeutic response in a transgenic mouse model of pancreatic cancer. *Int J Cancer*, 2016. **139**(3): 712–718.
101. Lin, Y. et al., Tumor characterization in small animals using magnetic resonance-guided dynamic contrast enhanced diffuse optical tomography. *J Biomed Opt*, 2011. **16**(10): 106015.
102. Kim, E.J. et al., In vivo tracking of phagocytic immune cells using a dual imaging probe with gadolinium-enhanced MRI and near-infrared fluorescence. *ACS Appl Mater Interfaces*, 2016. **8**(16): 10266–10273.
103. Bakalova, R. et al., Passive and electro-assisted delivery of hydrogel nanoparticles in solid tumors, visualized by optical and magnetic resonance imaging in vivo. *Anal Bioanal Chem*, 2016. **408**(3): 905–914.
104. Zhou, Z. et al., MRI detection of breast cancer micrometastases with a fibronectin-targeting contrast agent. *Nat Commun*, 2015. **6**: 7984.
105. Zhang, J. et al., In vivo tumor-targeted dual-modal fluorescence/CT imaging using a nanoprobe co-loaded with an aggregation-induced emission dye and gold nanoparticles. *Biomaterials*, 2015. **42**: 103–111.
106. Ale, A. et al., FMT-XCT: In vivo animal studies with hybrid fluorescence molecular tomography-X-ray computed tomography. *Nat Methods*, 2012. **9**(6): 615–620.
107. Houghton, J.L. et al., Site-specifically labeled CA19.9-targeted immunoconjugates for the PET, NIRF, and multimodal PET/NIRF imaging of pancreatic cancer. *Proc Natl Acad Sci USA*, 2015. **112**(52): 15850–15855.
108. Boyle, A.J. et al., MicroPET/CT imaging of patient-derived pancreatic cancer xenografts implanted subcutaneously or orthotopically in NOD-scid mice using ^{64}Cu-NOTA-panitumumab F(ab')$_2$ fragments. *Nucl Med Biol*, 2015. **42**(2): 71–7.
109. Chatalic, K.L. et al., A novel ^{111}in-labeled anti-prostate-specific membrane antigen nanobody for targeted SPECT/CT imaging of prostate cancer. *J Nucl Med*, 2015. **56**(7): 1094–1099.
110. Suzuki, A. et al., Analysis of biodistribution of intracranially infused radiolabeled interleukin-13 receptor-targeted immunotoxin IL-13PE by SPECT/CT in an orthotopic mouse model of human glioma. *J Nucl Med*, 2014. **55**(8): 1323–1329.
111. Plengsuriyakarn, T., J. Karbwang, and K. Na-Bangchang, Anticancer activity using positron emission tomography-computed tomography and pharmacokinetics of β-eudesmol in human cholangiocarcinoma xenografted nude mouse model. *Clin Exp Pharmacol Physiol*, 2015. **42**(3): 293–304.

112. Stellas, D. et al., Therapeutic effects of an anti-Myc drug on mouse pancreatic cancer. *J Natl Cancer Inst*, 2014. **106**(12): dju320.
113. Lin, L.T. et al., Evaluation of the therapeutic and diagnostic effects of PEGylated liposome-embedded 188Re on human non-small cell lung cancer using an orthotopic small-animal model. *J Nucl Med*, 2014. **55**(11): 1864–1870.
114. Pérez-Medina, C. et al., PET imaging of tumor-associated macrophages with 89Zr-labeled high-density lipoprotein nanoparticles. *J Nucl Med*, 2015. **56**(8): 1272–1277.
115. Cussó, L. et al., Combination of single-photon emission computed tomography and magnetic resonance imaging to track 111in-oxine-labeled human mesenchymal stem cells in neuroblastoma-bearing mice. *Mol Imaging*, 2014. **13**(10): 7290–8014.
116. Haeck, J.C. et al., Imaging heterogeneity of peptide delivery and binding in solid tumors using SPECT imaging and MRI. *EJNMMI Res*, 2016. **6**(1): 3.
117. Thorek, D.L. et al., Non-invasive mapping of deep-tissue lymph nodes in live animals using a multimodal PET/MRI nanoparticle. *Nat Commun*, 2014. **5**: 3097.
118. Penet, M.F. et al., Molecular imaging of the tumor microenvironment for precision medicine and theranostics. *Adv Cancer Res*, 2014. **124**: 235–256.
119. Hoffman, R.M. and M. Bouvet, Imaging the microenvironment of pancreatic cancer patient-derived orthotopic xenografts (PDOX) growing in transgenic nude mice expressing GFP, RFP, or CFP. *Cancer Lett*, 2015. **380**: 349–355.

Chapter 7 Quantum dots

Concepts, imaging, and therapeutic applications in cancer

Maria Sofia Fernandes,
Anabela Ferro,
Patrícia Carneiro,
Raquel Seruca, and
João M. Sanches

Contents

Introduction

Worldwide, cancer is still a major health burden due to its high incidence and associated mortality rates [1]. In clinical practice, early cancer detection and adequate therapies are key issues for effective and successful outcomes. In this context, biomarker evaluation using reliable imaging techniques is essential for an accurate diagnosis that will ultimately impact the selection of the best therapeutic strategy. At present, microscopy-based techniques using conventional dyes and fluorophores are used for biomedical molecular imaging as a complementary approach to standard diagnostic procedures. These have been used to analyze tumor markers and morphological features, in cells or tissues, for classification purposes as well as to better understand cell dynamics. However, and despite the many advances in the field, there are still many technical limitations that weaken such biomarker evaluation, mainly due to the intrinsic characteristics of the dyes and fluorescent probes. Therefore, in an attempt to overcome these drawbacks, the field of nanotechnology has gained much attention. Specifically, novel types of fluorescent nanoparticles, known as quantum dots, are considered promising tools for diagnosis purposes. Moreover, the potential medical applications of quantum dots are more widespread and could include imaging for guided surgery, targeted drug delivery, and photodynamic therapy [2,3].

A number of specific features characterize quantum dots and distinguish them from the conventional organic fluorophores and dyes [4]. Indeed, quantum dots are inorganic and semiconductor nanoparticles with diameters in the nanoscale range that have unique optical, physical, electrical, and magnetic properties [2,5]. More specifically, quantum dots improved features include broad excitation spectra, narrow emission spectra, tunable emission peaks, long fluorescence lifetimes, and diminished photobleaching that translate into a robust signal strength and allows the simultaneous detection of markers in a more accurate and reliable manner [3,4,6–8]. In addition, the size, structure, and composition of quantum dots are variable and can be modulated to obtain the desired emission properties, thus enabling the selection of molecules that are best suitable for each application [3]. Moreover, quantum dots can be conjugated with peptides, proteins, oligonucleotides, and other biomolecules for therapeutic approaches as drug delivery or imaging for surgical guidance, thus amplifying the range of applications in biomedicine [9]. Along with the improved characteristics of quantum dots, more reliable and reproducible data are generated and therefore novel processing tools should be made available for data analysis to automatically compute and integrate cell phenotypic and molecular characteristics. This is in contrast with organic fluorophores for which biomarker evaluation is mainly qualitative and operator dependent. Quantum dots are therefore a valuable tool with great potential that is being explored in many distinct areas including biomedicine.

Experimental procedures

The particularity of quantum dots relies mainly on its size-associated characteristics. With a large surface-to-volume ratio and nonplanar geometry, colloidal quantum dots are a challenge in the field of nanotechnology [10]. With a diameter of only about 2–10 nm, quantum dots behavior is distinct from that of the corresponding bulk semiconducting materials that are not at the nanometer scale [5,11]. Indeed, quantum dot nanocrystals possess the same crystal structure as the bulk semiconductor material but consist only of a few hundred to a few thousand atoms [9]. As a consequence, the energy levels of quantum dots are discrete similar to those observed in atoms, as compared to the continuum energy levels observed in bulk materials, thus explaining why quantum dots are also known as *artificial atoms* [12]. More specifically, the size of quantum dots is smaller than the Bohr exciton radius (the distance between the positive and the negative charges in the excited state of a material) of the bulk material, which means that the electrons in quantum dots are confined in a small region, the quantum box, and the distance between the discrete energy levels will depend on the size of the quantum dots, an effect termed as quantum confinement [2,13]. Accordingly, as the size of quantum dots decreases, the energy difference increases and the wavelength at which they fluoresce decreases therefore allowing modulation of quantum dots size and fine-tuning of emission wavelengths [5,14]. Such size-dependent feature enables the production, from the same material, of numerous fluorophores with distinct optical properties [5,14]. In addition, the extended fluorescence lifetime of quantum dots results from a large Stokes shift (the energy, and thus wavelength, difference between excitation and emission peaks), accounting for the stable and strong fluorescence without the interference of autofluorescence [5,8]. Moreover, a major advantage is that excitation of quantum dots can be performed using a unique light source allowing multiple quantum dots to be used simultaneously [3,8]. More specifically, the broad absorption bands of quantum dots enable the exposure of a single light source of short wavelength to generate narrow emission bands, thus creating distinct signals that are simultaneously detected with short cross talk [14]. Furthermore, the larger molar extinction coefficients of quantum dots allow very fast and large absorption excitation photons, resulting in an increased emission profile with brighter signals [5,14]. In addition to these intrinsic characteristics, quantum dot properties can be further modulated by other factors including composition of the quantum dot shell and surface coating, but core size and composition are known to affect the most [3]. Indeed, by modifying quantum dot size and composition, the emission of fluorescence light can be modulated from near the ultraviolet, through the visible, and into the infrared spectra [3,14]. Altogether, these unique characteristics render quantum dots as one of the most exciting and promising tools for biomedical applications. A detailed comparison of the properties of organic dyes *versus* quantum dots is available [4,7].

Regarding the composition of quantum dots, the typical structure includes a core of a semiconductor material that dictates the optical properties; a shell to increase the fluorescence efficiency and stability; and a coating for improved solubility and facilitate bioconjugation with targeting molecules [2,5,14,15] (Figure 7.1). In particular, the core is usually composed of elements from groups II and VI, with cadmium selenide (CdSe) crystals being the most common, or from groups III and V, and its size is the main factor determining the fluorescence emission spectra [2,14,15]. Enclosing the core, is a shell commonly of zinc sulphide (ZnS) that has a larger bandgap and a role in passivating surface defects eventually present in the crystal lattice of the core [2,5,15]. In addition, the shell serves to enhance quantum yields and photostability [2,5,15]. Notably, several methods can be used for the synthesis of quantum dot nanocrystals and importantly their synthesis determines surface termination and nanocrystal shape [10,16]. In particular, preparation

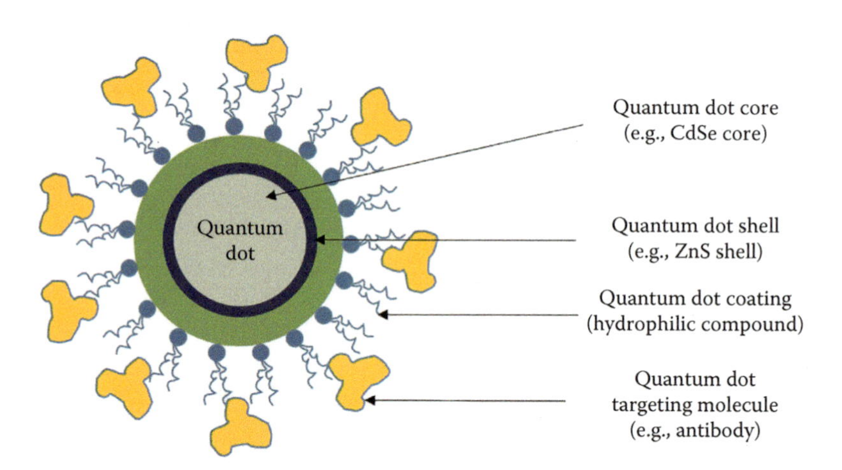

Figure 7.1 Illustrative basic structure of quantum dots. Quantum dots usually comprise a metal core typically of CdSe, an hydrophilic shell often composed of ZnS, an hydrophilic layer, and a targeting molecule (a functional ligand as an antibody, peptide, oligonucleotide, and a drug).

at high temperatures in organic solvents are often used and result in high-quality quantum dots [10,14]. Finally, in order to use quantum dots in live cells and other biological applications, an extra step is required, so quantum dots become hydrophilic and thus soluble in biological buffers. This can be achieved by modifications as silanization, surface exchange with bifunctional molecules, or encapsulation with molecules as phospholipids, micelles, and polymers [2,5,14]. Biofunctionalization is the last step of the process in which molecular moieties are conjugated to quantum dots for numerous biological applications. Peptides, proteins including avidin/streptavidin, albumin and antibodies, oligonucleotides, biotin, and other molecules can be conjugated to quantum dots through different strategies including electrostatic interaction, covalent cross-linking, adsorption, and mercapto-exchange [5,17]. Such plasticity is of major relevance as an unlimited number of possibilities are allowed. Both biocompatibility and biofunctionalization methods have associated advantages and disadvantages and the selection will depend on the properties of the quantum dots as well as on the molecules of interest and of the required applications [2]. In recent years, advances in the synthesis and modifications of quantum dots have led to the improvement of many of the strategies resulting in better efficiencies and outcomes.

Applications

At present, a wide range of biomedical applications has emerged using quantum dot technology, namely in the context of cancer. Indeed, the development of new strategies using quantum dots is continually expanding and, in the near future, it is believed they will have a major impact not only in terms of diagnosis and prognosis purposes but also in therapeutic management. However, there are still a number of technical limitations hampering the introduction of this technology in the routine of many clinical procedures. This is particularly observed in the field of biomedicine, in which several parameters have to be taken into consideration in terms of methodologies for the design and synthesis of the appropriate functional quantum dots. Indeed, the need to minimize quantum dots toxicity while maintaining quantum dots unique features as fluorescent yield, absorption and lifetime has been an ongoing challenge. Moreover, modification of the quantum dots for a specific target requires strategies that are able to maintain quantum dots characteristics. Although this process is still a challenge in the field of biomedical nanotechnology, there are already a number of approaches that have been successfully developed, namely in terms of biomedical imaging and therapeutic applications.

Bioimaging is indeed one of the most used and successful applications of quantum dots, either for research purposes or clinical applications. One example is cellular and tissue imaging for cancer detection. Specifically, the combination of specific antibodies with quantum dots enables an accurate and reliable detection of proteins further supporting the standard methods of diagnosis. Moreover, the improved features of quantum dots will certainly facilitate the detection of cancer at earlier stages of development and, subsequently allow the initiation of treatments at earlier phases of the disease. Ultimately, this will have a tremendous impact in the overall survival of cancer patients. A vast number of quantum dot-based approaches have been developed for distinct cancer cell types. For instance, in breast cancer patients, the detection of the cancer cell marker human epidermal growth factor receptor 2 (HER2) is important for the prognosis and therapy selection, but the current methods of detection have some limitations. In that context, the immunofluorescent labeling of HER2 and other cellular targets was successfully developed using semiconductor quantum dots and allowed improved labeling and specific and effective multiplexed detection [18]. More specifically, quantum dots were linked to immunoglobulin G (IgG) to label the HER2 on the surface of fixed and live cancer cells, to stain actin and microtubule fibers in the cytoplasm, and to detect nuclear antigens inside the nucleus [18]. Quantum dots conjugated with streptavidin were also evaluated as an alternative probe for HER2 detection and both strategies resulted in effective staining either in fixed and live cells as well as in tissue sections [18]. In a different study, quantum dots emitting at five distinct wavelengths (525, 565, 605, 655, and 705 nm) were directly conjugated to primary antibodies against HER2, estrogen receptor (ER), progesterone receptor (PR), epidermal growth factor receptor (EGFR) and mammalian target of rapamycin (mTOR) and the multicolor bioconjugates used for detection of the five clinically significant tumor markers [19]. Interestingly, simultaneous quantification was successfully observed in breast cancer cell lines but most importantly on paraffin-embedded clinical human tissue sections [19]. Moreover, comparison of quantum dots results with those using conventional techniques as immunohistochemistry (IHC) and fluorescence *in situ* hybridization (FISH) further confirmed the potential of quantum dots-based approaches [19]. In ovarian cancer, detection of the specific tumor marker CA125 was specifically and effectively achieved with conjugated quantum dots in different specimen types as fixed cells, tissue sections, and xenograft pieces and, in comparison with FITC, quantum dots signals were more stable, specific, and brighter [20]. In addition, quantum dots were shown to have superior quality in the detection of the prostate cancer marker prostate-specific antigen (PSA), in comparison to conventional IHC, even when multiplexed with androgen receptor [21]. In gastric cancer tissues, simultaneous quantum dot-based *in situ* detection of multiple biomarkers of stromal features, as infiltrating macrophages, tissue microvessels density, and neovessels maturity were used to predict clinical outcomes [22]. In a different study, quantification of membranous E-cadherin and EGFR by multiplexed quantum dots, in combination with other classical parameters such as age, gender, and grade, was shown to have greater predictive power than conventional IHC in respect to lymph node metastasis and prognosis in head and neck cancer [23]. Quantum dots have also been used in the detection of circulating tumor cells, as the number of these cells in the blood is very low and the establishment of novel methods of detection is necessary. More specifically, immunomagnetic separation of circulating tumor cells from body fluids was achieved with quantum dots conjugated with a specific epithelial cell adhesion molecule antibody (EpCAM) and monoclonal cytokeratin 19 antibody [24]. More recently, an efficient capture and simple quantification of circulating tumor cells approach was developed using sequential anti-EpCAM antibody-conjugated quantum dots and anti-IgG magnetic beads [25]. These strategies are of particular relevance these cells are considered a valuable measure for cancer diagnosis and an important tool in monitoring drug effects. In addition to these, a vast number of studies have now been published that focus on the identification of numerous tumor markers for other cancer types including pancreatic and colorectal cancer (CRC) [26,27].

In addition to cancer marker detection in fixed cells or tissues, quantum dots have proved to be a valuable tool for *in vivo* imaging, which is key to better understand the mechanisms of tumor development and progression. Notably, an advantage of nano-sized macromolecules is that they can undergo passive targeting through a phenomena known as *enhanced permeability and retention* effect meaning that macromolecules, such as quantum dots, accumulate preferentially at tumor sites, rather than at normal tissues, which is related to altered angiogenesis, abnormal blood supply, and a compromised lymphatic system [9,28]. This mechanism is key for solid tumor targeting and novel drug design. In contrast, this effect is not observed for low molecular weight drugs due to their rapid diffusion into the bloodstream and renal clearance [29,30]. Alternatively, active targeting takes advantage of alterations in cells, namely aberrant expression of a specific receptor in tumor cells [9]. The conjugation of quantum dots to molecules that specifically bind these receptors will subsequently increase the uptake of quantum dots, thus enabling the specific delivery of anticancer drugs to the tumor site. Noteworthy, passive targeting is also observed in active targeting but the specificity becomes greatly increased [9]. For this purpose, a variety of ligands have been conjugated with quantum dots including antibodies, peptides, and oligonucleotides. For instance, *in vivo* targeting studies of human cancer cells growing in nude mice have shown that quantum dot conjugates accumulate at the tumor sites by both the enhanced permeability and retention effect but also by antibody binding to cancer-specific cell-surface biomarkers [31]. In a different context, quantum dots labeled with anti-HER2 antibody were injected into living mice with HER2-overexpressing breast cancer to analyze the molecular processes of its mechanistic delivery to the tumor including: circulation within a blood vessel during extravasation, in the extracellular region, binding to HER2 on the cell membrane, moving from the cell membrane to the perinuclear region, and in the perinuclear region [32]. In another study, quantum dots linked to an alpha-fetoprotein (AFP) antibody, which is a biomarker for hepatocellular carcinoma, have shown good stability, specificity, and biocompatibility for ultrasensitive fluorescence imaging, both *in vitro* and *in vivo* [33]. The selective targeting of EGFR, which is overexpressed in many human cancers, was also achieved using quantum dots conjugated to EGF. Specifically, these quantum dot conjugates served as nanoprobes for noninvasive optical imaging of EGFR expression and were able to specifically bind EGFR *in vitro* and *in vivo* in human CRC xenografts in mice [34]. In a different study, *in vivo* magnetic resonance imaging (MRI) was able to show real-time imaging of CRC tumor on nude mice after intravenously injection of quantum dot conjugates [35]. These nanoprobes were found to be biocompatible *in vitro* and *in vivo* and are envisioned as promising agents for CRC cancer *in vivo* MRI diagnosis and *ex vivo* biopsy analysis [35]. An additional applicability of quantum dots is the surveillance of the metastasization process by evaluating sentinel lymph nodes (SLN), as SNL biopsies are important predictors of metastasis in cancer patients [36]. Indeed, several studies have evaluated the lymphatic system by using quantum dots. Studies reported the use of near-infrared (NIR) fluorescent quantum dots for intraoperative SLN mapping of the gastrointestinal tract [37], esophagus [38], and lung [39], constituting a powerful tool to aid in the identification and resection of SLN. However, evaluating and mapping lymph nodes is complex as each lymph node can receive lymphatic fluid from multiple afferent lymphatic channels [36]. More recently, a noninvasive *in vivo* approach was shown to enable the simultaneous visualization of several lymphatic flows in mice by using five different quantum dots injected at different sites, which could help understand the pathways for cancer metastasis into the lymph nodes [40]. Furthermore, and as above-mentioned, other molecular moieties can be conjugated with quantum dots such as peptides and oligonucleotides [9]. This is the case of the RGD peptide (arginine–glycine–aspartic acid) that was conjugated with quantum dots for *in vivo* targeting and imaging of tumor vasculature that may eventually aid in cancer detection and management including imaging-guided surgery [41].

However, the application of quantum dots is not limited to the detection, either *in vitro* or *in vivo*, of tumor biomarkers but has a rather important usefulness in terms of cancer therapy that relies on the application of quantum dots as nano-carriers for efficient targeted drug delivery. Indeed, many studies have evaluated several strategies to achieve the best possible outcome and the association of quantum dots with drugs to target cancer cells has been evaluated in many cancer types. For instance, by functionalizing the surface of fluorescent quantum dots with the A10 RNA aptamer, which recognizes the extracellular domain of the prostate specific membrane antigen (PSMA), a targeted quantum dot imaging system was developed capable of differential uptake and imaging of prostate cancer cells that express the PSMA protein [42]. Another example involved the assembly of quantum dots-small interfering RNA (siRNA) nanoplexes for gene silencing in tumor cells. Specifically, the nanoplexes were used to facilitate the intracellular delivery of siRNA molecules to pancreatic cancer cells, thereby inducing sequence-specific silencing of oncogenic KRAS mutations in pancreatic carcinoma [43]. An additional example involved the use of HER2 monoclonal antibody conjugated RNase-A-associated quantum dots for targeted imaging and therapy of gastric cancer [44]. In particular, subcutaneous gastric cancer nude mouse models and *in situ* gastric cancer mouse models were intravenously injected with the nanoprobes and the biodistribution and therapeutic effects evaluated. Interestingly, the nanoprobes could selectively kill gastric cancer cells, target subcutaneous gastric cancer cells, inhibit the growth of gastric cancer tissues, and extended the survival time of gastric cancer bearing mouse models [44].

Many other applications are known for quantum dots in the field of biomedicine. For example, quantum dots have been used in the identification of microorganisms as is the case of a recent report describing the rapid multiplexed detection of influenza A virus subtype H5 and H9 via quantum dot-based immunoassay [45]. For research purposes, the applications have expanded exponentially in the past few years. Indeed, these fluorescent tracers have been used to evaluate many cellular aspects and included colocalization and cell-trafficking studies to elucidate the dynamics of cellular processes. However, if successful, the impact of quantum dot-based technology in cancer detection and therapy can be enormous. From diagnosis to prognosis and therapy, the possible intervention steps are numerous. Among these are early cancer detection and follow-up procedures identifying primary, recurrent or metastatic disease, image-guided surgery, assist in the stratification of patients for therapy, target drug delivery and therapy monitoring, thus reinforcing why quantum dots are envisioned as a powerful biomedical tool.

Limitations

It is well established that quantum dot's physical, chemical, and fluorescent characteristics are unparalleled by any other conventional inorganic or organic molecules [46]. However, and particularly, in biomedical applications, there are several limitations and drawbacks associated with the use of quantum dots. One of the major concerns is the quantum dot's inherent toxicity due to the use of toxic materials in their basic composition [47]. Indeed, heavy metals such as Cd, which is one of the most used elements, is toxic for living cells and was shown to induce cell death in *in vitro* studies mediated by oxidative degradation of quantum dots and subsequent release of free Cd ions [48]. Coating of the quantum dots is one of the strategies used in the preparation of quantum dots, which is expected to increase stability and protect against toxicity [5]. However, this approach also has its associated risks for *in vivo* applications, as unstable surface coating could expose the heavy metals releasing metal ions from the quantum dot core; moreover, the coating itself can also present toxic effects [47]. As an alternative approach, nontoxic materials have been used in quantum dot preparation namely graphene quantum dots for cancer imaging applications. Indeed, in a recent study, graphene quantum dots were shown to have no obvious *in vitro* and *in vivo* toxicity, even under multidosing administration [49]. Still, other quantum dot-associated characteristics have been implicated in affecting *in vivo* toxicity including absorption, distribution, metabolism and excretion, and these will depend on many factors such as quantum dots size, charge, concentration, outer coating bioactivity, and stability [47]. In particular, quantum dots coating, to increase their solubility in aqueous buffers and functionalization through bioconjugation with targeting molecules, will result in quantum dots with increased size, thus possibly affecting fluorescence yield, stability, and ability to reach their targets within cellular compartments or molecular complexes [2]. Notably, the final size of quantum dots is comparable to that of large proteins due to the combination of the core/shell with a bioactive coating raising concerns on the influence of size on target functionality [50]. In addition, quantum dots are also susceptible to aggregation, which can be induced by external factors difficult to control as variations in pH and temperature [50].

In conclusion, before quantum dot's full potential can be reached for biomedical applications, improved understanding of the underlying toxicity/safety of the available quantum dots, as well as a more detailed knowledge for the development of alternative and safer quantum dots, has to be accomplished.

Future

Quantum dot-based technology has brought excitement into the field of biomedicine. Indeed, quantum dot nanoparticles are of particular interest due to their improved brightness and stability in comparison with standard organic dyes. Moreover, the plasticity in the modulation of quantum dot's features is of major relevance as it enables the generation of quantum dots suitable for a particular application. Currently, a number of quantum dot-based approaches have been established including bioimaging tools for cancer diagnosis and potential target delivery for medical therapy. Moreover, it has shown to have a tremendous impact for research purposes. However, the limitations and drawbacks of quantum dots as they exist, have hampered their use *in vivo*. Although research has been conducted in animal models, its use in humans is impaired mainly due to toxicity issues.

Therefore, the future in quantum dot-based technology aims at understanding and developing new and nontoxic quantum dots that will be safe to use not only for biomarker detection in cancer prognosis and diagnosis but also for real-time *in vivo* applications as guided surgery, target drug delivery and therapy monitoring. Moreover, making use of quantum dots for metastasis evaluation and subsequent drug delivery would have a major impact in clinical practice. Specifically, novel and less toxic synthesis methods should be made available to allow their use in biomedical applications but also to enable simple synthesis procedures to become routinely used and commercially available. Indeed, several types of quantum dots are already commercialized by many companies, which have facilitated their use by researchers. Still, a major goal in the future is the achievement of optimized protocols for the synthesis of quantum dots with good yield, which are stable and nontoxic, biocompatible, and biodegradable, while keeping their main advantageous characteristics.

In addition, as quantum dots enable the acquisition of a much more reliable and reproducible data, it calls for the urgent need to generate novel tools of analysis. These tools should be able to automatically compute and integrate phenotypic and molecular features but also assist in the development of programs to support data analysis from *in vivo* experiments. For instance, our group has previously developed an algorithm able to quantify E-cadherin expression from bioimaging analysis of *in situ* fluorescence microscopy images. In particular, we have shown that gastric cancer cells expressing mutant forms of E-cadherin display fluorescence profiles distinct from those of the wild-type cells reflecting the underlying molecular mechanism of trafficking deregulation [51]. This is of particular relevance in the context of hereditary diffuse gastric cancer (HDGC) in the identification of E-cadherin mutations that lead to deregulation and functional impairment of E-cadherin [51,52].

In summary, while far from using quantum dot-based approaches in cancer patients in routine clinical procedures, quantum dots already have clinical relevance in what concerns cancer biomarker detection. We envisage that in the future, we will witness a huge development in quantum dot-based technologies and these will be among the most promising tools for cancer detection and therapy.

Acknowledgments

This work was supported by FEDER Fundo Europeu de Desenvolvimento Regional funds through the COMPETE 2020—Operacional Programme for Competitiveness and Internationalisation (POCI), Portugal 2020, and by Portuguese funds through FCT—Fundação para a Ciência e a Tecnologia/Ministério da Ciência, Tecnologia e Inovação in the framework of the project *Institute for Research and Innovation in Health Sciences* (POCI-01-0145-FEDER-007274). This work was supported by FCT fellowships to M.S.F. (SFRH/BPD/63716/2009) and A. F. (SFRH/97295/2013). Salary support to P.C. was from NORTE-01-0145-FEDER-000029.

References

1. Torre LA, Bray F, Siegel RL, Ferlay J, Lortet-Tieulent J and Jemal A. Global cancer statistics, 2012. *CA: A Cancer Journal for Clinicians*, 2015, 65(2):87–108.
2. Azzazy HM, Mansour MM and Kazmierczak SC. From diagnostics to therapy: Prospects of quantum dots. *Clinical Biochemistry*, 2007, 40(13–14):917–927.
3. Pericleous P, Gazouli M, Lyberopoulou A, Rizos S, Nikiteas N and Efstathopoulos EP. Quantum dots hold promise for early cancer imaging and detection. *International Journal of Cancer*, 2012, 131(3):519–528.
4. Resch-Genger U, Grabolle M, Cavaliere-Jaricot S, Nitschke R and Nann T. Quantum dots versus organic dyes as fluorescent labels. *Nature Methods*, 2008, 5(9):763–775.
5. Mukherjee A, Shim Y and Myong Song J. Quantum dot as probe for disease diagnosis and monitoring. *Biotechnology Journal*, 2016, 11(1):31–42.
6. Barroso MM. Quantum dots in cell biology. *The Journal of Histochemistry and Cytochemistry: Official Journal of the Histochemistry Society*, 2011, 59(3):237–251.
7. Fang M, Peng CW, Pang DW and Li Y. Quantum dots for cancer research: Current status, remaining issues, and future perspectives. *Cancer Biology & Medicine*, 2012, 9(3):151–163.
8. Drummen GP. Quantum dots-from synthesis to applications in biomedicine and life sciences. *International Journal of Molecular Sciences*, 2010, 11(1):154–163.
9. Kamila S, McEwan C, Costley D, Atchison J, Sheng Y, Hamilton GR, Fowley C and Callan JF. Diagnostic and therapeutic applications of quantum dots in nanomedicine. *Topics in Current Chemistry*, 2016, 370:203–224.
10. Rosenthal SJ, McBride J, Pennycook SJ and Feldman LC. Synthesis, Surface studies, composition and structural characterization of CdSe, core/shell, and biologically active nanocrystals. *Surface Science Reports*, 2007, 62(4):111–157.
11. Yin Y and Alivisatos AP. Colloidal nanocrystal synthesis and the organic-inorganic interface. *Nature*, 2005, 437(7059):664–670.
12. Lei KW, West T and Zhu XY. Template-assembly of quantum dot molecules. *The Journal of Physical Chemistry B*, 2013, 117(16):4582–4586.
13. Michalet X, Pinaud FF, Bentolila LA, Tsay JM, Doose S, Li JJ, Sundaresan G, Wu AM, Gambhir SS and Weiss S. Quantum dots for live cells, in vivo imaging, and diagnostics. *Science*, 2005, 307(5709):538–544.

14. Smith AM, Dave S, Nie S, True L and Gao X. Multicolor quantum dots for molecular diagnostics of cancer. *Expert Review of Molecular Diagnostics*, 2006, 6(2):231–244.
15. Deerinck TJ. The application of fluorescent quantum dots to confocal, multiphoton, and electron microscopic imaging. *Toxicologic Pathology*, 2008, 36(1):112–116.
16. Rosenthal SJ, Chang JC, Kovtun O, McBride JR and Tomlinson ID. Biocompatible quantum dots for biological applications. *Chemistry & Biology*, 2011, 18(1):10–24.
17. Alivisatos AP, Gu W and Larabell C. Quantum dots as cellular probes. *Annual Review of Biomedical Engineering*, 2005, 7:55–76.
18. Wu X, Liu H, Liu J, Haley KN, Treadway JA, Larson JP, Ge N, Peale F and Bruchez MP. Immunofluorescent labeling of cancer marker Her2 and other cellular targets with semiconductor quantum dots. *Nature Biotechnology*, 2003, 21(1):41–46.
19. Yezhelyev MV, Al-Hajj A, Morris C, Marcus AI, Liu T, Lewis M, Cohen C et al. In situ molecular profiling of breast cancer biomarkers with multicolor quantum dots. *Advanced Materials*, 2007, 19(20):3146–3151.
20. Wang HZ, Wang HY, Liang RQ and Ruan KC. Detection of tumor marker CA125 in ovarian carcinoma using quantum dots. *Acta Biochimica et Biophysica Sinica*, 2004, 36(10):681–686.
21. Shi C, Zhou G, Zhu Y, Su Y, Cheng T, Zhau HE and Chung LW. Quantum dots-based multiplexed immunohistochemistry of protein expression in human prostate cancer cells. *European Journal of Histochemistry: EJH*, 2008, 52(2):127–134.
22. Peng CW, Tian Q, Yang GF, Fang M, Zhang ZL, Peng J, Li Y and Pang DW. Quantum-dots based simultaneous detection of multiple biomarkers of tumor stromal features to predict clinical outcomes in gastric cancer. *Biomaterials*, 2012, 33(23):5742–5752.
23. Hu Z, Qian G, Muller S, Xu J, Saba NF, Kim S, Chen Z et al. Biomarker quantification by multiplexed quantum dot technology for predicting lymph node metastasis and prognosis in head and neck cancer. *Oncotarget*, 2016, 7(28):44676.
24. Gazouli M, Lyberopoulou A, Pericleous P, Rizos S, Aravantinos G, Nikiteas N, Anagnou NP and Efstathopoulos EP. Development of a quantum-dot-labelled magnetic immunoassay method for circulating colorectal cancer cell detection. *World Journal of Gastroenterology*, 2012, 18(32):4419–4426.
25. Min H, Jo SM and Kim HS. Efficient capture and simple quantification of circulating tumor cells using quantum dots and magnetic beads. *Small*, 2015, 11(21):2536–2542.
26. Wang S, Li W, Yuan D, Song J and Fang J. Quantitative detection of the tumor-associated antigen large external antigen in colorectal cancer tissues and cells using quantum dot probe. *International Journal of Nanomedicine*, 2016, 11:235–247.
27. Li SL, Yang J, Lei XF, Zhang JN, Yang HL, Li K and Xu CQ. Peptide-conjugated quantum dots act as the target marker for human pancreatic carcinoma cells. *Cellular Physiology and Biochemistry*, 2016, 38(3):1121–1128.
28. Greish K. Enhanced permeability and retention (EPR) effect for anticancer nanomedicine drug targeting. *Methods in Molecular Biology*, 2010, 624:25–37.
29. Kobayashi H, Watanabe R and Choyke PL. Improving conventional enhanced permeability and retention (EPR) effects; what is the appropriate target? *Theranostics*, 2013, 4(1):81–89.
30. Iyer AK, Khaled G, Fang J and Maeda H. Exploiting the enhanced permeability and retention effect for tumor targeting. *Drug Discovery Today*, 2006, 11(17–18):812–818.

31. Gao X, Cui Y, Levenson RM, Chung LW and Nie S. In vivo cancer targeting and imaging with semiconductor quantum dots. *Nature Biotechnology*, 2004, 22(8):969–976.
32. Tada H, Higuchi H, Wanatabe TM and Ohuchi N. In vivo real-time tracking of single quantum dots conjugated with monoclonal anti-HER2 antibody in tumors of mice. *Cancer Research*, 2007, 67(3):1138–1144.
33. Chen LD, Liu J, Yu XF, He M, Pei XF, Tang ZY, Wang QQ, Pang DW and Li Y. The biocompatibility of quantum dot probes used for the targeted imaging of hepatocellular carcinoma metastasis. *Biomaterials*, 2008, 29(31):4170–4176.
34. Diagaradjane P, Orenstein-Cardona JM, Colon-Casasnovas NE, Deorukhkar A, Shentu S, Kuno N, Schwartz DL, Gelovani JG and Krishnan S. Imaging epidermal growth factor receptor expression in vivo: Pharmacokinetic and biodistribution characterization of a bioconjugated quantum dot nanoprobe. *Clinical Cancer Research: An Official Journal of the American Association for Cancer Research*, 2008, 14(3):731–741.
35. Xing X, Zhang B, Wang X, Liu F, Shi D and Cheng Y. An "imaging-biopsy" strategy for colorectal tumor reconfirmation by multipurpose paramagnetic quantum dots. *Biomaterials*, 2015, 48:16–25.
36. Kosaka N, McCann TE, Mitsunaga M, Choyke PL and Kobayashi H. Real-time optical imaging using quantum dot and related nanocrystals. *Nanomedicine*, 2010, 5(5):765–776.
37. Soltesz EG, Kim S, Kim SW, Laurence RG, De Grand AM, Parungo CP, Cohn LH, Bawendi MG and Frangioni JV. Sentinel lymph node mapping of the gastrointestinal tract by using invisible light. *Annals of Surgical Oncology*, 2006, 13(3):386–396.
38. Parungo CP, Colson YL, Kim SW, Kim S, Cohn LH, Bawendi MG and Frangioni JV. Sentinel lymph node mapping of the pleural space. *Chest*, 2005, 127(5):1799–1804.
39. Soltesz EG, Kim S, Laurence RG, DeGrand AM, Parungo CP, Dor DM, Cohn LH, Bawendi MG, Frangioni JV and Mihaljevic T. Intraoperative sentinel lymph node mapping of the lung using near-infrared fluorescent quantum dots. *The Annals of Thoracic Surgery*, 2005, 79(1):269–277.
40. Kobayashi H, Hama Y, Koyama Y, Barrett T, Regino CA, Urano Y and Choyke PL. Simultaneous multicolor imaging of five different lymphatic basins using quantum dots. *Nano Letters*, 2007, 7(6):1711–1716.
41. Cai W, Shin DW, Chen K, Gheysens O, Cao Q, Wang SX, Gambhir SS and Chen X. Peptide-labeled near-infrared quantum dots for imaging tumor vasculature in living subjects. *Nano Letters*, 2006, 6(4):669–676.
42. Bagalkot V, Zhang L, Levy-Nissenbaum E, Jon S, Kantoff PW, Langer R and Farokhzad OC. Quantum dot-aptamer conjugates for synchronous cancer imaging, therapy, and sensing of drug delivery based on bi-fluorescence resonance energy transfer. *Nano Letters*, 2007, 7(10):3065–3070.
43. Wang Y, Yang C, Hu R, Toh HT, Liu X, Lin G, Yin F, Yoon HS and Yong KT. Assembling Mn:ZnSe quantum dots-siRNA nanoplexes for gene silencing in tumor cells. *Biomaterials Science*, 2015, 3(1):192–202.
44. Ruan J, Song H, Qian Q, Li C, Wang K, Bao C and Cui D. HER2 monoclonal antibody conjugated RNase-A-associated CdTe quantum dots for targeted imaging and therapy of gastric cancer. *Biomaterials*, 2012, 33(29):7093–7102.
45. Wu F, Yuan H, Zhou C, Mao M, Liu Q, Shen H, Cen Y, Qin Z, Ma L and Song Li L. Multiplexed detection of influenza A virus subtype H5 and H9 via quantum dot-based immunoassay. *Biosensors & Bioelectronics*, 2016, 77:464–470.
46. Karakoti AS, Shukla R, Shanker R and Singh S. Surface functionalization of quantum dots for biological applications. *Advances in Colloid and Interface Science*, 2015, 215:28–45.

47. Hardman R. A toxicologic review of quantum dots: Toxicity depends on physicochemical and environmental factors. *Environmental Health Perspectives*, 2006, 114(2):165–172.
48. Kirchner C, Liedl T, Kudera S, Pellegrino T, Munoz Javier A, Gaub HE, Stolzle S, Fertig N and Parak WJ. Cytotoxicity of colloidal CdSe and CdSe/ZnS nanoparticles. *Nano Letters*, 2005, 5(2):331–338.
49. Chong Y, Ma Y, Shen H, Tu X, Zhou X, Xu J, Dai J, Fan S and Zhang Z. The in vitro and in vivo toxicity of graphene quantum dots. *Biomaterials*, 2014, 35(19):5041–5048.
50. Jaiswal JK and Simon SM. Potentials and pitfalls of fluorescent quantum dots for biological imaging. *Trends in Cell Biology*, 2004, 14(9):497–504.
51. Sanches JM, Figueiredo J, Fonseca M, Duraes C, Melo S, Esmenio S and Seruca R. Quantification of mutant E-cadherin using bioimaging analysis of in situ fluorescence microscopy. A new approach to CDH1 missense variants. *European Journal of Human Genetics*, 2015, 23(8):1072–1079.
52. Simoes-Correia J, Figueiredo J, Lopes R, Stricher F, Oliveira C, Serrano L and Seruca R. E-cadherin destabilization accounts for the pathogenicity of missense mutations in hereditary diffuse gastric cancer. *PLoS One*, 2012, 7(3):e33783.

Chapter 8 Selecting the appropriate *in situ* proximity ligation assay protocol

Doroteya Raykova,
Linda Arngården,
Axel Klaesson, Johan Heldin,
and Ola Söderberg

Contents

Introduction

The genetic composition of cells as well as the signals from their surroundings determines the phenotype of the cells. These signals are transmitted via protein–protein interactions, thus providing information on which genes should be expressed in response to environmental cues. Both tissues and cell cultures, particularly cancer cell lines and stem cell lines, display a remarkable heterogeneity and dynamics at a single cell level, which may be the key to understanding cell behavior in health and disease [1]. The current rapid development of large-scale *omics* methods offers abundant information on the quantity and type of transcripts and proteins found in the cell at a given time point [2,3]. However, these methods have the downside of failing to provide spatial information, which might be particularly important in the context of tissue profiling. Immunostaining has recently been used to map expression of all proteins in various types of tissues and diseases [4]. However, obtaining information on functional states of proteins, such as protein complexes and posttranslational modifications (PTMs), calls for other approaches. Methods that do allow spatial identification of protein–protein interactions are required to determine the signaling network profile of each cell, as this provides the foundation for analysis of cellular communication [5]. Förster resonance energy transfer (FRET) [6] can be used for such analysis. Nonetheless, although antibody-based FRET can be used to analyze protein–protein interactions [7], the method is mainly used by expressing pairs of fluorescent fusion proteins, which requires genetic manipulation of the cells.

An alternative is the *in situ* proximity ligation assay (*in situ* PLA) [8], which relies on affinity reagents for dual protein recognition and combines a requirement for proximal binding with a potent signal amplification to facilitate the use in formalin fixed paraffin-embedded tissue sections and cells [9]. *In situ* PLA can be used for highly specific detection of single proteins, protein–protein interactions, and PTMs such as phosphorylation [10,11] that is informative for determining the activation state of signaling cascades within the cell. Over the years, different protocols have been developed [12], and we herein describe several of them along with guidelines on choosing the protocol that suits the users' requirements and desired experimental outcomes.

Experimental procedures

In situ proximity ligation assays

Target recognition in *in situ* PLA is achieved via a pair of primary antibodies. Signal is generated by virtue of proximity probes—either the primary antibodies themselves [8], or secondary antibodies [11],—which are each conjugated to a short oligonucleotide (Figure 8.1). Whenever and only when the two proximity probes bind close to one another, such as when target proteins interact or form complexes, these oligonucleotides are brought in sufficient proximity, so as to serve as a template for hybridization and subsequent ligation of two additional single-stranded DNA molecules (called the long and the short circularization oligo) into a circle (Table 8.1). The circle can then be amplified via rolling circle amplification (RCA). RCA uses the oligonucleotide

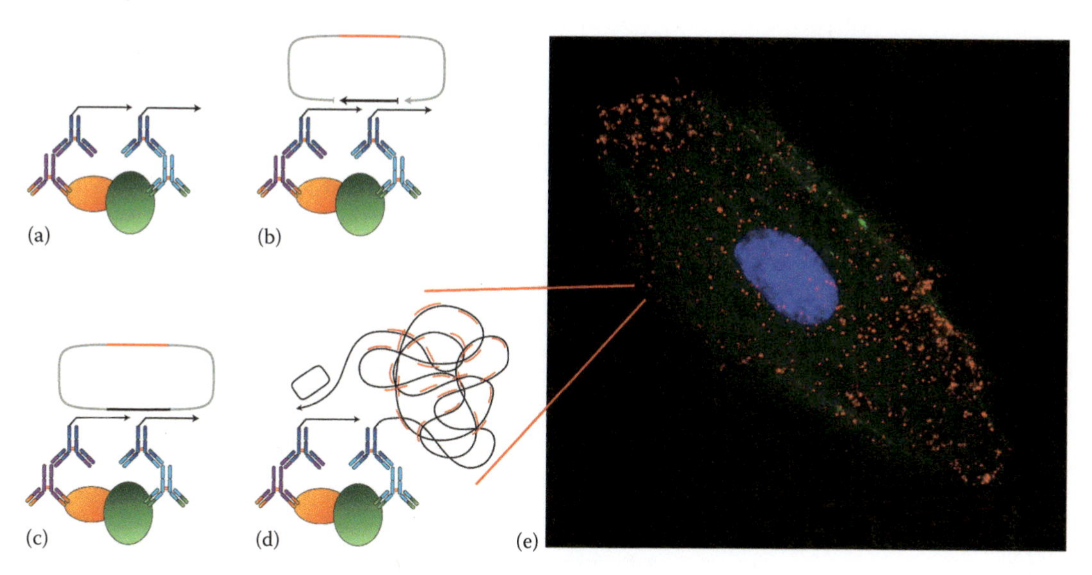

Figure 8.1 *In situ* proximity ligation assay (PLA) with secondary proximity probes. (a) Schematic image of two species–specific secondary antibodies (proximity probes) bound via primary antibodies to a protein complex. The arrowhead indicates a 3' end of an oligonucleotide. (b) Hybridization: added to the proximity probes are the long circularization oligonucleotide (LCO, in gray) and the short circularization oligonucleotide (SCO, black bolded arrow). The sequence for detection on the LCO is shown in red. (c) Ligation: The circularization oligonucleotides are joined together by ligation to form a complete DNA circle. (d) Rolling circle amplification (RCA): The circle is used as a template for amplification primed by one of the proximity probes. The resulting RCA product is a single-stranded DNA molecule containing several hundreds of repeated sequences for detection by a complementary fluorophore-labeled detection oligonucleotide (red). (e) Visualization: antibodies targeting PDGFR-β together with a pan-specific phospho-tyrosine were used to detect phosphorylated PDGFR-β in fibroblasts stimulated with PDGF-BB. The *in situ* PLA signals appear as red dots. The cytoplasm is visualized by phalloidin staining (green) and the nucleus by Hoechst staining (blue). The image is taken with a 63x objective.

attached to one of the probes as a primer, thus creating a long single-stranded product that remains attached to one of the proximity probes. In this way, the signal remains in vicinity of the protein(s) of interest. The RCA product is built-up by a repeated sequence: several hundred copies lined up in one molecule. It can be visualized by hybridization of fluorophore-labeled detection oligonucleotides that are reverse-complementary to a motif within the RCA product. As each RCA product will thereby contain several hundred fluorophores, the signal can be observed under a fluorescence microscope and appears as a bright discrete object with a diameter of up to 1 μm. The signal can then easily be quantified by automated software such as CellProfiler (CellProfiler, http://www.cellprofiler.org).

Table 8.1 Oligonucleotide Designs for PLA

Function	General PLA Protocol (Sequence: 5′ → 3′)	Modification	Figure
Proximity probe oligo 1	(A_N) GAC GCT AAT AGT TAA GAC GCT T (UUU)	5′-Aldehyde	8.1, 8.4
Proximity probe oligo 2	(A_N) AA ATA TGA CAG AAC TAG ACA CTC TT	5′-Aldehyde	8.1
Long circularization oligo	CTA TTA GCG TCC AGT GAA TGC GAG TCC GTC TAA GAG AGT AGT ACA GCA GCC GTC AAG AGT GTC TA	5′-Phosphate	8.1, 8.4, 8.5
Short circularization oligo (0A)	GTT CTG TCA TAT TTA AGC GTC TTA A	5′-Phosphate	8.1, 8.4, 8.5
Detection oligo	CAG TGA ATG CGA GTC CGT CT (UUU)	5′-Texas Red	8.1

Function	Multiplex PLA Protocol (Sequence: 5′ → 3′)	Modification	Figure
Proximity probe oligo 2 – Tag A	(A_N) AA ATA TGA CAG AAC CGG GCG ACA TAA GCA GAT ACT AGA CAC TCT T	5′-Aldehyde	8.4
Proximity probe oligo 2 – Tag B	(A_N) AA ATA TGA CAG AAC ATA CGG TCT CGC AGA TCG CTT AGA CAC TCT T	5′-Aldehyde	8.4
Proximity probe oligo 2 – Tag C	(A_N) AA ATA TGA CAG AAC GGA CGA TCA TCC AGC ACT AGT AGA CAC TCT T	5′-Aldehyde	8.4
Tag A	GTA TCT GCT TAT GTC GCC CG	5′-Phosphate	8.4
Tag B	AGC GAT CTG CGA GAC CGT AT	5′-Phosphate	8.4
Tag C	CTA GTG CTG GAT GAT CGT CC	5′-Phosphate	8.4
Detection oligo – Tag A	TA TCT GCT TAT GTC GCC CGG (UUU)	5′-Alexa488	8.3, 8.4
Detection oligo – Tag B	AGC GAT CTG CGA GAC CGT AT (UUU)	5′-Texas Red	8.3, 8.4
Detection oligo – Tag C	CTA GTG CTG GAT GAT CGT CC (UUU)	5′-Alexa647	8.3, 8.4

Function	Alternative Oligonucleotides (Sequence: 5′ → 3′)	Modification	Figure
Proximity probe oligo 1 (reversed)	(A_N) GAC GCT AAT AGT TAA GAC GCT T	3′-Aldehyde	8.5
Short circularization oligo (5A)	GTT CTG TCA TAT TAA AAA TAA GCG TCT TAA	5′-Phosphate	8.5
Short circularization oligo (10A)	GTT CTG TCA TAT TAA AAA AAA AAT AAG CGT CTT AA	5′-Phosphate	8.5
Short circularization oligo (20A)	GTT CTG TCA TAT TAA AAA AAA AAA AAA AAA AAA TAA GCG TCT TAA	5′-Phosphate	8.5
Compaction oligo	AGA GAG TAG TAC AGC AGC CGT AAA AGA GAG TAG TAC AGC AGC CGT (UUU)	–	8.2
Long circularization oligo – Tag A	CTA TTA GCG TCC AGT GAA TGC GAG TCC GTC TAA GTA TCT GCT TAT GTC GCC CGA AGA GTG TCT A	5′-Phosphate	8.3
Long circularization oligo – Tag B	CTA TTA GCG TCC AGT GAA TGC GAG TCC GTC TAA AGC GAT CTG CGA GAC CGT ATA AGA GTG TCT A	5′-Phosphate	8.3
Long circularization oligo – Tag C	CTA TTA GCG TCC AGT GAA TGC GAG TCC GTC TAA CTA GTG CTG GAT GAT CGT CCA AGA GTG TCT A	5′-Phosphate	8.3

(A_N) = Spacer of N Adenosines (UUU) = Mismatched 2′O-methyl RNA Uracils

Fine-tuning the fluorescent signal

Visualization of *in situ* PLA has the advantage of each initial target signal being amplified several hundred times. Hence, single molecules can be detected with standard microscopy. However, if signals are very close to one another and/or very abundant, it may be problematic to distinguish between and separate individual RCA products from each other, making automated counting difficult or even impossible. A way to bypass this problem is to add a compaction oligonucleotide (Table 8.1)

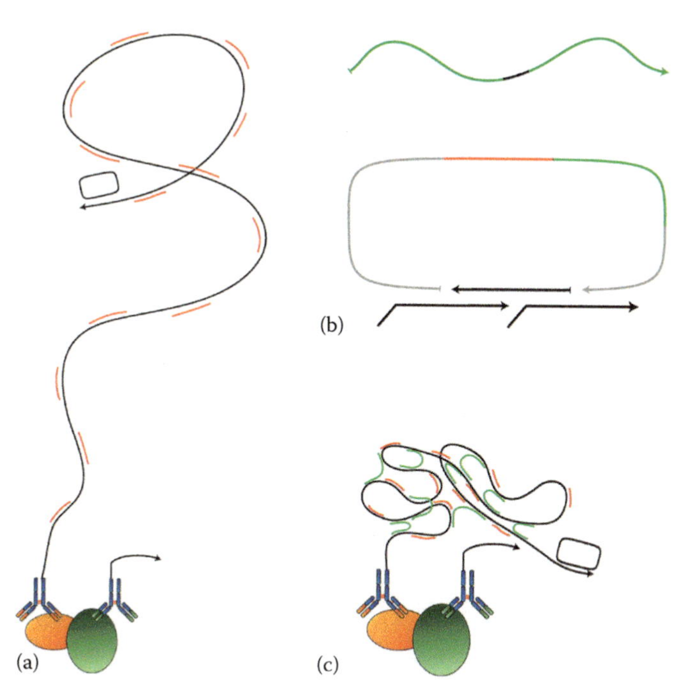

Figure 8.2 Reduction of RCA product size. (a) A regular RCA product will spread out into an object with a diameter of around 1 μm. Detection oligonucleotides hybridized to the RCA product are shown in red. (b) The compaction oligonucleotide design. The two identical sequences are shown in green and the spacer in between, in black. The same sequence (green) is also found in the LCO and will be amplified during RCA. (c) The two identical sequences of the compaction oligonucleotide (green) bind the RCA product and condense it down to an object with diameter of around 200 nm, which is labeled with the detection oligonucleotides (red).

that generates RCA products with a reduced size, a more compact structure, and a higher signal-to-noise ratio [13] (Figure 8.2). The compaction oligonucleotide is composed of two identical sequences, each 21 nucleotides long, which are separated by three bases. The whole sequence is protected from phi29 polymerase priming (which would render the RCA product double-stranded) and from phi29 degradation by three mismatched 2′O-methyl-RNA bases in the 3′ end. Each of the two identical sequences is reverse complementary to a site in each copy of the RCA product. This design enables the compaction oligonucleotide to pull close and distant regions of RCA product together, thus creating a smaller, more compact RCA product. Furthermore, Clausson, Arngarden et al. show that RCA products sometimes tend to disintegrate, resulting in one RCA product being counted several times or even invading other RCA products [13]. Such occurrences can generate false positive results or exclude existing RCA products during quantification. By introducing the

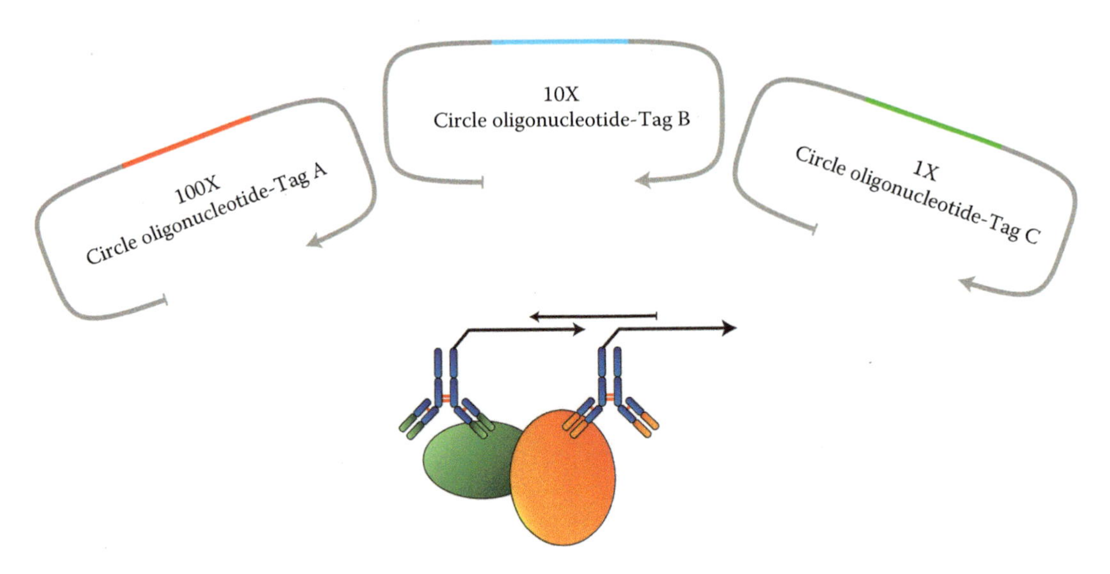

Figure 8.3 Increased dynamic range. Three LCOs with different tag (A, B, and C) sequences for the detection oligonucleotide are added at different concentrations: 100X, 10X, and 1X. These will be ligated with the SCO, generating DNA circles with alternative tags at the same ratio, which will be propagated into the RCA products and detected with unique detection oligonucleotides. By counting RCA products from a fluorescence channel that does not generate coalescent signals, the dynamic range can be extended.

compaction oligonucleotide during RCA, it is possible to increase the integrity of the RCA product, which excludes false positive detections and merged RCA products from the quantification, thereby providing higher accuracy of the final analysis.

An alternative approach to increase the dynamic range of *in situ* PLA is to use a mixture of long circularization oligonucleotides (LCOs) (Table 8.1), differing only in the sequence recognized by the fluorophore-labeled detection oligonucleotide (Figure 8.3). Ligation will thus create circles with different detection sites, which will be propagated into the RCA products. The amount of different RCA products will reflect the proportion of the different LCOs used. If three different LCOs are used in a 1:10:100 ratio, the amount of RCA products formed will have the same ratio. The different RCA products can be visualized with detection oligonucleotides labeled with unique fluorophores. Hence, if signals generated by the LCO with the highest concentration are so abundant that they coalesce, one can monitor an RCA product from an LCO present at a lower concentration, reporting at a different wavelength. By selecting the type of RCA products that generate discrete signals for each individual cell and multiplying the signals by the dilution factor, the dynamic range can be expanded by several orders of magnitude. This facilitates the analysis of heterogeneous cell populations with large differences in abundance of a protein or protein–protein interactions [14].

Choosing proximity probes: Multiplexing versus universality

If secondary antibodies are used as proximity probes, the possibility for observing multiple interactions at once becomes restricted, as that would require proximity probes that selectively bind a wide variety of primary antibody host species without any cross-reactivity. This limitation can be overcome by the use of directly conjugated primary antibodies as proximity probes (Table 8.1). Whenever antibodies are custom-conjugated, for example, by use of succinimidyl-6-hydrazino-nicotinamide (SANH), it is desirable that they come in a phosphate buffered saline (PBS) formulation. In addition, they need to be purified from admixtures such as bovine serum albumin (BSA), gelatin, sodium azide, and so on before the procedure. After conjugation, HPLC-based purification from unbound oligonucleotides and/or unconjugated antibody is recommended in order to decrease unspecific binding and background in the PLA assay. It is also possible that the conjugation disrupts the antibody's efficiency or ability to bind; therefore, antibodies need to be revalidated before they are used in a biological application.

In order to enable detection of multiple different interactions, the assay can be designed in such a way so that an additional, third piece of linear DNA called a tag is necessary for the creation of the circle (Figure 8.4; Table 8.1). The tag sequence is different for every proximity probe used. As the tag is incorporated in the ligated circle, it is propagated in the RCA product and can be visualized with detection oligonucleotides labeled with different fluorophores, generating signals in different wavelengths for each interaction pair [15]. In the design described herein, the tags are present in only one of the proximity probes, which allow interactions between one particular protein and several interaction partners to be monitored. It is, however, possible to use tags in both proximity probes, thus generating dual colored RCA products where the different colors will code for the proteins present in a complex.

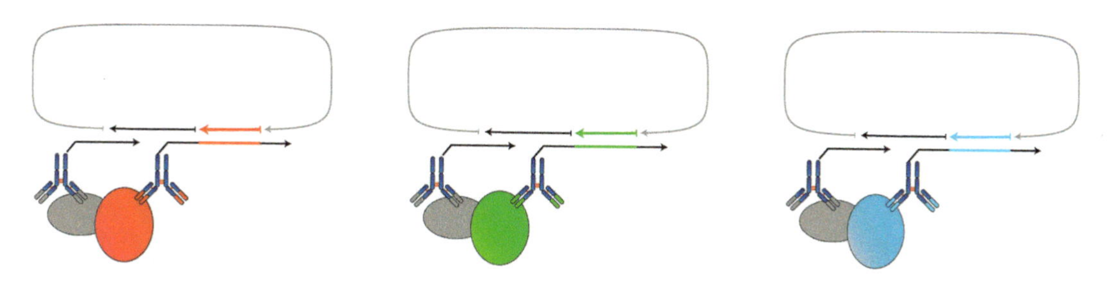

Figure 8.4 Multiplex *in situ* PLA. By equipping each proximity probe with a unique tag, representing the antibody used, alternative interaction partners can be assayed in parallel. Ligation of the complete DNA circle will incorporate both circularization oligonucleotides and the different tags. Each different protein complex will generate a distinctive RCA product that could be visualized by the fluorophore-labeled tag-specific detection oligonucleotides.

One downside of the multiplex method is that conjugation of the antibodies of interest has to be performed by the user, as opposed to the use of secondary proximity probes, which are commercially available through, for example, Sigma-Aldrich. Secondary antibody-based proximity probes have the additional advantage of being universal in the sense that they can be used in combination with any specific primary antibodies as long as they are reactive against the primary host (Figure 8.1).

Increasing or relaxing the threshold of interaction detection

A simple way to regulate the desired stringency of interaction detection is by allowing shorter or longer distance between the proximity probes. If only very closely positioned proteins are targeted or if there is just one protein of interest, minimal distance may be allowed, whereas if the proteins are part of a larger complex and might be positioned further away from each other, one could use longer oligonucleotides to allow for a larger distance between the proximity probes. The distance requirement in PLA is determined by the size of the affinity reagents and length of the oligonucleotides. Every nucleotide along a double helix is 0.34 nm, therefore a DNA strand comprised of 22 nucleotides is 7.5 nm long. Hence, the distance threshold for reporting an interaction is less than 10 nm measured between the points where the oligonucleotides are conjugated to the antibodies. Adding the size of an antibody, around 7–10 nm, the distance between targeted epitopes can become up to 25 nm for directly conjugated proximity probes and up to 40 nm for secondary proximity probes. This distance can be reduced or extended by using alternative affinity reagents of various sizes, or by adding a spacer region of a few adenosines in the proximity probe oligonucleotides, between the moiety used for conjugation to the antibodies, (e.g., the aldehyde group in Table 8.1) and the part that hybridizes to the circularization oligonucleotides. Another way to reduce distance is to change the length and orientation of the oligonucleotides used. Usually, in a proximity probe the oligonucleotides are attached to the antibodies at their 5′ ends. In order for RCA to start, one of the oligonucleotides needs to have a free 3′-OH end, which can be used by the phi29 polymerase as a primer. The other oligonucleotide that contributes to the formation of the circle is usually blocked at its 3′ end by mismatched 2′O-Methyl RNA bases that cannot be primed by phi29 DNA polymerase, in other words replication can only start from one origin. This kind of configuration can however be shrinked by conjugating one of the oligonucleotides by its 3′ end instead, and leaving the 5′ end free (Figure 8.5, Table 8.1). Phi29 polymerase cannot prime the free 5′ end, leaving only one option for priming, the free 3′ end. In this way, hybridization and ligation are only possible if the target proteins are very close together.

Alternatively, to regulate interaction distance different sizes of the DNA fragment can be used to act as a bridge between the proximity probe

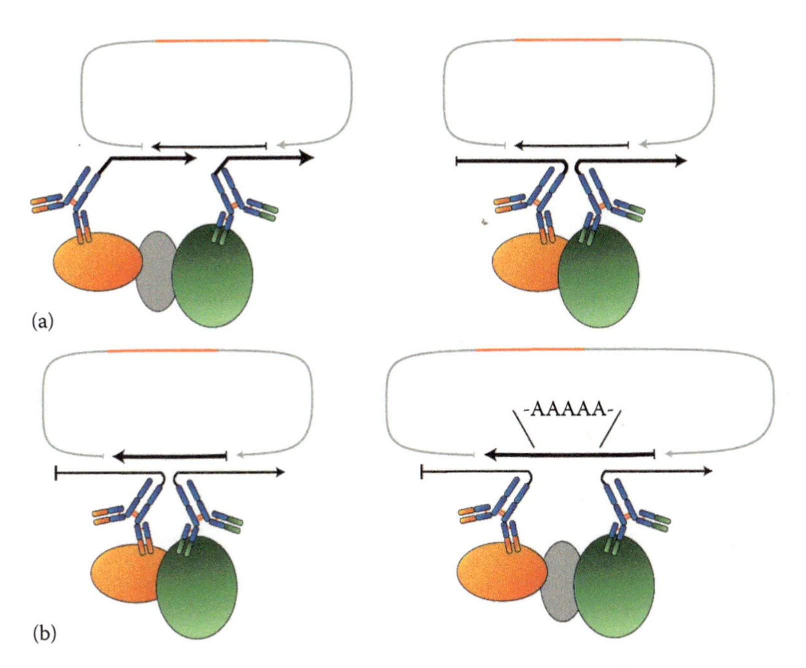

Figure 8.5 Stringency of interaction distance between epitopes. (a) High stringency: the distance threshold between epitopes can be decreased by conjugating one of the proximity probe oligonucleotides to the antibody by its 3′ end (right panel) instead of the 5′ end (left panel), as this forces the proximity probes to be closer to template ligation of the circularization oligonucleotides. (b) Lower stringency: instead of using a regular short circularization oligonucleotide (left panel), a poly(A) spacer can be added to increase SCO length (right panel) and hence increase the distance threshold required for the proximity probes to template ligation of the circularization oligonucleotides.

oligonucleotides—the short circularization oligonucleotide (Figure 8.5). By simply adding a stretch of adenosines to the part bridging the regions that are hybridizing to the proximity probes (Table 8.1), one can stretch and shorten the allowed distance between the interaction partners, much like playing the accordion.

Additional variants of *in situ* proximity ligation assay

In addition to the variants of *in situ* PLA described earlier, there are several others that are worth mentioning, although we will not provide protocols for these herein. In order to study protein–DNA interactions, it is possible to use a DNA molecule as a proximity probe, where one part hybridizes to the DNA assayed and the other part templates the ligation of circularization oligonucleotides. Alternatively, DNA can be used in place of just one of the proximity probes to template the ligation of circularization oligonucleotides [16]. For these types of assays, the targeted

DNA first needs to be rendered single-stranded by use of restriction enzymes and exonucleases. Hybridizing DNA proximity probes can also be used for detection of mRNA, where hybridization of a pair of probes enables a PLA reaction [17].

Another application is combining *in situ* PLA with padlock probes in order to simultaneously visualize protein–protein interactions/PTMs together with mRNA [18,19]. A padlock probe is a linear DNA molecule that upon hybridization brings the 5′ and 3′ end together, allowing the probe to be circularized by ligation. The ligated padlock probe can then be amplified by RCA, using the ligation template as primer. For analysis of mRNA, the method first requires reverse transcription to generate a cDNA molecule *in situ*, since ligation is ineffective with an RNA template [20].

Yet another modification is put to use a recently developed method called proximity-dependent initiation of HCR (proxHCR) [21]. It utilizes hybridization chain reaction (HCR) [22] for signal amplification. This method retains the proximity dependence of *in situ* PLA, while not requiring any enzymatic steps. It is solely built on invasion of metastable DNA hairpin molecules to liberate the initiation sequence that starts the chain reaction where fluorophore-labeled hairpins build up a long nicked double-stranded DNA molecule. The latter then acts as the surrogate marker of the targeted protein–protein interaction to which the proximity probes have bound.

Applications

Antibody conjugation (Optional step)

1. The final concentration of the antibody should be at least 2 mg/ml. If your antibody is concentrated enough, proceed directly to step 2
   a. Insert a concentration column such as the Amicon Ultra Centrifugal Filter 10K (Merck Millipore) in a microcentrifuge tube and prerinse by adding 500 μL 0.1 M NaOH. Incubate 5 min at room temperature (RT) and centrifuge at 1000X g for 5 min Discard the flow-through and reinsert in the tube.
   b. Add 500 μL 1X PBS and centrifuge at 1000X g for 5 min. Discard the flow-through and reinsert in the tube. Repeat (b).
   c. Load up to 400 μL of antibody onto the filter column and centrifuge at 10,000X g for 5 min. Discard the flow-through.
   d. Place the filter upside down in a new tube and centrifuge at 1200X g for 2 min. Collect the flow-though.
   e. Measure the relative antibody concentration in the flow-through using Nanodrop.

2. Proceed to SANH activation of the antibody as follows:
   a. Equilibrate SANH powder to RT before opening. Prepare 20 mM SANH solution in dimethyl sulfoxide (DMSO). Keep the solution in the dark and discard if unused after several days.
   b. Add SANH solution to the concentrated antibody in a 10–20-fold molar excess in favor of the conjugation reagent. Base your calculation on the antibody concentration determined at step 1(e).
   c. Mix well and incubate for 2 h at RT in the dark with gentle shaking.
3. Desalt, for example, using a Zeba Spin desalt column (ThermoFisher Scientific) and proceed as follows:
   a. Equilibrate the Zeba desalt column to RT. Remove the bottom, open the lid, and place into a 2 ml microcentrifuge tube.
   b. Centrifuge at 1500X g for 1 min. Discard the flow-through.
   c. Mark the upward-slanted side of the column and always place the column in the centrifuge with the mark outward.
   d. Equilibrate the Zeba column with 300 μL 1X conjugation buffer (see recipe). Centrifuge at 1500X g for 1 min and discard the flow-through. Repeat step (d) two more times.
   e. Insert the equilibrated column in a new microcentrifuge tube. Apply 30–130 μL sample directly on the filter. Note: If sample volume is smaller than 70 μL, add 15 μL 1X conjugation buffer (see recipe) to the sample.
   f. Centrifuge at 1500X g for 2 min and collect the flow-through.
   g. Recalculate sample concentration after desalting, assuming 5% loss.
4. Conjugate the SANH-activated antibody to an aldehyde-modified oligonucleotide*:
   a. Mix antibody with oligonucleotide in a molar ratio of 1:3 in 1X conjugation solution and 10 M aniline. Use a light-protective tube.
   b. Incubate for 2.5 h at RT in the dark with gentle shaking.
5. Use a Zeba Spin column to remove aniline. Repeat steps 3 (a–g), using 1X PBS instead of 1X conjugation buffer at (d and e).
6. HPLC-purification is recommended at this point. For long-term storage after purification, add BSA and sodium azide to the antibody and keep at 4°C.

* The aldehyde-modified oligonucleotides are chosen according to the intended method (see Table 8.1). For general *in situ* PLA: proximity probe oligo 1 and 2 (if allowing smaller distance of interaction detection is desired, proximity probe oligo 1 can be changed to the reversed proximity oligo 1). For multiplex *in situ* PLA: proximity probe oligo 1, in combination with either of the proximity probes oligo 2—tag A, B, or C.

In situ proximity ligation assay protocol

1. Sample preparation: this step should be optimized by the user and depends on the starting material, which can be cells, fresh frozen tissues, or formalin-fixed paraffin-embedded tissues. Sample preparation should include immobilizing/fixing the material of interest on a glass slide and permeabilization, for example, as described by [23].
2. Blocking.
   a. Incubate the cells with blocking buffer (see recipe or use other desired blocking buffer optimized for your biological material and antibodies) for 1 h at 37°C in a moisture chamber.
3. Primary antibodies.*
   a. Dilute your primary antibodies to working concentration (it must be optimized for each antibody and biological material) in blocking buffer or other desired antibody diluent optimized for your biological material and antibodies.
   b. Apply 40 μL antibody mix per well if using chamber slides, or enough to cover the cells/tissue. Incubate overnight (O/N) at 4°C in a moisture chamber.
   c. Wash 2 × 5 min in 1X TBS in a cuvette.
4. Secondary antibodies/proximity probes.
   a. Choose proximity probes corresponding to the hosts of the primary antibodies used in step 3. Note: Ensure that one of the proximity probes has a free 3′-OH end to the oligonucleotide that will be used for priming.
   b. Dilute the proximity probes to working concentration in blocking buffer or other desired antibody diluents optimized for your biological material and antibodies.
   c. Apply sufficient amount of antibody to cover the cells/tissue. Incubate with proximity probes for 1 h at 37°C in a moisture chamber.
   d. Wash 2 × 5 min in 1X TBS, 0.05% Tween-20 (TBST) in a cuvette.
5. Hybridization†.
   a. Prepare hybridization solution (see recipe) and apply on slide in a volume sufficient to cover the cells/tissue. Incubate for 30 min at 37°C in a moisture chamber.
   b. Wash 2 × 5 min in TBST.

---

* If multiplexing is desired, primary antibodies are conjugated to proximity oligonucleotides to create primary antibody-based proximity probes (see antibody conjugation protocol). In this case, skip step 4.

† Different circularization oligonucleotides are used depending on the method (see Table 8.1). For general *in situ* PLA use LCO and short circularization oligo SCO (0A). For multiplex *in situ* PLA use LCO, SCO (0A), Tag A, Tag B, and Tag C. For regulation of the desired stringency of interaction detection, use LCO together with either SCO 0A, 5A, 10A, or 20A. For *in situ* PLA with dynamic range detection use LCO-Tag A, LCO-Tag B, and LCO-Tag C together with SCO (0A).

6. Ligation.
   a. Prepare ligation solution (see recipe) and apply on slide. Incubate for 30 min at RT in a moisture chamber.
   b. Wash 2 × 5 min in TBST.
7. RCA* and detection†.
   a. Prepare RCA/detection mix (see recipe) and apply on slide. Incubate for 1.5 h at 37°C in a moisture chamber.
   b. Wash 3 × 10 min in 1X TBS.
8. Mounting and microscopy.
   a. Use SlowFade® Gold antifade reagent (ThermoFisher Scientific) or another mounting medium of your choice and cover slide with a cover glass. Seal with nail polish if needed. After this step, slides can be stored at −20°C for two months.
   b. Images of PLA signals can now be acquired using fluorescence microscopy. Quantification of the detected signals can be performed using the free software CellProfiler [24].

Recipes for buffers and solutions

NB: All ingredients given at final concentrations!

Blocking buffer. Prepare a solution of 20% sterile filtered goat serum and 0.0025 µg/µL salmon sperm DNA in 1X TBS.

Conjugation buffer 10x. Prepare 10x conjugation buffer consisting of 1 M $NaHPO_4$ and 1.5 M NaCl with pH 6.0.

Hybridization solution. Prepare a mix of 0.25 mg/ml BSA, 1X T4 ligation buffer (ThermoFisher Scientific), 0.1% Tween-20, 0.125 µM short circularization oligonucleotide (phosphorylated), and 0.125 µM long circularization oligonucleotide (phosphorylated) in water. If you are using the multiplex *in situ* PLA protocol, add the appropriate tag-specific oligonucleotide(s) at a final concentration of 0.125 µM to the mix.

Ligation solution. Mix 1X T4 ligation buffer (ThermoFisher Scientific) and 0.025 U/µL T4 DNA ligase (ThermoFisher Scientific) in water. Note that the T4 ligation buffer may contain a white precipitate of dithiothreitol; if so, heat at 37°C for 10 min and vortex before use.

RCA/detection mix. Prepare a solution containing 0.25 mg/ml BSA, 1X phi29 polymerase buffer (ThermoFisher Scientific), 250 µM dNTP mix, 0.5 U/µL phi29 polymerase (ThermoFisher Scientific), 0.025 µM detection oligonucleotide, and 1X Hoechst (optional) in water. If you are using the compaction oligonucleotide, add it to a final concentration of 0.025 µM.

* For smaller and more compact RCA products add the compaction oligonucleotide (see Table 8.1) to a final concentration of 25 nM.

† The choice of detection oligonucleotides depends on the desired method. For general *in situ* PLA with or without regulation of the desired stringency of interaction detection use detection oligo (see Table 8.1), for multiplex *in situ* PLA and *in situ* PLA with dynamic range detection use detection oligo A, B, and C (see Table 8.1).

References

1. Huang, S., Non-genetic heterogeneity of cells in development: More than just noise. *Development*, 2009, **136**(23): 3853–3862.
2. Saliba, A.E. et al., Single-cell RNA-seq: Advances and future challenges. *Nucleic Acids Res*, 2014, **42**(14): 8845–8860.
3. Wang, D. and S. Bodovitz, Single cell analysis: The new frontier in 'omics'. *Trends Biotechnol*, 2010. **28**(6): 281–290.
4. Uhlen, M. et al., Proteomics. Tissue-based map of the human proteome. *Science*, 2015. **347**(6220): 1260419.
5. Koos, B. et al., Next-generation pathology—surveillance of tumor microecology. *J Mol Biol*, 2015, **427**(11): 2013–2022.
6. Kenworthy, A.K., Imaging protein-protein interactions using fluorescence resonance energy transfer microscopy. *Methods*, 2001, **24**(3): 289–296.
7. Bastiaens, P.I. and T.M. Jovin, Microspectroscopic imaging tracks the intracellular processing of a signal transduction protein: Fluorescent-labeled protein kinase C beta I. *Proc Natl Acad Sci USA*, 1996, **93**(16): 8407–8412.
8. Soderberg, O. et al., Direct observation of individual endogenous protein complexes in situ by proximity ligation. *Nat Methods*, 2006, **3**(12): 995–1000.
9. Conze, T. et al., MUC2 mucin is a major carrier of the cancer-associated sialyl-Tn antigen in intestinal metaplasia and gastric carcinomas. *Glycobiology*, 2010, **20**(2): 199–206.
10. Koos, B. et al., Platelet-derived growth factor receptor expression and activation in choroid plexus tumors. *Am J Pathol*, 2009, **175**(4): 1631–1637.
11. Jarvius, M. et al., In situ detection of phosphorylated platelet-derived growth factor receptor beta using a generalized proximity ligation method. *Mol Cell Proteomics*, 2007, **6**(9): 1500–1509.
12. Koos, B. et al., Analysis of protein interactions in situ by proximity ligation assays. *Curr Top Microbiol Immunol*, 2014, **377**: 111–126.
13. Clausson, C.M. et al., Compaction of rolling circle amplification products increases signal integrity and signal-to-noise ratio. *Sci Rep*, 2015, **5**: 12317.
14. Clausson, C.M. et al., Increasing the dynamic range of in situ PLA. *Nat Methods*, 2011, **8**(11): 892–893.
15. Leuchowius, K.J. et al., Parallel visualization of multiple protein complexes in individual cells in tumor tissue. *Mol Cell Proteomics*, 2013, **12**(6): 1563–1571.
16. Weibrecht, I. et al., Visualising individual sequence-specific protein-DNA interactions in situ. *N Biotechnol*, 2012, **29**(5): 589–598.
17. Frei, A.P. et al., Highly multiplexed simultaneous detection of RNAs and proteins in single cells. *Nat Methods*, 2016, 13(3): 269–275.
18. Weibrecht, I. et al., Simultaneous visualization of both signaling cascade activity and end-point gene expression in single cells. *PLoS One*, 2011, **6**(5): e20148.
19. Weibrecht, I. et al., In situ detection of individual mRNA molecules and protein complexes or post-translational modifications using padlock probes combined with the in situ proximity ligation assay. *Nat Protoc*, 2013, **8**(2): 355–372.
20. Larsson, C. et al., In situ detection and genotyping of individual mRNA molecules. *Nat Methods*, 2010, **7**(5): 395–397.
21. Koos, B. et al., Proximity-dependent initiation of hybridization chain reaction. *Nat Commun*, 2015, **6**: 7294.
22. Dirks, R.M. and N.A. Pierce, Triggered amplification by hybridization chain reaction. *Proc Natl Acad Sci USA*, 2004, **101**(43): 15275–15278.
23. Maity, B., D. Sheff, and R.A. Fisher, Immunostaining: Detection of signaling protein location in tissues, cells and subcellular compartments. *Methods Cell Biol*, 2013, **113**: 81–105.
24. Carpenter, A.E. et al., CellProfiler: Image analysis software for identifying and quantifying cell phenotypes. *Genome Biol*, 2006, **7**(10): R100.

Chapter 9 Cell detection and joint shape tracking using local image filters

Tiago Esteves and
Pedro Quelhas

Contents

Introduction

The majority of automatic image analysis approaches applied to microbiology are based on some type of segmentation or machine learning applied to image analysis [1–3]. Although these approaches can obtain good results, they are in many cases difficult to parameterize or require large amounts of annotated data for good results. This means that once an analysis tool is deployed in the laboratory, the data under analysis must not change in a way not anticipated during the tool's development.

A variety of image segmentation techniques have been used in the scope of cell analysis. When there is a high contrast between cells and background, detection can be performed by image thresholding [4,5]. However, this approach is not able to discriminate touching cells, as the spatial relations are not embedded in image pixel levels. To perform segmentation in images where cells may touch the watershed transform is widely used, which treats image intensity as a topological surface [3]. Watersheds, or crest lines, are built when the water rise from different minima, or seeded locations to avoid oversegmentation [3,6]. All pixels associated with the same catchment basin are assigned to the same label [4].

A more robust family of cell segmentation approaches applied to cell detection are deformable models, which find object boundaries by evolving contours or surfaces guided by internal and external forces [7–9]. Level-set deformable models have become increasingly popular for cell segmentation because they do not require any explicit contour parameterization and have no topological constraints [5,10,11], and enable the use of shape priors [12,13]. However, any method based on contour evolution suffers from dependency on finding starting shapes and encounter problems when objects are clustered spatially [11,14].

However, almost all segmentation methods have parameters, which are fragile when variations to image quality occur, and have very specific optimal parameter values for different contrast and noise conditions. As such, most parameters must be readjusted based on image-specific properties. This reduces the method's robustness to imaging conditions.

To analyze cells in microscopy images with robustness and increased performance we present the use of local image filters. These filters are based in local image characteristics such as local image pixel appearance [15–17] or local image gradient convergence [18]. The local nature of these filters gives them inherent invariance to image illumination variations, enable greater robustness to noise, and low contrast [18]. This has been shown to work well on multispectral fluorescence microscopy image [19], single channel laser scanning confocal microscopy, [20,21] and brightfield microscopy images [19]. In addition, the parameters of local image filters are directly related to the size and shape of cells allowing for easy reparameterization when needed.

The previously presented cell-detection approaches have widely been extended to the automatic analysis of microscopy time-lapse images for cell mobility analysis, which is fundamental for cell migration analysis that is used for tissue development, tissue repair [22], and disease analysis [23]. The properties of cell motion and morphological changes are closely related to the biomechanical properties of the surrounding environment [24]. To evaluate cell mobility, biologists established *in vitro* assays, recording time-lapse microscopy videos, which lead to a large amount of data. This large amount of data does not lend itself to easy analysis due to the tedious work it represents and the possibility of subjective bias from which such analysis would suffer. Therefore, the development of automated quantitative analysis methods for accurate quantification is a key for proper understanding of the complex mechanobiology of cell migration and development [25–27].

In the study of cancer cells specifically, experiments are conducted under different conditions to dissect the associated molecular mechanisms and to validate possible therapies for cancer regulation [28]. Cyanobacteria in brightfield images are also a case study where the analysis of mobility in response to a light is fundamental for understanding how cells sense the position of a light source [29]. Bise et al. also performed cell migration analysis in phase-contrast microscopy images of a wound healing assay [22]. Harder et al. also applied a tracking approach for cell nuclei in fluorescent images, to quantify cell migration and proliferation [30].

To automatically analyze cell mobility both detection and identity propagation is required [31]. This is performed by automatic object tracking methods, which can be classified into two categories: detection association [32,33] and model estimation/evolution [25,34]. In tracking by detection association, cell are detected in each frame and then cell detections are associated between each adjacent frames, based on spatial location and image features [32,33,35,36]. The use of cell tracking based on a detection–association approach has the advantage of simplicity but is limited by the initial detection.

To achieve higher accuracy, detection and tracking can be unified, assuming some prior knowledge about cell dynamics, in an approach that uses temporal information to estimate the cells' location and even shape in the following frames in a more robust and time-consistent manner, even with missing data.

This is achieved by representing cell position, dynamics, and shape as a state variable for which a likelihood can be estimated [37,38]. State modeling approaches that assume linear dynamics and Gaussian noise in the tracking estimation can make use of the Kalman filter [11,39,40]. Kalman filtering was applied by Huth et al. for the automatic tracking of pancreatic cancer cells in differential interference contrast (DIC) microscopy image sequences [39]. Li et al. used the Kalman filter to perform online tracking of migrating and proliferating cells in phase-contrast microscopy images [11]. However, in real biological applications more complex models may be required, which may not be linear or Gaussian, invalidating the use of the Kalman filter.

Particle filter-based tracking is applied when modeling nonlinear dynamics, as they are less restrictive in their assumptions [41–43]. Chowdhury et al. used particle filters for muscle satellite cells tracking, in brightfield microscopy videos [41]. Khan et al. used particle filters for the tracking of interacting targets that can be applied in cell tracking and deals with nonlinear motion [44]. The authors describe a joint particle filter that deals with interacting targets influenced by the proximity and behavior of other targets. Chenouard et al. also presented several particle-tracking methods that range from simple particle detection–association (nearest neighbor) to more complex approaches such as the assumption of motion models applied in particle detection in simulated image data of particles [45]. Ortiz-de-Solorzano developed several tracking methods for cell tracking, such as detection–association based on Euclidean distance and level sets where cell detections are used as initialization for the next frame [46].

Finally, it is known that cell morphology plays an important role on cell mobility more precisely in the directionality and randomness of the cell movement. The use of morphology to improve the tracking, and subsequent mobility analysis of cells was proposed by Smith et al. [47], using the cell nuclei's shape to estimate motion direction. Based on that we explored an approach that is capable of modeling cell morphology and cell mobility jointly [48,49]. By modeling cell motion together with cell morphology, it is possible to improve the tracking accuracy.

In the next section, we will introduce the detection of cells using local image filters and their benefits in comparison with some classical approaches to cell detection. We will then introduce the extension of the use of local image filters to tracking of cells in time-lapse microscopy images and how to explore the shape estimation for better tracking results.

Experimental procedures

To explore use of local image filters for microscopy image analysis we first describe their use in cell detection, with benchmark comparison against baseline segmentation methodologies. We then explore cell tracking using local image filters and explore the relation between cell shape and mobility in joint automatic cell tracking.

Cell detection

As baseline methods, we consider standard image segmentation approaches widely used in cell image analysis [4]: Otsu [50], watershed [6], and level sets [4]. For illustration of typical results for the presented approaches as well as local image filters, we explore the task of *Arabidopsis thaliana* root nuclei cell segmentation [18].

Otsu thresholding selects an adequate threshold of gray level for extracting objects from their background. The basis for this segmentation algorithm is the assumption that the image histogram is bimodal, one mode being related to the objects and the other to the background. The main drawbacks of this technique are that it cannot handle touching/overlapping objects and that results are too tightly coupled with the thresholds used. This result in the frequent need to redefine optimal parameters for specific image conditions, making this method semisupervised at best (Figure 9.1c).

Watershed tries to solve the problem of separating touching objects through an immersion simulation approach. The input image is considered as a topographic surface, which is flooded by water starting from regional minima or given seed locations [6,18]. Watershed lines are formed on the meeting points and all pixels associated with the same catchment basin are assigned to the same label [6,51]. The difficulties

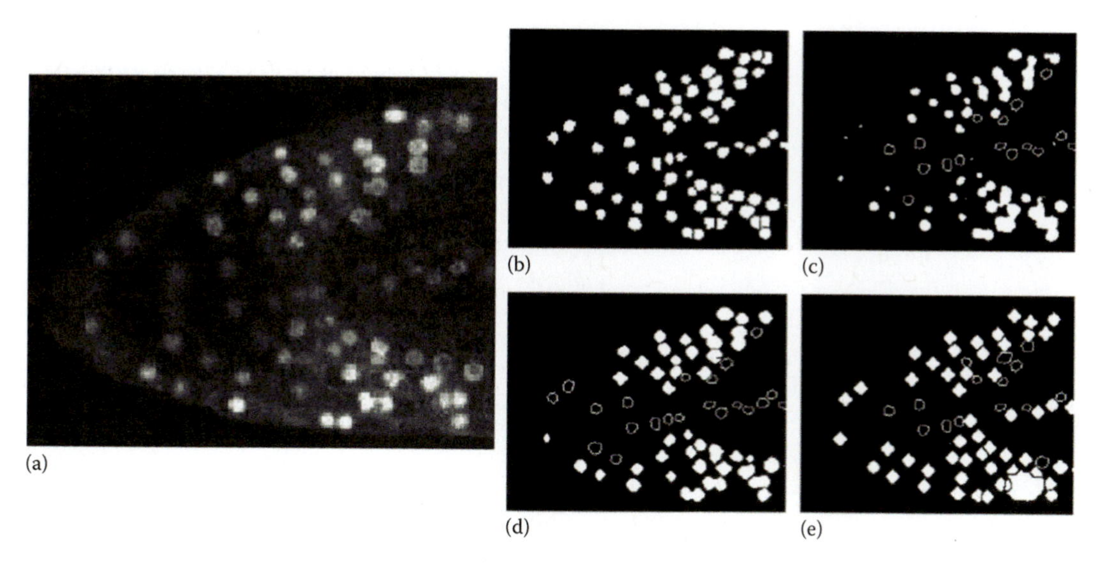

Figure 9.1 Segmentation results for the baseline methods applied to the problem of in *in vivo* cell nuclei detection: (a) input image of the *Arabidopsis thaliana* plants root tip, (b) expert hand annotation, (c) Otsu thresholding, (d) level-sets segmentation (the specific approach for level-sets segmentation used in our work is that proposed by Li et al. [38].), and (e) watershed segmentation. (From Esteves, T. et al., *MVAP*, 23, 623–638, 2012.)

in robust ways to find initialization of seeds in each image make this method highly dependent on parameter (Figure 9.1e).

Level Sets Deformable models based on level sets do not require explicit contour parameterization nor suffer from topology constraints [4,5,11,38,52]. However, these methods require an initial contour from which to start the evolution of the contour curve, which in many cases is not easy to obtain [4]. Level-set contour evolution is capable of adapting to a variable number of objects while preserving spatial coherence. However, it is unable to solve the problem of touching cells that exhibit weak or no edges across their touching boundaries, see Figure 9.1d. Level-set methods also allow for the introduction of priors into the segmentation framework, which is performed by analyzing the statistics of known shapes [12,13]. This has been widely applied in medical image segmentation of anatomical regions [53]. However, the application of such methods to cell segmentation is not trivial as it relies on a trained prior for each individual cell [14].

Local Image Filters have a high level of robustness to illumination changes and noise. This is due to their locality and their local prior, which prevents false positive detection even with high sensitivity. When applied to cell detection, local filter parameters are related to the characteristics of the cells in a way that is intuitive, enabling easy reparameterization and avoiding false detections that do not fit the expected cell characteristics [17,18]. Local image filters can also deal with touching and overlapping cells as long as the cell shape is still within the expected shape by the filter [3,28]. They can also be used to tune their own parameters to best fit the image data in an iterative way [17].

Although there is a large corpus of research on local image filtering for a wide range of applications, we focus on those that are related to the approximated blog such as appearance of cells (or cell nuclei), more relevant for microscopy image analysis.

Scale-normalized Laplacian of Gaussian filter is based on the image scale-space representation to enhance the blob-like structure as introduced by Lindeberg [54]. Given an input image $I(x, y)$, the Gaussian scale-space representation at a certain scale t is

$$L(x,y,t) = g(x,y,t) * I(x,y) \text{ where } g(x,y,t) = \frac{1}{2\pi t} e^{-\left(\frac{x^2+y^2}{2t}\right)}$$

The scale-normalized Laplacian of Gaussian (LoG) operator is then defined as

$$\nabla^2 L(x,y,t) = t^2 \left(L_{xx}(x,y,t) + L_{yy}(x,y,t) \right)$$

where L_{xx} and L_{yy} are the second derivatives of the input image in x and y, respectively, and t^2 is the scale normalization parameter. The scale of the filter, t, is set to the expected range of cell radii (Figure 9.2a). We perform detection of cells by finding local maxima of LoG response for the input image as shown in Figure 9.2(b). The detected maxima enable us to estimate the position and radius ($r = 1.5 \times t$) of cells (Figure 9.2c). By using the LoG filter for cell detection, there is an underlying assumption of a circular shape.

Local convergence filters By assuming a convex shape and a limited range of sizes for cell areas, we can use local convergence filters (LCF) for cell detection and shape estimation. This is possible because LCF detect local gradient convergence in the image that is usually an indication of cell membrane locations [18]. LCF are based on the maximization

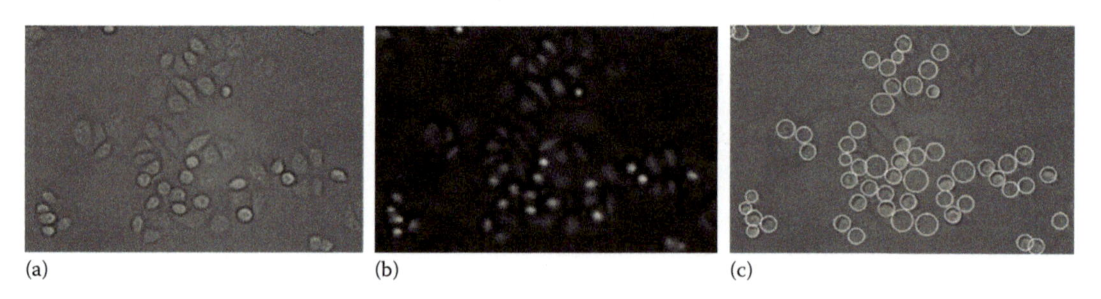

(a) (b) (c)

Figure 9.2 LoG-based cell detection: (a) brightfield frame with cancer cells, (b) LoG response, and (c) detections overlaid in the image.

of the convergence index (CI) at each image spatial coordinate (x, y). The CI within the support region (SR), which is the region that defines each specific filter, is estimated by the cosine between the polar direction θ_i and the image gradient for coordinate (x, y, θ_i, m):

$$CI(x, y, i, m) = \cos(\theta_i - \alpha(x, y, \theta_i, m))$$

where (θ_i, m) are polar coordinates within the filter's SR. The overall convergence is obtained by summing over the specific SR of each filter. We present in the next paragraphs a range of possible LCF filters suitable for nuclei detection.

COIN filter The COIN filter (CF) assumes a circle with variable radius as SR. The value of the radius is varied in search for maximum convergency [18]. The CF formulation is given by the diagram and formulas in Figure 9.3a, where r is the radius of the circle of the SR that

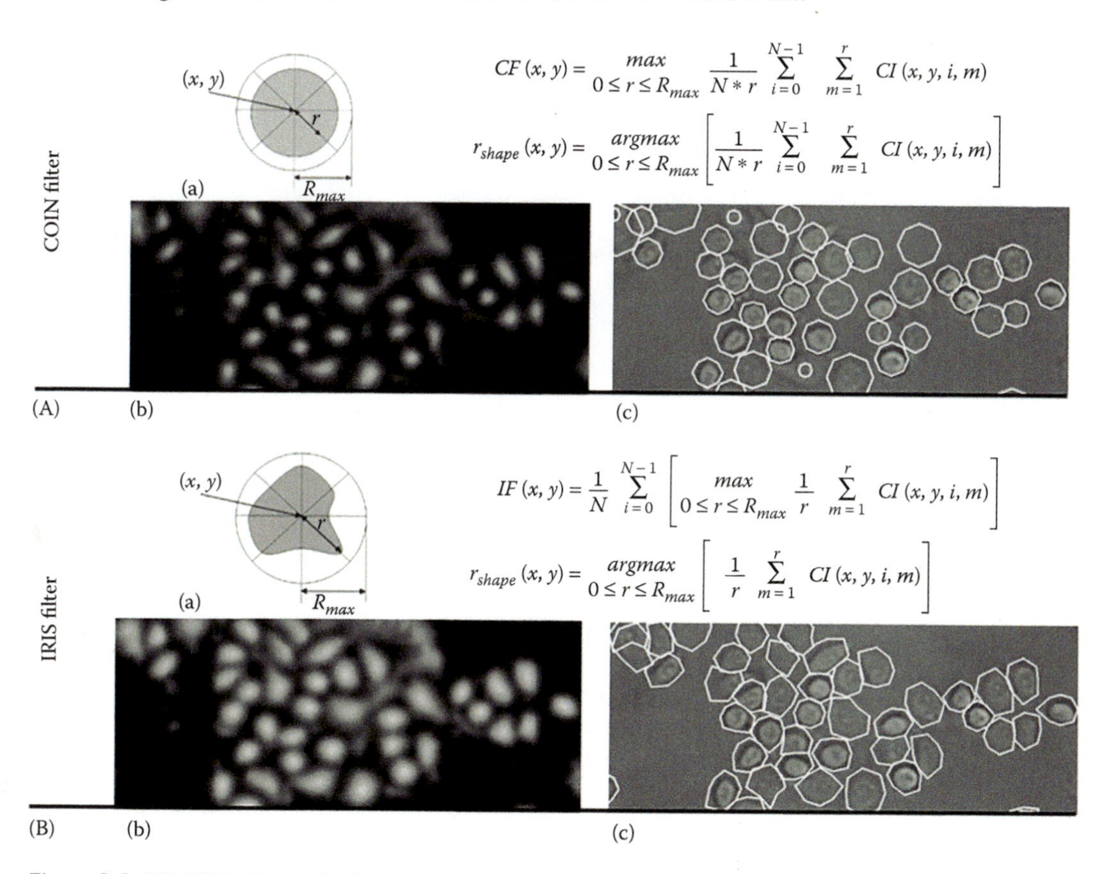

Figure 9.3 (A) COIN filter: (a) schematics of the filter's support region (gray) with $N = 8$, (b) CF response from input image, and (c) detections overlaid in the input image. (B) IRIS filter: (a) schematics of the filter's support region (gray) with N = 8, (b) IF response from input image, and (c) detections overlaid in the input image.

(*Continued*)

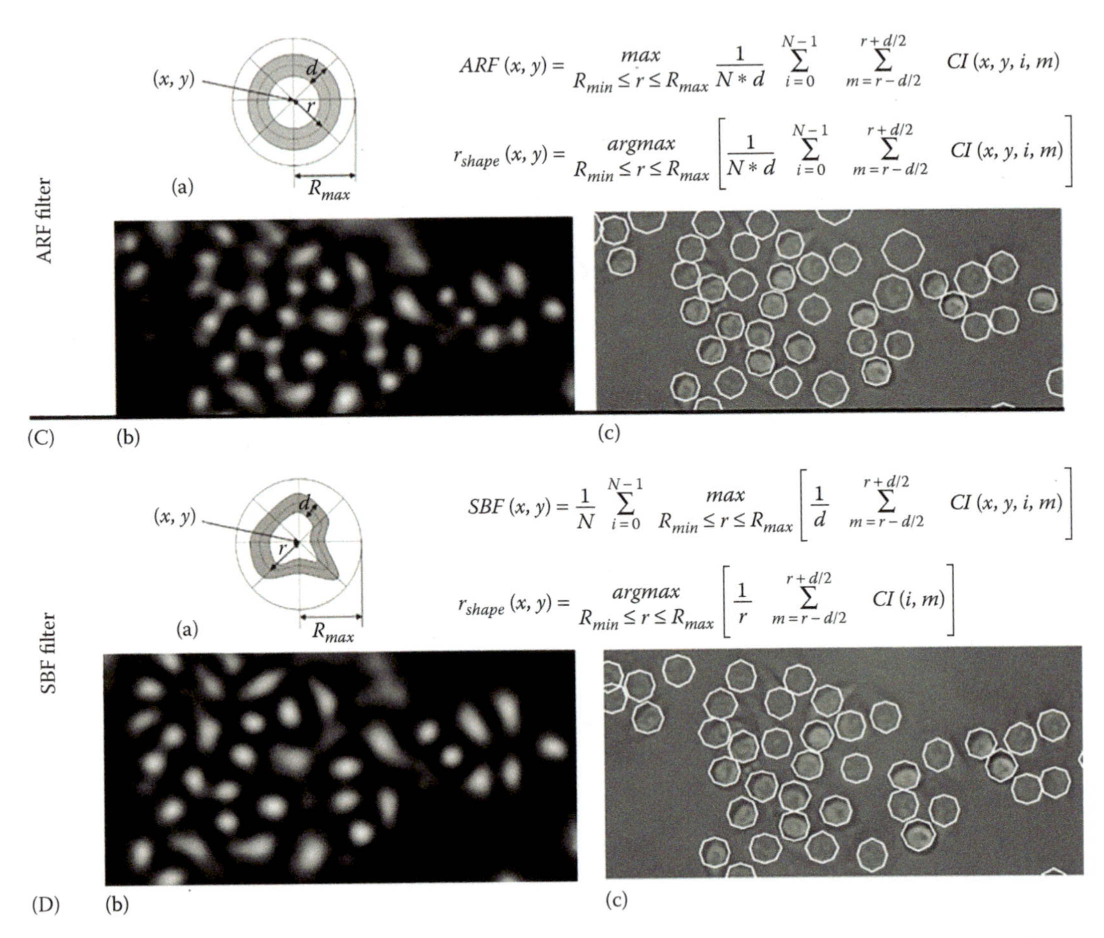

Figure 9.3 (Continued) (C) ARF filter: (a) schematics of the filter's support region (gray) with $N = 8$, (b) ARF response from the input image, and (c) detections overlaid in the input image. (D) SBF filter: (a) schematics of the filter's support region (gray) with $N = 8$, (b) SBF response from—a input image, and (c) detections overlaid in the input image.

varies from 0 to $R_{\max}$, N is the number of radial directions for which convergence is evaluated. The *SR* of this filter corresponds to N half-lines that radiate from the point (x, y), where we calculate the filter's response. The maxima of such response indicate the locations of interest. For each location we obtain the radius of the *SR* at that location through the $r_{\text{shape}}(x, y)$ equation, shown in Figure 9.3c, overlaid on the input image.

IRIS filter The Iris filter (IF) adapts the radius of its *SR* for each of the N directions, maximizing convergence for each radial direction independently [18]. IF is not restricted to circular shapes. Figure 9.3a shows the filter's *SR* and formulation. Although filter maxima still indicates possible object centers, the shape estimation is defined by N independent radii (support points [SP]). The sets of SP corresponding to the image maxima lead to the shapes detected in the image (Figure 9.3c).

Adaptive ring filter The adaptive ring filter (ARF) defines a ring-shaped convergence region [55]. The size of the ring used as *SR* is varied searching for the radius of maximum convergence as given by ARF equation from Figure 9.3a where d is the ring width. Similarly to the process performed when dealing with the CF the shape estimation is performed by searching for the radius of the ring *SR* for the location in the image given by equation from Figure 9.3a. This filter tends to provide a tighter fit than the CF while having the same final estimated shape (circle).

Sliding Band Filter The sliding band filter (SBF) combines the ideas of IF and ARF by defining a *SR* formed by a band of fixed width, with varying radius in each direction [18]. The SBF formulation derives from ARF and IF convergence estimation and is given by the SBF equation from Figure 9.3a where d corresponds to the width of the band, which is moved between R_{min} and R_{max}. The shape estimation of the SBF filter is similar to that of the IF (Figure 9.3c). This filter combines both the shape flexibility of the IF with the limited band search of the ARF.

Cell tracking

To track cells in time-lapse microscopy image some detection method must be applied to each image and a tracking method is used to obtain a consistent identity track of the cells across frames. These methods can be classified into two categories: tracking by detection association [32,46] and tracking by model estimation/evolution [25,34]. For cell detection and shape estimation, we explore local image filters as their advantages in cell detection are even more important for tracking.

Tracking by detection association Cells are detected in each frame and then detections in consecutive frames are associated based on spatial and feature-based distance measures [36,56]. Although this approach is heavily dependent on cell detection, a process that is prone to errors, it does have the upside of being simple and easy to implement.

Considering a detection–association approach based solely on spatial distance, the tracking of cells is performed by measuring the smallest distance between cells and detections in consecutive frames according to

$$D\left(X_{\text{cell}}, X_{\text{detection}}\right) = \left\| X_{\text{cell}} - X_{\text{detection}} \right\|$$

where X_{cell} correspond to the (x, y) coordinates of the cell being tracked in the current frame and $X_{\text{detection}}$ correspond to the (x, y) coordinates of a candidate for correspondence in the next frame [49].

The simplicity of this method has a drawback: if two cells are near each other, they can be incorrectly associated to each other's detections in the next frame. To improve the accuracy of this association process we consider the use of cell appearance descriptors.

Cell appearance descriptors for tracking Tracking by detection association can be extended to include features in the association distance, for improved characterization of cells across frames. We explored cross correlation and local scale invariant feature transform (SIFT) descriptors but other features could be easily used [57]. Pixel information can be directly used in tracking by using image cross correlation [58]. As each cell detection, obtained using local image filters, has a known location and scale, it is easy to define a template region to compare between cell detection (normalized for size) [59]. The template size is parameterized based on the cell radius (template size = $4 \times$ cell radius). The cross-correlation coefficient ranges from 1 to -1 and correlates directly with the quality of the possible match, indicating the similarity between cell detection [57].

Another feature that easily fits the use of local image filters is the SIFT descriptors [60,61]. These features have shown good performance in a large variety of applications from multiview geometry to generic image classification [60,61]. A SIFT descriptor is based on the local image gradient magnitude and orientation information, which is accumulated into orientation histograms of the local image texture/structure [60,61]. A descriptor is extracted for each detected cell, using the detection's scale, and the Euclidian distance between such descriptors can be used to weigh the association metric [57].

For any feature used in this process, the detection–association formulation, using descriptors and spatial distance, is defined by Esteves et al.[62]

$$D\left(C_i^f, \hat{C}_j^{f+1}\right) = \left\| \hat{X}_i^{f+1} - X_j^{f+1} \right\| + kD_F\left(F\left(c_i^f\right), F\left(\hat{c}_i^{f+1}\right)\right)$$

where C_i^f is the cell i in frame f and $\hat{C}_j^{f+1}$ is a candidate cell detection in frame $f+1$. The estimated position of the cell i in the frame $f+1$ is identified by coordinate vector $\hat{X}_i^{f+1}$ and X_j^{f+1} corresponds to the location of the j candidate cell detection in frame $f+1$. The function $D_F\left(F(c_i^f), F(\hat{c}_i^{f+1})\right)$ is the distance between feature descriptor $F(c_i^f)$ of cell i and the feature descriptor $F(\hat{c}_i^{f+1})$ of candidate cell j in the next frame $f+1$, which can be given by the cross-correlation coefficient or by the SIFT descriptors distance [62]. The k parameter is a distance weighting value that controls the influence of each term in the formula.

Although dynamics can be included in tracking by detection association by defining a distance function that favors spatial association of detections in accordance to a specific motion, detection of cells in each frame is still independent of motion prediction. However, the use of motion dynamics for cell-location prediction is valuable for its detection, and can even be used to correct detections in cases of missing data [37,46]. Next, we explore the use of state modeling-based tracking where detection and tracking are integrated in a way that detection can become more robust and shape information can be easily integrated into the tracking process.

State modeling tracking: Tracking by detection–association fails in the case of a missing cell and may also fail when two cells cross trajectories, where the cell identity may change, or if the cells are not well separated. For better robustness, we explore state modeling-based tracking approaches, in specific with the use of Kalman and Particle filtering. Through these approaches, it is possible to assume prior knowledge on the nature of the cell motion in order to overcome the problems related with the detection–association-based tracking.

Kalman filter is an efficient filter that estimates the state of a dynamic system from a series of incomplete and noisy measurements, minimizing the mean of the squared error. The solution is based on the assumption that the system is linear and the probability density function at each state follows a Gaussian distribution [42,43]. The Kalman filter works with two alternating operations: first, it predicts the new state and its uncertainty, and then it corrects with a given new measurement:

$$\text{Prediction: } X_k^- = AX_{k-1} + Bu_{k-1} + \omega_{k-1}$$

$$\text{Measurement: } Z_k = HX_k + v_k$$

where:

X_k^- is the predicted state vector at frame k

A is the dynamics matrix that relates the state X_{k-1} at frame $k-1$ with the state in the current frame

k and B relate the optimal control input u_{k-1} to the state, being u a direct input on the state

ω_{k-1} and v_k represent the prediction and measurement Gaussian noise, respectively

H is the measurement matrix that relates the state vector to the measurement Z_k

The state vector X in frame k is defined by the position of the cell, velocity, and radius: $X_k = [x_k, y_k, vx_k, vy_k, r_k]$ [40,62].

The Kalman filter uses a linear dynamic model assumption, which is problematic when the state has nonlinear dynamics; such as when shape and motion relationships are used or object interaction (grouping or avoidance behavior). The use of particle filters relaxes such assumptions on state dynamics.

Particle filter is a technique for implementing recursive Bayesian filter by Monte Carlo (MC) sampling. It is used to estimate the parameters of Bayesian models and are sequential analogs of Markov chain Monte Carlo (MCMC) batch methods. The key idea is to represent the required posterior density function by a set of random samples with associated weights and to compute estimates based on these samples and weights. As the number of samples becomes very large, this MC characterization becomes an equivalent representation to the usual functional description of the posterior probability density function, and the particle filter approaches the optimal Bayesian estimate [42,43]. On account of the Markov assumption, the probability of the current state given the previous one is conditionally independent of earlier states:

$$p(X_k | X_{k-1}, X_{k-2}, \ldots, X_0) = p(X_k | X_{k-1})$$

State X is an unobserved Markov process, and the measurements Z are the observations of a hidden Markov model (HMM). The state vector X_k comprises the position (x_k, y_k), velocity (vx_k, vy_k), and radius (radius_k) [62].

Similarly, the measurement at the $k-th$ timestep is dependent only on the current state, so is conditionally independent of all other states given the current state:

$$p(Z_k | X_k, X_{k-1}, \ldots, X_0) = p(Z_k | X_k)$$

Using these assumptions, the probability distribution over all the states can be written as

$$p(X_0, \ldots, X_k, Z_1, \ldots, Z_k) = p(X_0) \prod_{i=1}^{k} p(Z_i | X_i)\, p(X_i | X_{i-1})$$

The use of particle filters enables the assumption of more complex state models than the Kalman filter, enabling the modeling of cell mobility and morphology jointly to improve tracking [62].

Joint cell shape and mobility modeling

In experimental *in vitro* assays, biological researchers have found evidence that supports the existence of a relationship between cell morphology and cell mobility, more precisely, the directionality and randomness of the cell movement are dictated by cell morphology [11,13,33,63]. With the aim of verifying such relationship, we set out to analyze representative time-lapse data. We focus our analysis on hand annotated brightfield images containing adenocarcinoma-derived colon cancer cells (RKO) that were seeded on native surfaces or on surfaces coated with extracellular matrix components [28,48,57], see Figure 9.4.

First, we can observe that cells have initially similar appearance, rounded and high contrast (Figure 9.4a), but become more irregular and have lower contrast as time into the experiment increases (Figure 9.4c). This correlates to a decrease in mobility due to greater interaction with the surface's extracellular matrix components [44]. As such, cells with high contrast are observed to have higher mobility (Figure 9.4d: red), whereas low-contrast cells tend to remain static (Figure 9.4d: dotted green).

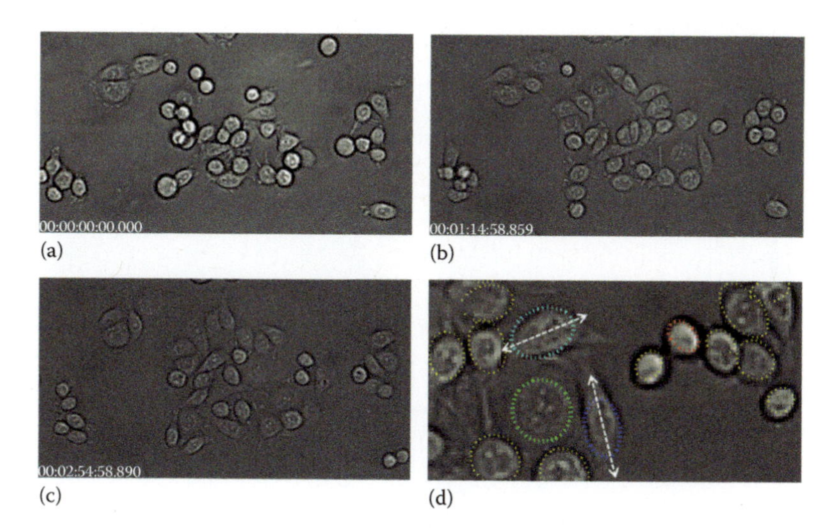

Figure 9.4 Time-lapse images at different time-points (a–c) from colon cancer cells obtained through brightfield microscopy; (d) example of a typical cancer cell time-lapse movie frame (cropped), with detection of final motility phenotypes overlaying the initial brightfield image. Different colors indicate different morphologies (eccentricity) where the cell marked with the dotted green ellipse have reduced mobility, whereas the cells marked with the dotted bright and dark blue ellipses have more mobility.

In addition, cells with elongated shapes are mobile but mostly along their major eccentricity direction (Figure 9.4d: blue).

For better quantification of this relationship, we studied the relationship between shape and mobility on 6 manually annotated time-lapse videos of distinct experimental conditions. We estimated the persistence for each cell, based on its mean-square displacement for several time intervals (1, 2, 4, 8, 16, 32, 64, and 128 frames) [64]. A cell with motion in a predominant direction will have a persistence curve with high slope, whereas a random moving or static cell will exhibit a low slope persistence curve [64].

To validate the relationship between cell appearance and mobility we grouped cells based on their persistence curve slope and obtained exemplar images for such clusters (Figure 9.5b), where we can observe consistent morphology. The elongated shape with some contrast is consistent with a directional motion and is confirmed by the relatively high slope of the persistence curves (Figure 9.5a). On the contrary, clusters of cells with a low persistence curve slope (see Figure 9.6a), tend to have lower contrast and low eccentricity (Figure 9.6b).

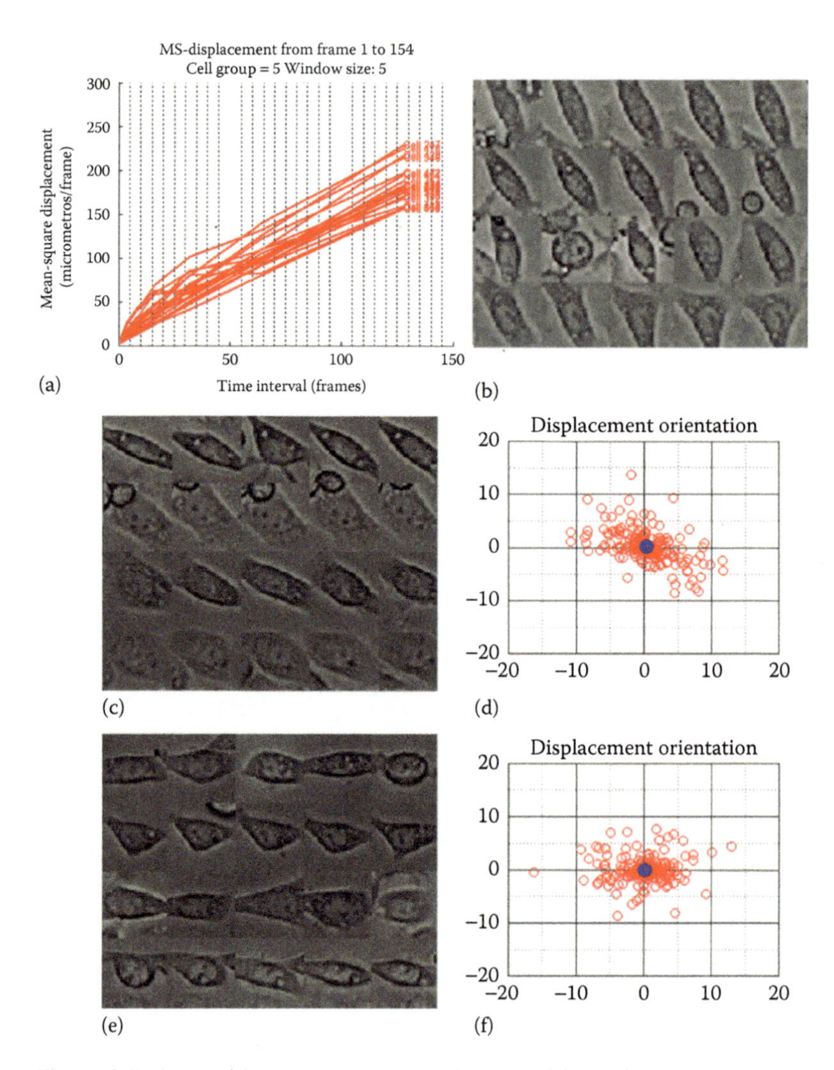

Figure 9.5 Cells with the same cell persistence: (a) persistence cluster with strait persistence lines with an high score indicating that cells are moving always with the same directionality; (b) the most frequent cell shape examples; (c, e) cells with the same shape type, in this case elongated cells; and (d, f) displacement of each cell where it is visible that displacements are performed mainly according to the main axis of eccentricity of the cell shape.

We also analyzed the relationship between cells' shapes and their dominant motion, by grouping cells based on appearance and computing the interframe displacement. In Figure 9.5c and e, we observe that ellipsoid/elongated cells tend to exhibit displacement in the same direction of the main cell axis of eccentricity (Figure 9.5d and f). However, when cells have lower contrast and eccentricity (Figure 9.6c and e), they have a

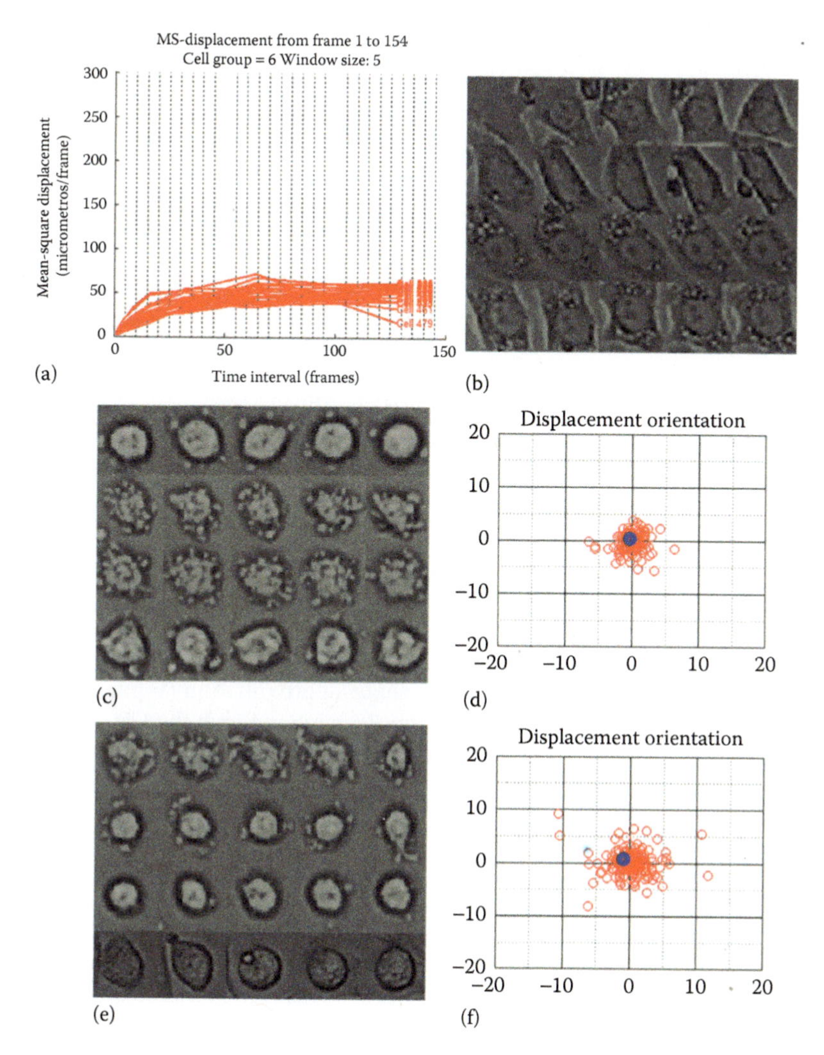

Figure 9.6 Cells with the same cell persistence: (a) persistence cluster with persistence lines with a low slope that indicates that cells tend to stay in the same position or at least move around its own position not moving to distant places; (b) the most frequent cell shape examples from that cluster; (c, e) cells with the same shape type, in this case mainly circular ones; and (d, f) displacement of each cell where it is visible that displacements are performed randomly and toward all possible directions.

more random and isotropic motion with lower displacement magnitude (Figure 9.6d and f).

From this analysis, we can conclude that there is in fact a correlation between cell shape and its mobility, which validates the initial premise. With this in mind, we explore how this can be integrated into the tracking process with the objective of improving results.

Tracking by association using shape information We extended the tracking by association methodology to include shape in its association distance. Cell morphology is in this case modeled by the cell's detection eccentricity and association is favored along the cells largest eccentricity axis [48,49]. The cell's eccentricity is estimated using the local image gradient Hessian matrix. The new distance formula is

$$D(X_{\text{cell}}, X_{\text{detection}}) = \left[X_{\text{cell}} - X_{\text{detection}}\right]^T \mathcal{H}(X_{\text{cell}})^{-1} \left[X_{\text{cell}} - X_{\text{detection}}\right]$$

$$+k\left\|\nabla^2 L(X_{\text{cell}}) - \nabla^2 L(X_{\text{detection}})\right\|$$

where X_{cell} correspond to the (x, y, θ) coordinates and scale of the cell being tracked in the current frame, $X_{\text{detection}}$ correspond to a candidate for correspondence in the next frame, and $\mathcal{H}$ is the local image Hessian matrix defined [54] as

$$\mathcal{H}(x,y,\theta) = \begin{bmatrix} L_{xx} & L_{xy} \\ L_{yx} & L_{yy} \end{bmatrix}$$

From the eigenvalue of $\mathcal{H}$, we obtain the orientation and eccentricity values for the respective ellipsoid shape approximation (Figure 9.4d). The second part of the distance formula is related with the LoG filter's response magnitude of the cells under analysis, which is related to the cell's local aspect (Figure 9.2b). The k parameter is a distance weighting value that controls the influence of this information in the distance formula [48,49].

Particle filter joint position and shape tracking Shape and motion are not governed by linear relationships and as such, the joint tracking of cells and shape estimation process is better suited to be tackled using inference based on particle filters. The cell shape and position are represented by a state vector

$$X = \left[x_k, y_k, r_k, \Sigma_k\right]^T$$

where:

(x_k, y_k) is the cell's position
r_k is its radius
Σ_k is the covariance matrix that describes the cell shape (ellipsoid approximation)

Due to the increase in the size of X and the complexity of the state space, we apply partitioned sampling to improve the performance of the particle filter [65]. This makes the necessary number of samples (particles) increase linearly with the number of partitions ($O(N)$), instead of increasing exponentially with the number of variables ($O(N^2)$). The space was partitioned by first estimating the cell position, followed by its radius and finally its shape. With our chosen state partitioning, we were able to factorize the state's posterior sampling likelihood into multiple sequential sampling steps, each for a given partition:

$$
\begin{aligned}
&P(X_k|Z_k) \rightarrow \\
&\qquad \sim h_1(X'_k|X_{k-1}) \rightarrow q(Z_k|X'_k) \\
&\qquad \sim h_2(X''_k|X''_{k-1}, X'_k) \rightarrow q(Z_k|X''_k, X'_k) \\
&\qquad \sim h_3(X'''_k|X'''_{k-1}, X'_k) \rightarrow q(Z_k|X'''_k, X'_k) \\
&\rightarrow P(X'''_k|Z_k)
\end{aligned}
$$

where the $\sim$ symbol denotes resampling, $h_j, j = 1,2,3$ is the dynamics, and $q(Z|X^i), i = ', '', '''$ denotes for the *weighting* process of each subspace. The dynamic was based on the assumption of *random motion* ($\mathcal{N}(X^i_k, \Sigma)$) or *constant velocity* ($\mathcal{N}(X^i_k + v^i_k, \Sigma)$) models where Σ corresponds to shape eccentricity influence in the dynamics and v^i_k is the velocity. The distribution $P(X_k|Z_k)$ is approximated using a *weighted* particle set $\{X^i_k, \pi^i_k\}^N_{i=1}$.

Initialization of the state of each cell (position and shape estimation) is performed given user selection of cells' approximate positions and radii range, and an initial 500 particles burn-in run in the same frame is used to estimate the initial cell shape. Sampling of particles for state X^i_k is performed with distribution $\mathcal{N}(X^i_k, \sigma)$, at (x, y), with *radius* r_k, and covariance Σ (Figure 9.7b). Each particle weight is estimated, using convolution, with a local LoG image filter defined by the state's values. The state estimate $\hat{X}_k$ (Figure 9.7c) is obtained by the weighted sum over all particles (after particle normalization):

$$\hat{X}_k \approx \sum_{i=1}^{N} \pi^i_k X^i_k$$

Given the set of best particles from the initial burn-in procedure, we estimate the cell position in the next frame, corresponding to the state partition X'. This estimation is based on sampling particles using diffusion and dynamics according to the eccentricity Σ, see Figure 9.7d. Particles will be diffused with higher probability along the main axis of the particle shape distribution. The particles (visible in Figure 9.7e) are then weighted according to similarity information between particles in consecutive frames, using Euclidean distance between SIFT descriptors of each predicted particle i at frame $k(\pi^i_k)$ and the SIFT descriptor of the same particle in the previous frame (π^i_{k-1}) [57].

The next step in partitioned sampling deals with the cell size *radius*. We sample particles from the previous state distribution $\{X'^i_k, \pi^i_k\}^N_{i=1}$, where each X'^i_k is selected with probability π^i_k obtaining a new set of particles with equal weight $\{X''^i_k, 1/N\}^N_{i=1}$. We apply diffusion and dynamics to the scale information of each particle. The likelihood (weight) for each

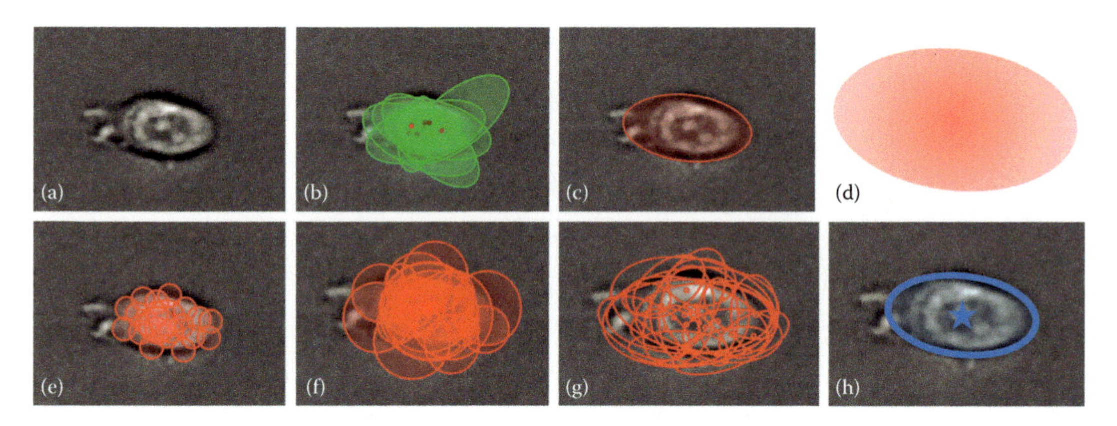

Figure 9.7 Process of acquiring the initial shape of the cell to be tracked (burn in process of particle filter): (a) original image with the cell shape; (b) burn in process where several particles in different positions, with different sizes and shapes are created; (c) initial cell shape acquisition result as a mean of the different weighted particles (considering position, size, and shape); (d) shape distribution according to the previously estimated cell shape from which particles are sampled and diffused; (e) particles diffused on top of the cell being tracked and weighted (different circle sizes means different weights) where it is visible that higher weight is given to the particles in the center of the cell where the SIFT descriptor is probably more similar to the SIFT descriptor of the cell being tracked; (f) different cell sizes are considered (through the diffusion process) and each particle is weighted; (g) different cell shapes are considered (through the diffusion process); and (h) estimated cell shape according to the particle weights.

particle is estimated using a LoG filter on the particle's predicted position (x, y), using the scale information for that particle (Figure 9.7f).

The final part of the partitioned sampling proposed is responsible for estimating the cell shape. Another particle sampling is performed. Diffusion is applied to the covariance matrix Σ from the previous particle shape information at frame $k-1$. Likelihood is estimated using an anisotropic LoG filter based on the particle's orientation and eccentricity information (Figure 9.7g). The final shape, size, and position for each cell can be estimated by the weighted sum over all particles (Figure 9.7h).

Tracking with models for cell interaction The interaction between cells can also induce changes in their mobility. Therefore, we extend our model to handle cell interactions, based on the assumptions that cells do not overlap and isolated cells do not influence other cells.

In order to introduce the interaction model to the particle filter we need to model the properties of all cells in a single state variable. However, this would lead to a very large dimensionality of the state space. As such, we reduce interaction modeling to cells near other cells with a graph-based Markov random field (MRF) constructed for each individual frame. Assuming the pairwise MRF, we can add it to the Bayes filter formulation [44]:

$$p(X_k|X_{k-1}) \propto \prod_{i=1}^{N} p(X_k^i|X_{k-1}^i) \prod_{ij \in E} \Psi(X_k^i, X_k^j)$$

where the $\Psi(X_k^i, X_k^j)$ are pairwise interaction potentials (close cells) of the MRF. This enables us to specify how to control the joint behavior of interacting targets. At the same time, the low weight (absence) of an edge in the MRF encodes the domain knowledge that targets do not influence each other's behavior. Thus, given a potential interaction, we consider the Gibbs distribution to express $\Psi(X_k^i, X_k^j)$ [44]:

$$\Psi(X_k^i, X_k^j) \propto \exp(-g(X_k^i, X_k^j))$$

with

$$g(X_k^i, X_k^j) = \frac{1}{D(X_k^i, X_k^j)}$$

where $D(X_k^i, X_k^j)$ is the spatial distance between X_k^i and X_k^j particles. If D is small (close particles) then $g(X_k^i, X_k^j)$ will be high, meaning a high penalty by $\Psi(X_k^i, X_k^j)$ [44]. By using this cell, proximity information particles are penalized if they overlap with other particles and we expect that this leads to improved tracking results.

Applications

We now explore how the different local image filter-based methodologies perform in a set of tasks related to each method.

Cell detection

In order to test the use of local image filters when applied for cell nuclei detection, we present an objective study on a database of 32 fluorescence images and compare with baseline methods against the respective expert manually labeled ground-truth segmentation [18]. Evaluation measures were obtained over the images using k-folds cross-validation (Table 9.1). In the validation process, the measure maximized was accurate for all methods.

Regarding the total number of errors (false positives and false negatives), there are significantly better results for convergence filters than for baseline methods ($p < 0.05$), with the exception of the CF. Although the LoG detector obtains a better level of false negatives, this is accompanied by an increase in false positives that decreases the performance. Relative to

Table 9.1 Comparison between Nuclei Detection Performance for All Discussed Methods. Results for All Cells in the 32 Fluorescence Images of *Arabidopsis thaliana* Plants. All Numbers are Averaged Results Obtained Over All Images

Method	F_1-score	False Positives	False Negatives	True Positives	Accuracy
Otsu	0.71	0.22	0.70	0.33	0.27
Level Sets	0.77	0.11	0.43	0.60	0.53
Watershed	0.77	0.19	0.42	0.63	0.52
LoG	0.61	0.25	0.34	0.71	0.55
CF	0.72	0.18	0.40	0.65	0.52
IF	0.73	0.16	0.37	0.69	0.56
ARF	0.74	0.16	0.37	0.69	0.56
SBF	0.74	0.14	0.38	0.68	0.56

Source: Esteves, T. et al., *MVAP*, 23, 623–638, 2012.

the F_1-score, both level set and watershed algorithms obtained the highest performance, surpassing all convergence filters based on gradient orientation. In the case of detection accuracy, all convergence filters performed better than baseline methods, with the exception of the CF filter ($p < 0.05$). The LoG detector obtained an accuracy similar to that of the local convergence filter, while not being better; this indicates the advantage of nuclei detection based on local image filters and maxima detection. The main disadvantage of the LoG filter was its simpler shape adaptation, which can be improved based on the Hessian matrix information of the local image gradient [18]. From these results, it is clear that local image filtering approaches for nuclei detection lead to good results that in many cases surpasse de performance of the most widespread baseline methods.

We further evaluated the performance of the LoG filter and local convergence filters on 90 brightfield images from one time lapse video for the task of adenocarcinoma-derived colon cancer cells (RKO) detection, presented in Table 9.2. From the results, we can observe LoG led to the best results, indicating that this is the most adequate methodology for cell detection on these data. By using the LoG

Table 9.2 Cancer Cell Detections and Shape Fit Performance Evaluation on 90 Brightfield Images from Time-Lapse Video

Method	F_1-score	False Positives	False Negatives	True Positives	Accuracy
LoG	0.80	0.10	0.13	0.83	0.78
LoG Hessian	0.81	0.10	0.13	0.83	0.78
CF	0.78	0.13	0.17	0.79	0.72
IF	0.78	0.11	0.19	0.77	0.73
ARF	0.78	0.13	0.17	0.79	0.72
SBF	0.78	0.11	0.17	0.79	0.73

filter together with the local image Hessian information, we were able to increase the cell shape estimation, increasing the F_1-score to 0.81 [28]. It must be emphasized that the same methodologies performed well for both fluorescence brightfield microscopy images, indicating that these methodologies can be applied over a wide range of imaging modalities.

We also applied the LoG filter to electron microscopy images for the detection of immunogold particles. The immunogold labeling technique is used to permit the study of different types of cell wall growth and to understand if there are relevant differences on their composition and if their transport capabilities are similar [62]. The performance achieved was above 85% of both precision and recall. Immunogold particle detection is an extreme time-consuming task where a single image containing almost a thousand particles can take several hours to annotate. The application of the LoG filter to perform this task improved significantly the work of scientists, resulting in prototype software where user need to only specify the range of the size of particles to detect [62].

Tracking by detection–association

Extending the local image filter approach to tracking, we started by evaluating the use of a detection–association approach on 156 brightfield images containing 81 adenocarcinoma derived colon cancer cells (RKO). Cells were detected using the LoG filter and associated by minimum spatial distance [57,62]. The results were obtained by considering 30 randomly selected cells. The obtained accuracy was around 90% and we tracked correctly in the entire time-lapse video around 36% of the selected cells. A simple example of a cell tracked considering the proposed methodology is available in Figure 9.8a.

We also applied automatic tracking by detection–association using the LoG filter for detection of cells in the 3rd edition of the *Cell Tracking Challenge* organized by the *IEEE International Symposium on Biomedical Imaging* (ISBI 2015) in Brooklyn [66]. The challenge included images from different microscopy techniques such as phase contrast and differential interference contrast microscopy as well as brightfield and fluorescence microscopy [66]. The datasets are based on sequences of nuclei or cells, fluorescently stained, moving on top of substrates and microscopy videos of phase contrast and differential interference contrast images of cells moving on a flat substrate of varying rigidity. The videos have examples of a large range of cell types and image quality (different spatial and temporal information resolution and different levels of noise). We were among the top ranked participants of the challenge and one example of a tracking result is available in Figure 9.8b where the cells from the dataset Fluo-N2DH-GOWT1 were tracked with an accuracy of 88%. The fact that we were able to obtain a high level of results with the

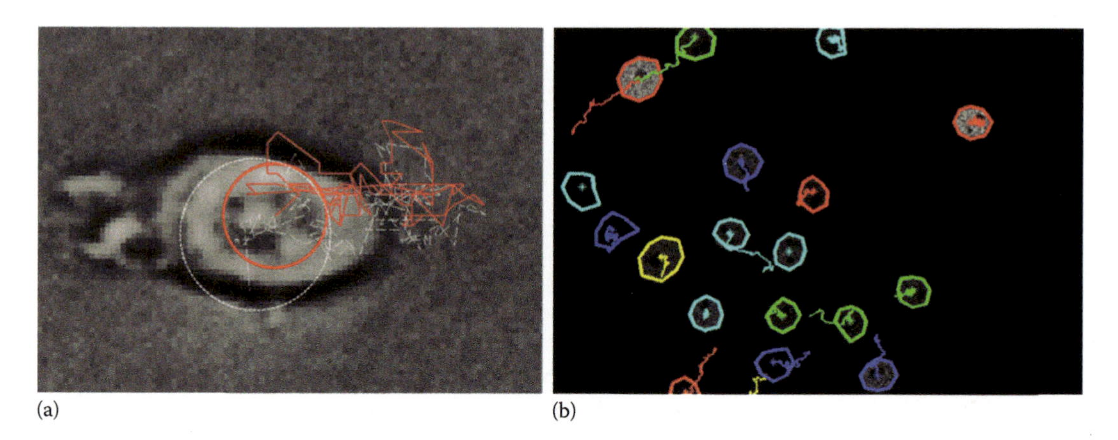

Figure 9.8 Example tracking results: (a) tracking result of one cell considering the tracking by association approach (red continuous line) and the respective manual annotation (white dash line) and (b) tracking result for one dataset (Fluo-N2DH-GOWT1) from the cell tracking challenge.

same approach for all data types, while other participants tailored their approach to each dataset, indicates the robustness of local image filters for microscopy image analysis.

This methodology was also easy to apply to another brightfield problem of automatic tracking of unicellular *cyanobacterium Synechocystis* sp. In this problem, researchers want to quantify bacteria's mobility in response to light, in order to understand how cells sense the position of a unidirectional light source [29]. This work resulted in a user-friendly application, named BacteriaMobilityQuant software, to perform the automatic quantification of cyanobacteria mobility in time-lapse videos.

Cell appearance descriptors for tracking

We tested the detection–association approach with the addition of local descriptors, which are extracted at cell locations based on the known scale of cells. The scale of cells and cell detection is trivial to obtain when using local image filters. The tracking results obtained in the same 30 colon cancer cells (RKO) (randomly selected) in one time-lapse video with 156 frames [57]. Although we maintained the accuracy of the method in relation to the detection–association approach (around 90%), we were able to increase the number of the cells correctly tracked in the entire time-lapse video from 36% (detection–association approach) to 40% as well as a higher average number of frames where cells were correctly identified. Thus, the cell appearance characterization allowed us to improve association of cells between frames.

Kalman filter

To add dynamics into the tracking analysis we applied the Kalman filter to the tracking of each cell, using the LoG filter for detection in each frame.

Table 9.3 Tracking Results Obtained by the Particle Filter and Kalman Filter without Using the Cell Shape Information and Interaction Models

Method	Correct	Coverage	Accuracy
Kalman filter	50,0	75,5	89,8
Particle filter	53,3	80	97,3

Assuming a *constant position* dynamics the use of the Kalman filter led to a tracking accuracy of more than 90%, an increase when compared with detection–association [40]. It was possible to obtain more than 50% of cells completely well tracked in the entire time-lapse video, improving in this way the results achieved by the tracking approach based on the cell appearance descriptors (Table 9.3). Assuming a constant velocity, the results were not improved and less cells were correctly tracked in the entire time-lapse video (around 27%) as well as less frames, where cells were correctly identified (around 50%) [62]. We concluded that the use of this information alone is not adequate for tracking performance improvement.

Particle filter

For a better dynamics modeling we applied particle filter-based tracking on the same data. Using particle filters with LoG filter-based likelihood estimation of cell location and local image cells descriptors, it was possible to achieve a tracking accuracy of 97%, exceeding the performance obtained by detection–association and Kalman filter. In addition, a higher number of cells completely well tracked in the entire time-lapse video was obtained (53%) as well as a higher number of frames in which cells were correctly tracked (80%, which represents 125 frames) (Table 9.3).

Tracking by detection association using shape

To integrate shape into the tracking process we started by evaluating the use of tracking by detection–association with the cell shape eccentricity information, based on Hessian modified LoG filter detection. Tracking was performed on 21 brightfield images from one time-lapse video, containing 81 cancer cells (RKO) manually annotated. For the applied distance formula containing the spatial distance and shape information, we also tested the addition of the LoG filter response information because we assume that cell morphology changes are not abrupt between frames, and so the LoG response should be similar for the same cell in consecutive frames [62].

From the obtained results (Table 9.4), we can observe that, by using Euclidean distance-based association, the obtained accuracy was 74%. By using our local eccentricity-based distance, we were able to improve the tracking by association results, from 74% to 76% (considering $k = 0$), which corresponds to 34 less errors. By adding the LoG response to the

Table 9.4 Performance Evaluation of the Tracking Results (Percentage values), Considering a Detection–Association Approach Using Only the Spatial Euclidean Distance and Considering Associations Using Eccentricity and LoG Response Information. The *k* Parameter Controls the Influence of the LoG Response Information. The Results were Obtained on 21 Brightfield Images Containing 81 Colon Cancer Cells

Detection-Association Approach	Correct	Coverage	Accuracy
Euclidean distance	30	62	74
Eccentricity-based distance	Correct	Coverage	Accuracy
($k = 0$)	31	63	76
($k = 1$)	31	61	77
($k = 3$)	27	60	79
($k = 5$)	24	54	75
($k = 10$)	25	48	74

distance formula, by changing the *k* parameter to 3, the tracking accuracy value over the 21 brightfield images became higher (79%), leading to 85 more cells being correctly identified.

We concluded that by using local anisotropy clues given by the analysis of the Hessian matrix of the image gradient information it was possible to correctly estimate the cell's eccentricity and using that information to improve the tracking results. The results were improved further by using the LoG filter's response magnitude values, as we were able to track more cells correctly for the full movie's length.

Particle filter-joint location and shape tracking

In order to further integrate shape into our tracking approach we developed a particle filter based on the joint motion and shape model, where the shape likelihood of cells is also estimated with the state sampling [62]. We used the same brightfield data, with 30 selected colon cancer cells (RKO) as before.

In Table 9.5, we can observe the results of considering the joint motion and shape model to perform cell tracking considering the partitioned sampling technique in comparison with the classical particle filter. We also obtained the results from applying or not applying the interaction model. In conclusion, we found that the use of the particle filter with partitioned sampling, and considering both shape model and the interaction model, achieved the best results. Although maintaining a high-tracking accuracy (*Tracking accuracy* value around 96%), we were able to obtain a higher number of cells tracked correctly in the entire time-lapse video (*Correct* value of 70% that corresponds to 21 cells correctly tracked) and a higher number of frames, where cells were correctly tracked (*Coverage* value of 93% that corresponds to 145 frames approximately). In comparison with

Table 9.5 Tracking Results Obtained by the Particle Filter Considering Motion and Shape Joint Tracking. The Results were Obtained for 156 Brightfield Images of One Time-Lapse Video Where 30 Cells were Automatically Tracked. The Use of Partitioned Sampling was Compared With the Use of the Classical Particle Filter and With and Without Interaction Modeling

Particle Filter + Shape	Interaction	Correct	Coverage	Accuracy
Without partitioned sampling	Not using	50,0	88,5	90,5
	Using	56,7	87,7	92,1
With partitioned sampling	Not using	60,0	89,4	95,9
	Using	70,0	93,3	95,7

the traditional particle filter, the results improved by considering the partitioned sampling approach and by adding the interaction model.

In Table 9.3, we present the performance results of using the traditional Kalman and Particle filters in the same data for comparison (tracking of 30 cells on 156 brightfield images of one time-lapse video). It is clear that by considering the joint motion and shape model the results improved, specifically for the number of frames where cells were correctly tracked (coverage) and also the total number of cells that were tracked correctly in the entire time-lapse video (correct). As in the proposed methodology for cell tracking, the predicted state for both Kalman and Particle approaches is measured using the LoG filter and by computing SIFT descriptors for similarity analysis between cells.

Limitations

Although local image filters have an overall good performance across many applications and data modalities, they have a high computational complexity when compared with some more classical approaches. With the increase in complexity for each method, the processing time also increases, taking more time to obtain the results.

In the case of using shape in a joint tracking framework, it is clear that the more complex the tracking the more processing time will be required for data analysis (Table 9.6) [62].

The processing time of the tracking procedure considering cell appearance descriptors is comparable to the simple detection–association approach.

The processing time of using the particle filter is higher when compared with the Kalman filter and the detection–association approach. However, the results of using the particle filter were better and the processing time is not only the most important evaluation parameter. Since we are performing the analysis of cell mobility off-line and not in real time, this is

Table 9.6 Processing Time Comparison for All the Tested Methods in the Same Dataset. The Total Processing Time was Annotated for Comparison (seconds)

					Particle Filter Joint Location and Shape Tracking	
Method	**Tracking by Detection-Association**	**Cell Appearance Descriptors for Tracking**	**Kalman Filter**	**Particle Filter**	**Without Partitioned Sampling**	**With Partitioned Sampling**
Time (s)	85,6	118,7	537,8	1573,1	225000	9000

not an important issue in most applications. The researchers take several hours to prepare the experiments and get the time-lapse videos and they do not require an online processing of the data.

Finally, the processing time of using the joint motion and shape model also increased compared with the remaining approaches, but again, this is not an issue because the researchers do not want an online processing.

Summary

We present an overview of the application of local image filters for the problems of cell detection and tracking in microscopy images, and also extend their use to the joint tracking of motion and shape of cells in time-lapse videos. The overall performance and easy parameterization of these methods allow for the development of easy to use and adapt software for lab environment use by a biology researcher. Local image filters also show advantages in the presence of noise and varying image quality, where through their prior on cell shapes, they are able to maintain high-performance levels with no reparameterization.

We investigated the correlation between cell mobility and morphology in cancer cells present in *in vitro* assays and we were able to explore such relationship to create an accurate cell-tracking methodology.

By using cell eccentricity information, through the analysis of the Hessian matrix of the image gradient information, we were able to improve the tracking by detection–association results from 74% to 76%. Using particle filters for joint motion and shape tracking, we were able improve accuracy to 90% and successfully track 50% of cells over the entire time-lapse video. By exploiting partitioned sampling, we further improved the tracking accuracy to 96% and tracking 60% of all cells throughout the complete video. By also modeling cell interaction, tracking further improved to a final accuracy level of 96% and 70% of all the cells tracked correctly in the entire time-lapse video. From the tracking results obtained, it is clear about the importance of the cell shape information incorporation in the tracking process as well as the existence of the correlation between cell shape and mobility.

Overall, this improved analysis favors biological accuracy regarding cell mobility toward external stimuli, such as promotile or, proinvasive factors as well as therapeutics agents. In addition, it may also provide quantitative information to biologists regarding morphological alterations along cell movement.

References

1. E. Meijering, O. Dzyubachyk, and I. Smal, Methods for cell and particle tracking. In *Imaging and Spectroscopic Analysis of Living Cells*, P. Michael Conn (Ed.) 2012, **504**: 183–200.
2. Q. Chen et al., Cell classification by moments and continuous wavelet transform methods. *International Journal of Nanomedicine*, 2007, **2**(2): 181–189.
3. M. Marcuzzo et al., Automated Arabidopsis plant root cell segmentation based on SVM classification and region merging. *Computers in Biology and Medicine*, 2009, **39**(9): 785–793.
4. G. Xiong, X. Zhou, and L. Ji, Automated segmentation of *Drosophila* RNAi fluorescence cellular images using deformable models. *IEEE Transactions on Circuits and Systems I*, 2006, **53**(11): 2415–2424.
5. G. Xiong et al., Segmentation of *Drosophila* RNAI fluorescence images using level sets. In *Proceedings IEEE International Conference on Image Processing*. 2006, pp. 73–76.
6. A. Bleau and L.J. Leon, Watershed-based segmentation and region merging. *Computer Vision and Image Understanding*, 2000, **77**(3): 317–370.
7. P. Moore and D. Moore, A survey of computer-based deformable models. *International Machine Vision and Image Processing Conference*, 2007, p. 55–66.
8. M. Hu, X. Ping, and Y. Ding, A new active contour model and its application on cell segmentation. *Proceedings of Control, Automation, Robotics and Vision Conference*, 2004, pp. 1104–1107.
9. P. Yan, X. Zhou, M. Shah, and S. Wong, Automatic segmentation of high-throughput RNAi fluorescent cellular images. *IEEE Transactions on Information Technology in Biomedicine*, 2008, **12**(1): 109–117.
10. F. Bunyak, K. Palaniappan, S.K Nath, T. Baskin, and G. Dong, Quantitative cell motility for in vitro wound healing using level set-based active contour tracking. *Proceedings of the IEEE International Conference on Biomedical Imaging (ISBI)*, 2011, pp. 1–4.
11. K. Li et al., Cell population tracking and lineage construction with spatiotemporal context. *Medical Image Analysis*, 2008, **12**: 546–566.
12. W. Fang and K.L. Chan, Incorporating shape prior into geodesic active contours for detecting partially occluded object. *Pattern Recognition*, 2007, **40**: 2163–2172.
13. M. Leventon, W. Grimson, and O. Faugeras, Statistical shape influence in geodesic active contours. In *Proc. IEEE Conf. on Computer Vision and Pattern Recognition (CVPR)*, 2000, pp. 316–323.
14. K.R. Mosaliganti, A. Gelas, A. Gouaillard, R. Noche, N. Obholzer, and S. Megason, Detection of spatially correlated objects in 3D images using appearance models and coupled active contours. Medical image computing and computer-assisted intervention: MICCAI. *International Conference on Medical Image Computing and Computer-Assisted Intervention*, 2009, 12(2): 641–648.

15. S. Eom, R. Bise, and T. Kanade, Detection of hematopoietic stem cells in microscopy images using a bank of ring filters. In *Biomedical Imaging: From Nano to Macro, 2010 IEEE International Symposium on*, 2010, pp. 137–140.
16. L. Ferro, P. Leal, M. Marques, J. Maciel, M. Oliveira, M. Barbosa, and P. Quelhas, Multinuclear cell analysis using Laplacian of Gaussian and Delaunay graphs. In *Iberian Conference on Pattern Recognition and Image Analysis*, 2013, pp. 441–449.
17. P. Leal, L. Ferro, M. Marques, S. Romão, T. Cruz, A. Tomás, H. Castro, and P. Quelhas, Automatic assessment of Leishmania infection indexes on in vitro macrophage cell cultures. *Lecture Notes in Computer Science* (including subseries Lecture Notes in Artificial Intelligence and Lecture Notes in Bioinformatics), 2012, 7325 LNCS(2): 432–439.
18. T. Esteves, P. Quelhas, A.M. Mendonça, and A. Campilho, Gradient convergence filters for cell nuclei detection: A comparison study with a phase based approach. *MVAP*, 2012, **23**(4): 623–638.
19. P. Quelhas, M. Marcuzzo, A.M. Mendonça, and A. Campilho, Cell nuclei and cytoplasm joint segmentation using the sliding band filter. *IEEE Transactions on Medical Imaging*, 2010, **29**(8): 1463–1473.
20. M. Marcuzzo, P. Quelhas, A. Mendonça, and A. Campilho, Evaluation of symmetry enhanced sliding band filter for plant cell nuclei detection in low contrast noisy fluorescent images. *Proceedings of the International Conference on Image Analysis and Recognition*, 2009, Springer **LNCS 5627**: 824–831.
21. P. Quelhas, A.M. Mendonca, and A. Campilho, 3D cell nuclei fluorescence quantification using sliding band filter. In *International Conference on Pattern Recognition* (*ICPR*), 2010, pp. 2508–2511.
22. R. Bise et al., Automatic cell tracking applied to analysis of cell migration in wound healing assay. In *33rd Annual International Conference of the IEEE EMBS*, 2011, pp. 6174–6179.
23. P. Friedl and D. Gilmour, Collective cell migration in morphogenesis, regeneration and cancer. *Nature Review Molecular Cell Biology*, 2009, **10**: 445–457.
24. P. Friedl and D. Alexander, Cancer invasion and the microenvironment: Plasticity and reciprocity. *Cell*, 2011. **147**: 992–1009.
25. E. Meijering, O. Dzyubachyk, I. Smal, and W. van Cappellen, Tracking in cell and developmental biology. *Seminars in Cell Developmental Biology*, 2009, **20**(8): 894–902.
26. P. Quelhas, M. Marcuzzo, M.J. Oliveira, A.M. Mendonça, and A. Campilho, Cancer cell detection and invasion depth estimation in brightfield images. *BMVC*, 2009: 7–10.
27. M. Usaj et al., Cell counting tool parameters optimization approach for electroporation efficiency determination of attached cells in phase contrast image. *Journal of Microscopy*, 2010, **241**(3): 303–314.
28. T. Esteves, M.J. Oliveira, and P. Quelhas, Cancer cell detection and morphology analysis based on local interest point detectors. In *Iberian Conference on Pattern Recognition and Image Analysis*, 2013, pp. 624–631.
29. N. Schuergers, T. Lenn, R. Kampmann, M. Meissner, T. Esteves, M. Temerinac-Ott, J. Korvink, A. Lowe, C. Mullineaux, and A. Wilde, Cyanobacteria use micro-optics to sense light direction. *eLife*, 2016, **5**: e12620.
30. N. Harder et al., Large-scale tracking for cell migration and proliferation analysis and experimental optimization of high-throughput screens. In *Microscopic Image Analysis with Applications in Biology, MIAAB 2011–Heidelberg*, 2011, pp. 1–6.
31. P. Kostelec, L. Carlin, and B. Glocker, Learning to detect and track cells for quantitative analysis of time-lapse microscopy image sequences. In *Engineering in Medicine and Biology Society*, 2015.

32. R. Bise, Z. Yin, and T. Kanade, Reliable cell tracking by global data association. *IEEE International Symposium on Biomedical Imaging: From Nano to Macro*, 2011, pp. 1004–1010.
33. K. Smith, A. Carleton, and V. Lepetit, General constraints for batch multiple-target tracking applied to largescale videomicroscopy. In *Proceedings of the IEEE Conference on Computer Vision and Pattern Recognition (CVPR 2008)*, 2008, **1**: 1–8.
34. K. Rohr, W. Godinez, N. Harder, S. Wörz, J. Mattes, W. Tvaruskó, and R. Eils, Tracking and quantitative analysis of dynamic movements of cells and particles. *Cold Spring Harbor Protocols*, 2010, **6**: pdb.top80.
35. D. House, M.L. Walker, Z. Wu, J.Y. Wong, and M. Betke, Tracking of cell populations to understand theirs spatial temporal behavior in response to physical stimuli. In *Proceedings of IEEE Conference on Computer Vision and Pattern Recognition*, 2009, **1**: 186–193.
36. T. Kanade et al., Cell image analysis: Algorithms, system and applications. *IEEE Computer and Information Science*, 2011: 374–381.
37. N.N. Kachouie et al., Probabilistic model-based cell tracking. *International Journal of Biomedical Imaging*, 2006, 2006: 1–10.
38. C. Li et al., Level set evolution without re-initialization: A new variational formulation. In *IEEE International Conference on Computer Vision and Pattern Recognition (CVPR)*, 2005, pp. 430–436.
39. J. Huth et al., Significantly improved precision of cell migration analysis in time-lapse video microscopy through use of a fully automated tracking system. *BMC Cell Biology*, 2010, **11**(24): 1–12.
40. T. Esteves, M.J. Oliveira, and P. Quelhas, Cancer cell tracking using a Kalman filter. In *Proceedings of RecPad*, 2013, pp. 4–5.
41. A. Chowdhury et al., Semi-automated tracking of muscle satellite cells in brightfield microscopy video. In *Image Processing (ICIP), 2012 19th IEEE International Conference on*, 2012, pp. 2825–2828.
42. F. Gustafsson et al., Particle filters for positioning, navigation and tracking. *IEEE Transactions on Signal Processing*, 2002, **50**(2): 425–437.
43. M. Arulampalam et al., A tutorial on particle filters for online nonlinear/non-Gaussian Bayesian tracking. *IEEE Transactions on Signal Processing*, 2002, **50**(2): 174–188.
44. Z. Khan, T. Balch, and F. Dellaert, MCMC-based particle filtering for tracking a variable number of interacting targets. *IEEE Transactions on Pattern Analysis and Machine Intelligence*, 2005, 27: 1805–1819.
45. N. Chenouard et al., Objective comparison of particle tracking methods. *Nature Methods*, 2014, **11**: 281–289.
46. C.O.D. Solorzano, A benchmark for comparison of cell tracking algorithms. *Bioinformatics*, 2014, **30**(11): 1609–1617.
47. K. Smith, A. Carleton, and V. Lepetit, General constraints for batch multiple-target tracking applied to largescale videomicroscopy. In *Proceedings of the IEEE Conference on Computer Vision and Pattern Recognition (CVPR 2008)*, 2008, **1**: 1/8.
48. T. Esteves, M.J. Oliveira, and P. Quelhas, Local interest detector based cancer cell mobility and morphology joint analysis. *International Symposium in Applied BiomImaging*, 2012, pp. 92.
49. T. Esteves, M.J. Oliveira, and P. Quelhas, Local interest detector based cancer cell mobility and morphology joint analysis. *RecPad*, 2012, pp. 1–2.
50. N. Otsu, A threshold selection method from gray level histograms. *IEEE Transactions on Systems, Man and Cybernetics*, 1979, **9**(1): 62–66.
51. M. Marcuzzo, T. Guichard, P. Quelhas, A.M. Mendonça, and A. Campilho, Cell division detection on the Arabidopsis thaliana root. In *Proceedings of the 4th Iberian Conference on Pattern Recognition and Image Analysis (IbPRIA)*, 2009.

52. T. Chan and L. Vese, Active contours without edges. *IEEE Transactions on Image Processing*, 2001, **10**(2): 266–277.
53. X. Bresson, P. Vandergheynst, and J. Thiran, A priori information in image segmentation: Energy functional based on shape statistical model and image information. In *Proceedings of International Conference on Image Processing, ICIP*, 2003, p. 428.
54. T. Lindeberg, Scale-space theory: A basic tool for analysing structures at different scales. *Journal of Applied Statistics*, 1994, **21**(2): 224–270.
55. J. Wei, Y. Hagihara, and H. Kobatake, Detection of cancerous tumors on chest X-ray images candidate detection filter and its evaluation. In *Proceedings of International Conference on Image Analysis and Processing (ICIP)*, 1999, pp. 397–401.
56. R. MacLachlan, Tracking moving objects from a moving vehicle using a laser scanner. Carnegie Mellon University, 2005: pp. 21–27.
57. T. Esteves, M.J. Oliveira, and P. Quelhas, Cancer cell detection and tracking based on local interest point detectors. *International Conference Image Analysis and Recognition*, 2013, pp. 434–441.
58. C.R. Gonzalez and R.E. Woods, *Digital Image Processing (3rd Edition)*, 2006, Upper Saddle River, NJ: Prentice Hall.
59. J. Lewis, Fast template matching. *Vision Interface*, 1995, **95**: 15–19.
60. R.M. Jiang et al., Live-cell tracking using SIFT Features in DIC microscopic videos. *IEEE Transactions on Biomedical Engineering*, 2010, **57**(9): 2219–2228.
61. D. Lowe, Distinctive image features from scale-invariant keypoints. *International Journal of Computer Vision*, 2004, **60**(2): 91–110.
62. T. Esteves, M.J. Oliveira, and P. Quelhas, Cell mobility and morphology joint analysis in Biology assays. PhD Thesis. Faculty of Engineering of the University of Porto, 2016.
63. T. Roberts et al., Estimating the motion of plant root cells from in vivo confocal laser scanning microscopy images. *Machine Vision and Applications*, 2010, **21**: 921–939.
64. L. Li, S. Narrelykke, and E. Cox, Persistent cell motion in the absence of external signals: A search strategy for eukaryotic cells. *PLoS One*, 2008, **3**(5): e2093.
65. J. MacCormick and M. Isard, Partitioned sampling, articulated objects, and interface-quality hand tracking. *Lecture Notes in Computer Science*, 2000, **1843**: 3–19.
66. C. Solorzano, E. Meijering, and A. Muñoz-Barrutia, *12th IEEE International Symposium on Biomedical Imaging (ISBI 2015)*, 2015. *http://www.codesolorzano.com/celltrackingchallenge/Cell_Tracking_Challenge/Welcome.html.*

Section II

Molecular and cellular applications

Chapter 10 Studying subcellular signaling events in living microglial cells using fluorescence resonance energy transfer-based nanosensors

Renato Socodato,
Camila C. Portugal,
Paula Sampaio, and
João B. Relvas

Contents

Introduction

The development of green fluorescent protein (GFP) variants has provided genetically encoded tags for labeling molecules in biological systems [1]. This single fact has spurred great interest in the development of imaging techniques based on Föster resonance energy transfer (FRET) to study protein functioning in single living cells with high subcellular resolution [2]. FRET is a short-range phenomenon in which energy emitted by an excited donor fluorophore is transferred to and excites an acceptor molecule via a nonradioactive mechanism. When donor and acceptor fluorophores are genetically attached to different cellular components, such as proteins, and two labeled proteins are separated by less than 10 nm, the two fluorophores come in close proximity to each other and intermolecular FRET occurs, which can be interpreted as protein–protein interaction. Likewise, by attaching donor and acceptor fluorophores to the same protein, intramolecular FRET is used to monitor conformational changes and can predict biochemical activity changes. Therefore, FRET can be employed to directly study and quantitate molecular interactions and changes in biochemical activities in living biological systems [3]. With the development of biosensors, FRET imaging allowed studying of the subcellular topography of protein conformational changes, protein–protein interactions, and posttranslational modifications (determination of activation states, binding events, etc.) in real time [4], which would not be possible using traditional fluorescence-based approaches in fixed samples or biochemical methods. Ratiometric-based FRET sensors are highly efficient and are useful for comparing relative FRET changes in single living cells [3,5] given that donor and acceptor molecules are expressed in an equimolar ratio within the same chimeric protein. However, this method does not allow comparing data in terms of real FRET efficiencies without proper correction of bleed through artifacts [6].

Microglia are a population of myeloid resident cells in the CNS. It is believed that during early embryogenesis, yolk sac-derived macrophages populate the developing CNS giving rise to microglial cells found in the mature nervous system [7]. Emerging roles of microglia have now established that these cells actively patrol the neuronal parenchyma, monitor synapse functioning, and control neuronal activity [8]. Classically, microglia are known to modulate immune responses and to regulate several branches of the inflammatory process within the neuronal parenchyma, which might also be linked with onset/progression of neurodegenerative disorders [9]. Besides, it is known that specific modulation of microglial function under inflammatory conditions might alleviate neurodegeneration in the CNS. Therefore, the study of signaling pathways in living microglia might be paramount to better understand microglial biology in the healthy and diseased CNS, raising the possibility to delay the onset or halt the progression of neuropathologies.

In this chapter, we describe and exemplify image-processing implementations for single-chain ratiometric FRET biosensors to study microglial function. We show that such ratiometric FRET approach is extremely sensitive for routine inspection of signaling events in real time with high spatio-temporal resolution in living microglia.

Microglial cultures

Microglial cultures are one of the most used experimental paradigms in studies concerning microglial biology. Some authors, however, criticize the use of this model because cultured microglia are generally considered to be in a semiactivated state and are not as quiescent as *in vivo*. Moreover, microglia sense and interact with the environment in which they are inserted in, and for this reason, several authors also question whether the responses obtained with isolated microglia, without the input from other CNS cells, are indeed representatives of the complexity of the CNS environment. On the other hand, studies using isolated cells are also a powerful tool to identify and dissect cell-autonomous signaling mechanisms. We believe that FRET-based approaches will enable us to investigate in real time and with unprecedented detail the spatiotemporal changes in different biochemical processes and protein signaling cascades occurring in living microglial cells. Here we give a brief description of the methods we routinely use in our microglial FRET experiments.

Primary cortical microglial cell cultures

This is one of the most used microglial culture models. It has the benefit to be directly derived from an animal; the microglial cells obtained are not transformed with cancer genes, viruses, or other modifications, as are cell lines, that are a lesser representative model of the *in vivo* situation. However, two experimental drawbacks of this model are the low yield and the low efficiency of transfection when compared with cell lines.

Cultivation and experimental procedure Primary microglial cell cultures are obtained from either Wistar rat or C57BL/6 mouse pups. Animals are sacrificed at P0–P2 and their brain cortices are dissected, including removal of meninges, in Hanks' balanced salt solution (HBSS), pH 7.2 on ice. After that, cortices are digested in a mixture containing 100 μl of Trypsin 0.025% *per* pup and 5 μg/mL (w/v) DNAse *per* pup in 10 mL of DMEM GlutaMAX™-I for 15 min. Next, tissues are centrifuged at 453 g for 10 min, the pellet is dissociated in DMEM GlutaMAX™-I supplemented with 10% FBS and 100 U/mL penicillin and 100 μg/mL streptomycin and filtered in a 105 μm cell strainer. Cells are plated on poly-D-lysine-coated T-flasks (75 cm^2) in a density of two cortices *per* flask (Wistar rat cultures) or four cortices *per* flask (mouse cultures) diluted in 12 mL of supplemented culture media. Cultures are kept at 37°C and 95% air/5% CO_2 in a humidified incubator. Culture media are partially changed every 3 days up to 10 days. After at least 10 days, culture flasks are subjected to orbital shaking at 200 rpm for 2 h and the supernatant enriched in microglia is centrifuged at 453 g for 5 min at RT. Then, the supernatant is discarded and the pellet, containing microglia, is resuspended in DMEM F12 GlutaMAX I supplemented with 10% FBS, 100 U/mL penicillin and 100 μg/mL streptomycin. The plated cells are maintained at 37°C, 95% air and 5% CO_2 in a humidified incubator. Cortical microglia are cultured for 5 days. The shaking and harvesting of microglia can be repeated once a week for three consecutive weeks. Microglial purified cultures obtained from the first and third shakes are no different from each other in all tests performed in response to LPS stimulation (iNOS expression, latex beads phagocytosis, and Iba-1 expression). Purity of these cultures is determined with CD11b immunocytochemistry and varies between 95% and 99% relative to total cells stained with DAPI. In order to perform FRET experiments, we plate freshly isolated primary cortical microglia on uncoated plastic-bottom culture dishes (μ-Dish 35 mm, iBidi).

CHME3 human microglial cell line

The human microglial cell line CHME3 was obtained from primary cultures of human embryonic microglial cells by transfection with a plasmid encoding for the large T antigen of SV40 [10]. These cells were previously used in microglial studies demonstrating similar behavior relative to primary cultures [10]. We also opted for using this cell line because CHME3 cells are of human origin, are comparable with primary cultures in terms of LPS responsiveness, and are much easier to transfect. When the target protein under study is not conserved between rodent and humans, it is important to corroborate the results obtained in human cells.

Cell culture and experimental procedure: These cells are cultured in DMEM GlutaMAX™-I, 100 U/mL penicillin, and 100 μg/mL streptomycin supplemented with FBS. The FBS supplementation is the key factor for their expansion. When using frozen aliquots, immediately after thawing we cultivate them in DMEM supplemented with 10% FBS, but changed them to 0.5% FBS when performing experimental work. Cells are kept at 37°C, 95% air, and 5% CO_2 in a humidified incubator. As for primary cells, CHME3 cells are plated on plastic-bottom culture dishes (μ-Dish 35 mm, iBidi) for FRET experiments.

Microglial transfection

In order to perform FRET experiments, both primary cortical microglial cells and human CHME3 microglia, previously plated on plastic-bottom culture dishes (μ-Dish 35 mm, iBidi), have to be transfected with the desired FRET probe. To transfect microglial cells, we normally use commercially available formulations, such as Jetprime from Polyplus, according to manufacturer's instructions. Briefly, we first wash cell cultures three times with the corresponding supplemented culture media, add 2 mL of the same culture medium, and return the plates to the incubator. In the meantime, we prepare the transfection cocktail as follows: we add 200 μL of transfection buffer (available from the commercial kit) to a tube containing 1 μg DNA of the FRET probe, vortex for 10 sec, and then add 2 μL of transfection reagent (Jetprime transfection reagent from Poliplus) to the mixture, vortex again for 10 sec, and incubate for 10 min at room temperature (RT). Then, we add the transfection cocktail dropwise to the cells, gently rock the plates, and return them to the incubator for not more than 4 h before replacing the transfection media with normal supplemented media. Expression of the FRET probe can be seen after 18 h in a standard fluorescent microscope equipped with appropriate GFP/FITC filter. As primary cells have a slower metabolism compared to CHME3 cells, we usually wait for 48 h to evaluating the probe expression in these cells.

Hardware setup

We routinely use in our FRET experiments a full-motorized microscope DMI6000 (Leica Microsystems) equipped with excitation and emission filters for cyan fluorescent protein (CFP) (BP 427/10) and yellow fluorescent protein (YFP) (BP 504/12) working with specific dichroic (CG1 440/520) mounted into a microscope filter carrousel (Leica fast filter wheels). Cells are observed with a PlanApo 63X 1.3NA glycerol immersion objective with correction ring adjusted for best image quality. An ORCA-Flash4.0 V2 Digital CMOS camera (Hamamatsu) coupled to the microscope by a 1.3x c-mount adapter is used to acquire images. The LAS X software (Leica Microsystems) controls all microscope systems. A small-stage incubator (iBidi) on the microscope is used to control environmental conditions (temperature, CO_2, and O_2) in the culture dish.

Time-lapse routine

In our experimental conditions, images are usually acquired every 30–60 sec time intervals. In primary cortical microglia, we always acquire images using 4 × 4 binning on the full field of the ORCA-Flash4.0 V2 CMOS camera. Primary cortical microglia are difficult to transfect and even in cells transfected with high efficiency, the signal-to-background fluorescence is low. Besides, microglia are very sensitive to blue light-induced photodamage, which is not compatible with the intense bright light coming from the illumination source for acquiring donor and FRET images. The light source we use is a mercury metal halide bulb equipped with a fast internal shutter, online-controlled by the acquisition software, coupled to a light attenuator, which decreases both photodamage and photobleaching. However, optimizing the light source is not enough for good quality images of FRET biosensors in microglia. We found that binning was an essential necessity to achieve meaningful data when using primary cortical microglia given that the 4 × 4 binning operation greatly reduces both the intensity and exposure time of sample illumination while keeping a reliable signal-to-noise ratio, principally in very dim regions near the microglial cell edge or in filopodia-like structures. We always use the autocontrast function during fluorescence inspection in live mode and normally values ranging from 1200–2500 maximum field pixel values in the look-up ramp in the donor configuration (CFP/CFP) are acceptable to respect the dynamic range of the probe using a 16 bit intensity range CMOS camera. The CHME3 human microglial cell line suffer from the same photosensitivity problem of primary cortical microglia but are much easier to transfect and display an excellent signal-to-background ratio of the biosensor, which allow using 4 × 4 binning with reduced light intensity. These settings allow us to acquire donor and FRET images from the CHME3 cell line with reliable spatial resolution while minimizing photodamage.

The first automated step in the FRET experiment is to capture as many cells as possible within the 30–60 s time intervals between two consecutive frames. For standardization and reproducibility reasons, we always acquire donor and acceptor images from at least five random microglial cells in a given culture dish. This is possible using a full-motorized microscope coupled to the control software that allows automating the acquisition in multiple XYZ positions. In our case, we use the mark/find function in the LAS X software from Leica Microsystems. Another important function implemented for live-cell imaging acquisition in the Leica Z-stage is the adaptive focus control. We noticed that time-lapse experiments using microglial cells without proper focus correction directly affect final data interpretation. A hardware adaptive focus control allows constant focus correction to maintain the distance between objective and coverslip during the time-lapse. The LAS X software has to be adjusted for the proper Z-position within the stack in live data mode before running the time-lapse routine.

After choosing appropriate microglial cells to image and setting all illumination parameters for image acquisition, the uneven field illumination is corrected online using a shading correction routine implemented for the LAS X software. Uneven illumination is an inherent system problem even when using plan apochromatic objectives such as the PlanApo 63X 1.3NA glycerol immersion objective that we use. The online correction routine of LAS X requires the acquisition of images of the donor and acceptor channels in a sample region devoid of cells. These images should be acquired with the same illumination parameters and exposure setup to be used in the experimental time-lapse. Images of donor and acceptor channels are divided by their respective shaded image during every frame of the time-lapse. LAS X implements ratio calculations over the entire time course using floating point's decimal places to minimize rounding errors to shade-correct donor and FRET images. For the sake of reproducibility, time-lapse files are saved with all raw parameters preserved and with the metadata incorporated in the work file to allow proper off-line quantifications.

Background subtraction

The first off-line operation in the FRET time-lapse image setting is to subtract background from each raw channel image (donor/donor and donor/

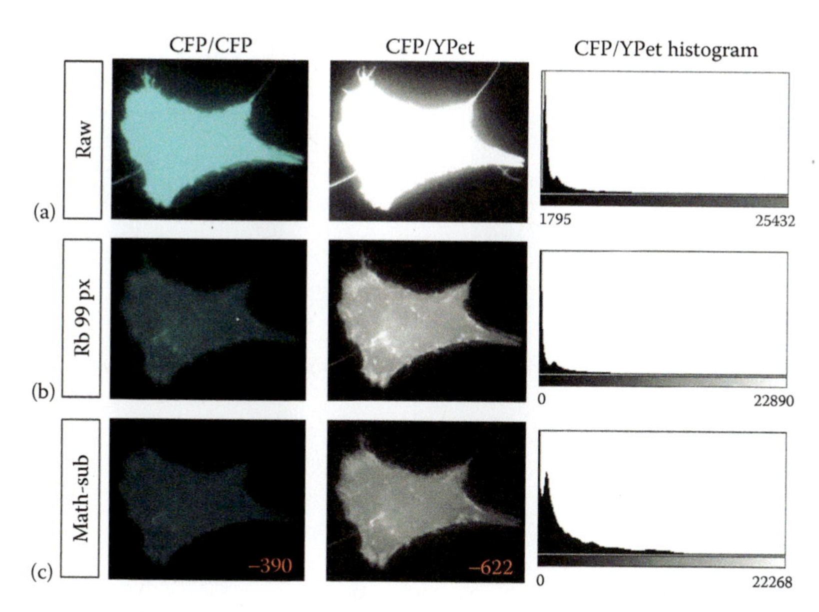

Figure 10.1 Background subtraction operation in CHME3 microglia expressing the c-Src FRET biosensor: (a) Raw image of donor (CFP/CFP) and acceptor (CFP/YPet) channels, (b) donor and acceptor images after subtracting the background with the roller-ball algorithm with the radius set at 99 pixels, and (c) donor and acceptor images after additional pixel subtraction (390 pixels in the donor and 622 pixels in the acceptor).

acceptor; Figure 10.1a). There are several ways by which different software can operate background subtraction routines. All off-line routines described in this chapter are performed using the open source Java-based software FIJI [11]. We employ dynamic background subtraction for the entire time-lapse in individual channels using FIJI's *subtract background function* and a 99-pixel roller ball radius is used for the initial assessment (Figure 10.1b). Later adjustments using the subtract function from the process/math menu, using a hatched background area in the image as reference, can be implemented to refine further the contrast in dim regions near the cell edge (Figure 10.1c). Although the difference after the first background operation is notorious (compare Figure 10.1a and b), the second operation is much harder to appreciate when only comparing photomicrographs (compare Figure 10.1b and c). Differences however become apparent by observing the pixel distribution histogram for the CFP/YPet channel after running a final math operation (compare Figure 10.1b and c).

Image filtering

Filtering of donor and acceptor channels is essential to balance the steep ends of pixels distribution histograms and to reduce noise. We found that stacks of microglial images in both donor and acceptor configurations suffered from flickering noise between consecutive frames due

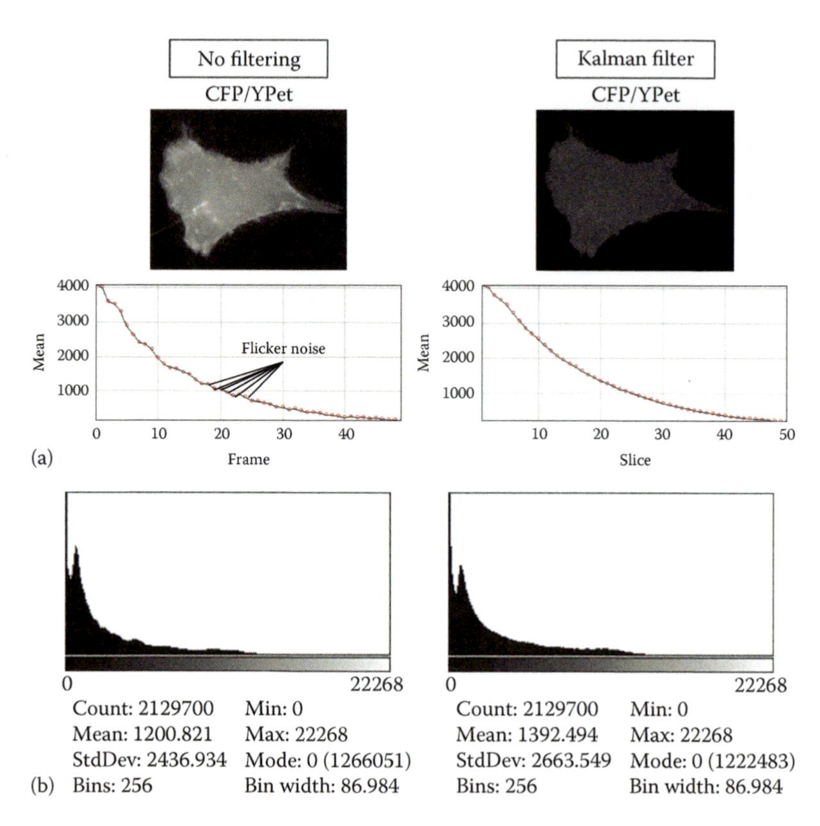

Figure 10.2 Image filtering: (a) Acceptor (CFP/YPet) image before (left) and after (right) application of a Kalman stack filter. Mean pixels intensities over the time frames display the flicker noise in the nonfiltered time-lapse. The pixel distribution histograms of both configurations are displayed in (b).

to the oxidation-based imaging processing method within the CMOS camera microchip. We implemented a Kalman stack filter [12] to minimize the flickering noise between frames and to reduce acquired noise throughout the recorded time-lapse for both donor and acceptor channels (Figure 10.2a). It is interesting to note that such filtering operation does not modify the pixels count in the image or alter the min and max pixel values, but largely reshape the pixel distribution histogram (Figure 10.2b).

Thresholding and masking

Generation of a binary reference image for masking donor and acceptor channels requires image thresholding. Proper thresholding is the most critical step in this ratiometric FRET imaging quantification. Sloppy threshold operation usually leads to masked donor and/or acceptor images lacking complete pixel information, which may generate false negative results or obliterate important spatial information in a given microglial subcellular compartment and ultimately compromise experimental reproducibility. FIJI offers a plethora of global and local thresholding

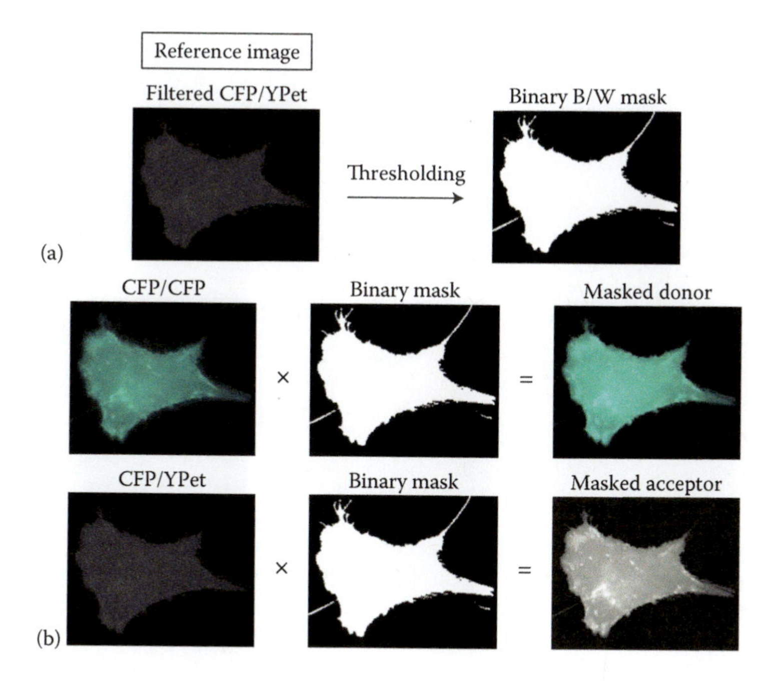

Figure 10.3 Image masking: (a) The filtered acceptor (CFP/YPet) image was used as reference to generate a binary black and white (B/W) mask using the Phansalkar local threshold and (b) math operation applied to donor and acceptor channels to produce masked images from the binary reference.

methods, which may be further modified in their source codes if one finds it necessary. We apply the Phansalkar local threshold method [13] and either donor or acceptor image may serve as the reference. The operation is automated in FIJI and generates a binary black background and white foreground image by default (Figure 10.3a). Binary masks can be further edited to contain only the desired spatial information.

The next step is to individually mask donor and acceptor images according to the binary mask, generating images where relevant pixels are preserved and unwanted pixels are turned to zero. FIJI performs this operation using the *Image Calculator* function. It is worth mentioning that the new generated image must be in 32 bit float points. Multiplying donor and acceptor images by the binary mask produces masked donor and acceptor images (Figure 10.3b).

Registration

After masking donor and acceptor channels, images must be corrected for the pixel drift during the time-lapse routine. This drift is an inherent problem of a filter wheel-based FRET imaging system and thus cannot be eliminated. To equalize drift, images have to be registered at subpixel resolution. FIJI offers several registration routines, which can be adjusted according to

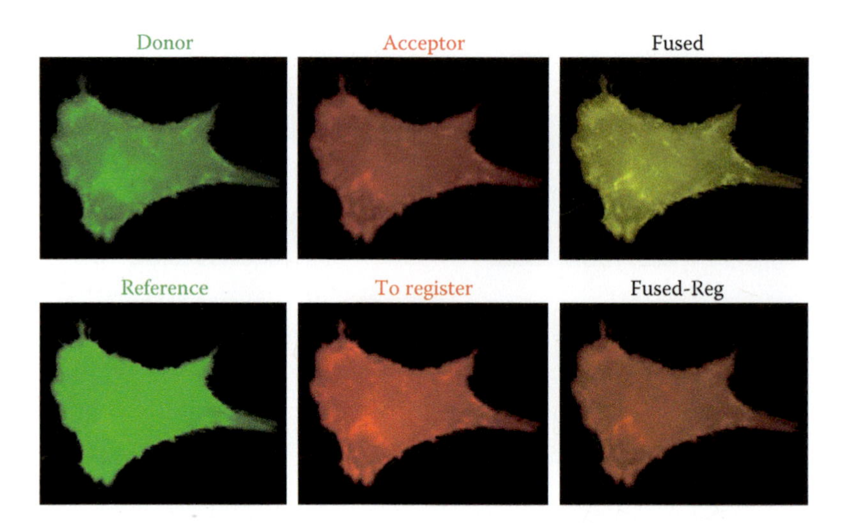

Figure 10.4 Donor (green) and acceptor (red) channels were pseudocolored and image registration on the acceptor pixel grid was performed using the donor channel as the template. Fused (overlaid donor to acceptor) images before and after registration are also shown.

the experimental purpose, but we mainly use the *Descriptor-based series registration* (*2d/3d+t*) plug-in [14]. For ratiometric FRET images, we register the acceptor channel using the donor channel as template. Drift over time in donor image can be disregarded because relevant experimental information is determined in the form of donor-to-acceptor ratio. Therefore, we only equalize the drift in the acceptor according to the drift in the donor channel using the CFP/CFP image as reference (Figure 10.4) at each time point. After registration, merged donor–acceptor image might differ a lot from the fused unregistered counterpart (Figure 10.4), which reflects the huge drift in both donor and acceptor channels due to the constant exchange of filter cubes (CFP/CFP—CFP/YFP) within the filter wheel.

Photobleach correction and generation of ratiometric image

Live imaging experiments using GFP variants usually suffer the bleaching effect due to continuous exposure of the fluorescent protein to intense bright light. Although fine control of excitation light intensity plus the use of UV and IR blocking filters helps minimizing photobleaching, FRET biosensors are very prone to bleach over time and thus time-lapse images must be corrected off-line for an appropriate ratio FRET image to be generated. Quick inspection of mean intensity values on individual donor and acceptor channels over time is sufficient to detect whether or not the biosensor underwent photobleaching during the experiment and whether should or should not be bleach-corrected.

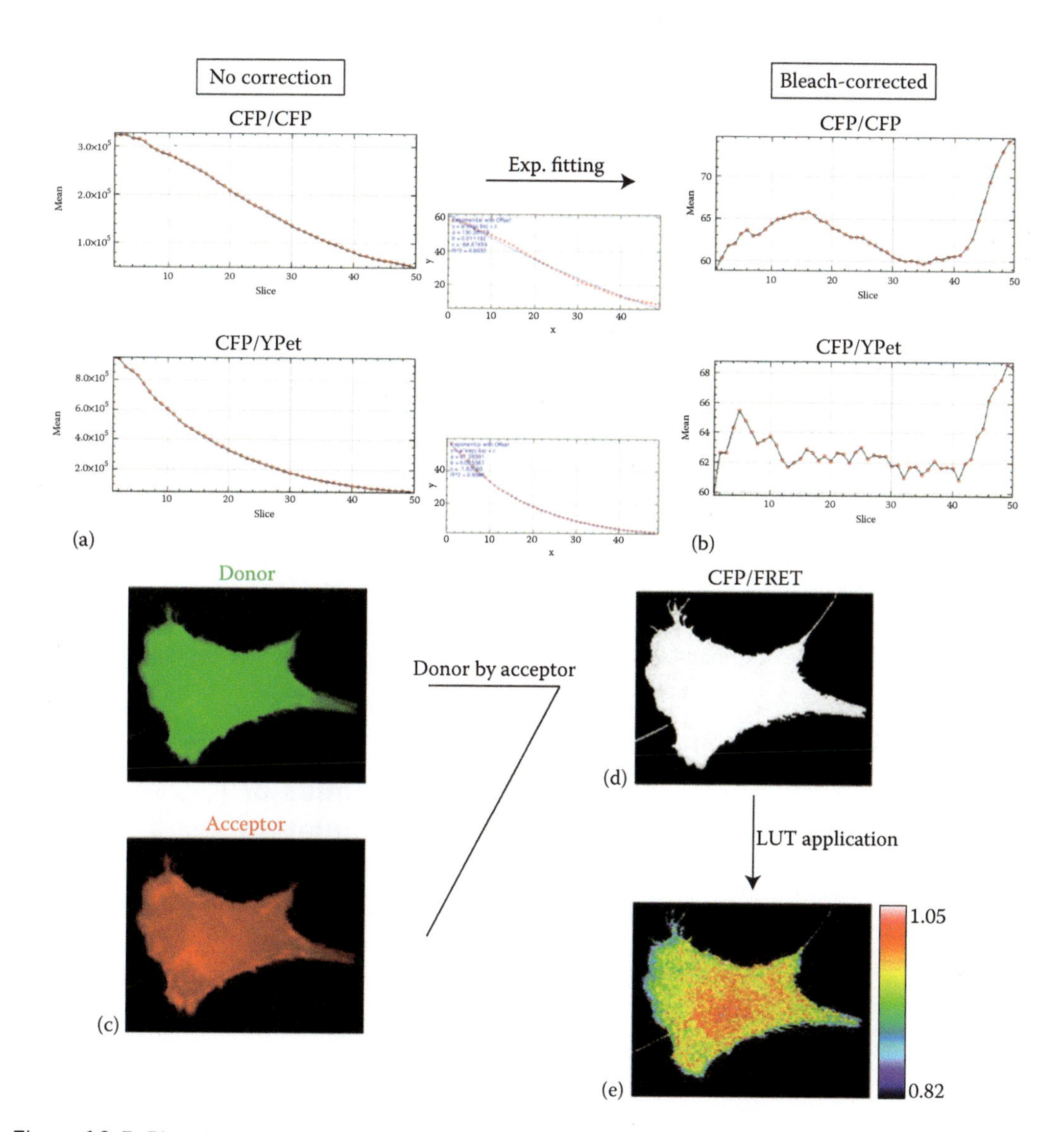

Figure 10.5 Photobleach correction and ratio image generation: (a, b) Mean pixels intensities over the time frames for donor and acceptor are shown in (a). After fitting these curves with a one-exponential decay function, the bleach-corrected mean intensities over time for both donor and acceptor are generated (b), (c, e) bleach-corrected donor image stack is divided by the bleach-corrected acceptor image stack (c) to generate a gray scale donor/acceptor image (d). Min and max contrast calibration (using the brightness/contrast menu) and LUT application produce a final ratio FRET image for the c-Src FRET biosensor (e).

In the following example, we deliberately increased the light intensity over our sample to provoke photobleach in the time series but we fixed a low exposure time to avoid photodamage. Note that donor and acceptor channels bleached at different rates (Figure 10.5a). Photobleach in ratiometric FRET biosensors are usually expected to decay exponentially as a function of the GFP-based fluorophore [15] and fitting the mean intensity values of individual channels over time with an exponential decay function produces reasonable bleach-corrected intensity curves for both donor and acceptor in microglia (Figure 10.5b). FIJI presents different exponential fitting options using one or two decay functions as the PixBleach plug-in [16] or even different macros based on FRAP that can also be used to correct the bleaching effect. The routine can be further complemented with more elaborated corrections in different subregions of the cell or in different parts of the decay curve as documented for Rho-based FRET biosensors using MATLAB® [17].

After correcting the photobleach (when appropriate) ratiometric image is ready to be generated. The *Image Calculator* function in FIJI is used to divide donor by acceptor images (Figure 10.5c) or acceptor by donor (when convenient), generating a final ratio image in 32 bit float points (Figure 10.5d). A ratio image with native pixel values in gray scale is highly noninformative and subcellular differences only become apparent when min and max saturation ramps are adjusted and a rainbow look-up table (LUT) is applied to the ratio image (Figure 10.5e).

Subcellular activation dynamics of c-Src during microglial stimulation

We exemplify the use of this ratiometric FRET approach in the subcellular activation of the tyrosine kinase c-Src in live microglia exposed to LPS, a component of bacterial cell wall classically known to trigger a proinflammatory response on these cells [18]. For this, we used a specific c-Src FRET biosensor, namely KRas Src YPet chimera [19], in the CHME3 human microglia. We observed that LPS elicited fast c-Src activation (increase in the CFP-to-FRET ratio of the biosensor) on several different subcellular compartments of microglia such as at the cell front, at the cell rear, at c-Src-associated focal adhesions, and at lamellipodia (Figure 10.6). The LPS-mediated c-Src activation displays different time-dependent signatures and amplitudes according to the microglial subcellular domain studied (Figure 10.6). Of note, c-Src is a tyrosine kinase claimed to control microglia proinflammatory polarization during inflammation [20] and to mediate microglia-associated excitotoxic cell death of embryonic neurons [21]. Therefore, understanding the kinetics of c-Src activation in different microglial subcellular domains may help explain microglia-associated neuronal impairment during chronic neurodegeneration [22].

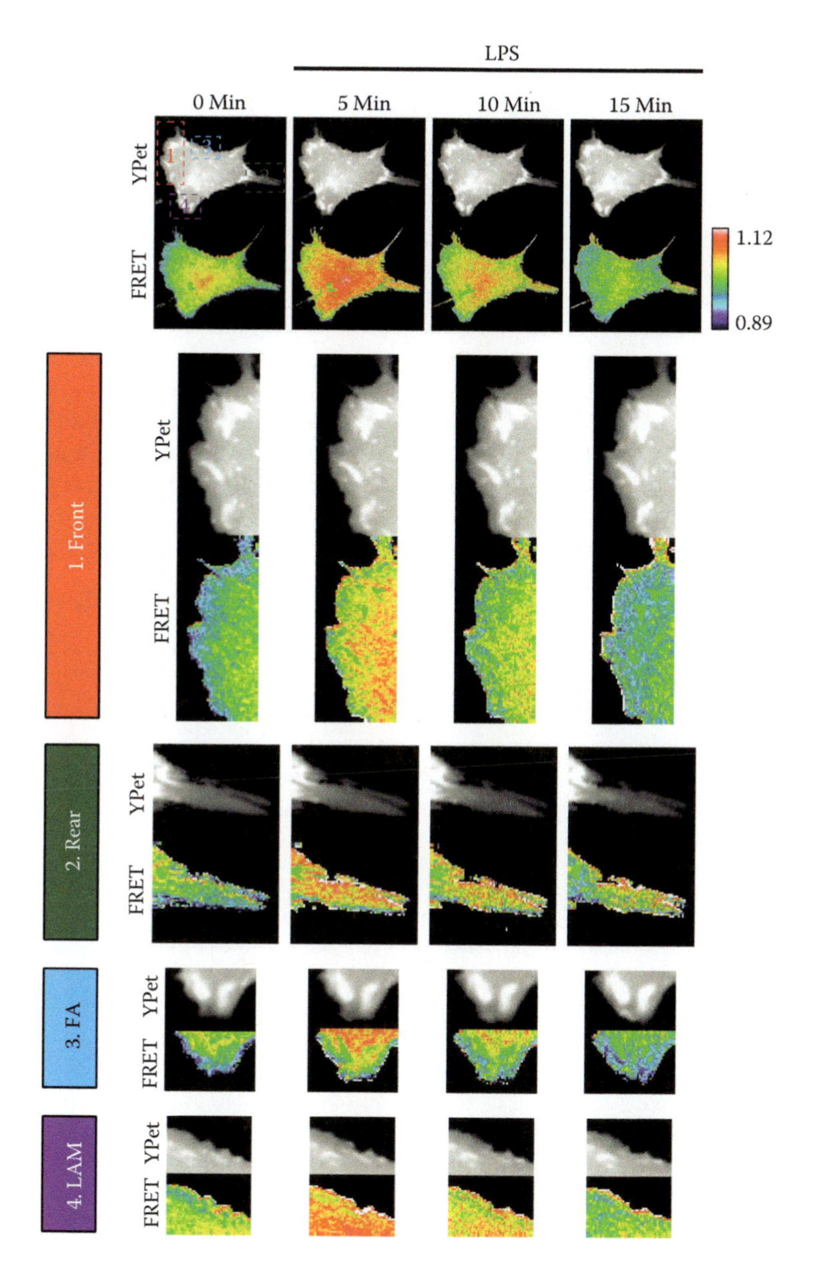

Figure 10.6 Subcellular activation dynamics of c-Src in microglia. CHME3 human microglia expressing KRas Src YPet quimera (c-Src FRET biosensor) were recorded in the presence of LPS (1 μg/mL) for 15 min (1 min between frames). Time-lapse CFP/FRET images are shown. Pseudocolor ramps represent the mean with min/max ratios. FA = focal adhesion; LAM = lamellipodia.

Concluding remarks

FRET-based time-lapse imaging microscopy is a very powerful technique to study signaling events in living cells. Microglia are highly dynamic cells from the CNS and usually respond very quickly and characteristically to extracellular cues. The implementations portrayed herein allow studying subcellular dynamic events during the activation of living microglia with high spatio-temporal resolution. The routines described for the c-Src ratiometric FRET biosensor can be perfectly accommodated for other FRET-based ratiometric nanosensors with only minimal adjustments and can also be scaled-up with batch modes to allow FRET-based mid-to-high throughput drug screenings in microglia. Overall, FRET-based live-cell imaging will largely contribute to the study of microglial biology and may help pave the way for better understanding the signaling mechanisms governing microglial homeostasis in the healthy or diseased CNS.

References

1. Cubitt, A.B. et al., Understanding, improving and using green fluorescent proteins. *Trends Biochem Sci*, 1995. **20**(11): 448–455.
2. Tsien, R.Y., B.J. Bacskai, and S.R. Adams, FRET for studying intracellular signalling. *Trends Cell Biol*, 1993. **3**(7): 242–245.
3. Sekar, R.B. and A. Periasamy, Fluorescence resonance energy transfer (FRET) microscopy imaging of live cell protein localizations. *J Cell Biol*, 2003. **160**(5): 629–633.
4. Giepmans, B.N. et al., The fluorescent toolbox for assessing protein location and function. *Science*, 2006. **312**(5771): 217–224.
5. Giuliano, K.A. et al., Fluorescent protein biosensors: measurement of molecular dynamics in living cells. *Annu Rev Biophys Biomol Struct*, 1995. **24**: 405–434.
6. Periasamy, A. et al., Chapter 22: Quantitation of protein-protein interactions: confocal FRET microscopy. *Methods Cell Biol*, 2008. **89**: 569–598.
7. Prinz, M. et al., Heterogeneity of CNS myeloid cells and their roles in neurodegeneration. *Nat Neurosci*, 2011. **14**(10): 1227–1235.
8. Salter, M.W. and S. Beggs, Sublime microglia: expanding roles for the guardians of the CNS. *Cell*, 2014. **158**(1): 15–24.
9. Block, M.L., L. Zecca, and J.S. Hong, Microglia-mediated neurotoxicity: Uncovering the molecular mechanisms. *Nat Rev Neurosci*, 2007. **8**(1): 57–69.
10. Janabi, N. et al., Establishment of human microglial cell lines after transfection of primary cultures of embryonic microglial cells with the SV40 large T antigen. *Neurosci Lett*, 1995. **195**(2): 105–108.
11. Schindelin, J. et al., Fiji: An open-source platform for biological-image analysis. *Nat Methods*, 2012. **9**(7): 676–682.
12. Gürellidot, M.I. and L. Onural, A class of adaptive directional image smoothing filters. *Pattern Recognition*, 1996. **29**(12): 1995–2004.
13. Phansalkar, N., et al. Adaptive local thresholding for detection of nuclei in diversity stained cytology images. in *Communications and Signal Processing (ICCSP), 2011 International Conference on*, Calicut, India. 2011. IEEE.
14. Preibisch, S. et al., Software for bead-based registration of selective plane illumination microscopy data. *Nat Methods*, 2010. **7**(6): 418–419.

15. Sinnecker, D. et al., Reversible photobleaching of enhanced green fluorescent proteins. *Biochemistry*, 2005. **44**(18): 7085–7094.
16. Wustner, D. et al., Selective visualization of fluorescent sterols in Caenorhabditis elegans by bleach-rate-based image segmentation. *Traffic*, 2010. **11**(4): 440–454.
17. Hodgson, L. et al., Imaging and photobleach correction of Mero-CBD, sensor of endogenous Cdc42 activation. *Methods Enzymol*, 2006. **406**: 140–156.
18. Cunningham, C., Microglia and neurodegeneration: the role of systemic inflammation. *Glia*, 2013. **61**(1): 71–90.
19. Ouyang, M. et al., Determination of hierarchical relationship of Src and Rac at subcellular locations with FRET biosensors. *Proc Natl Acad Sci USA*, 2008. **105**(38): 14353–14358.
20. Socodato, R. et al., c-Src function is necessary and sufficient for triggering microglial cell activation. *Glia*, 2015. **63**(3): 497–511.
21. Socodato, R. et al., c-Src deactivation by the polyphenol 3-O-caffeoylquinic acid abrogates reactive oxygen species-mediated glutamate release from microglia and neuronal excitotoxicity. *Free Radic Biol Med*, 2015. **79**: 45–55.
22. Gomez-Nicolà, D. and V.H. Perry, Microglial dynamics and role in the healthy and diseased brain: A paradigm of functional plasticity. *Neuroscientist*, 2015. **21**(2): 169–184.

Chapter 11 Membrane trafficking under the microscope, what new imaging technologies have brought to light

Julia Fernandez-Rodriguez
and Tommy Nilsson

Contents

Introduction

The secretory pathway

Golgi organization is closely tied to the mechanism(s) of membrane traffic and it is essential for protein sorting and transport; understanding how the Golgi is constituted and how it works is therefore of fundamental importance. Despite more than six decades of productive Golgi research, the underlying question as to how biosynthetic material is sorted and transported through the secretory pathway remains unclear. Various models for protein transport have been put forward, each with merits and limitations, and we are yet unable to resolve which model correctly describes the true nature of the Golgi apparatus. At present, no single model explains all available data. As such, we have to consider multiple possible mechanisms of Golgi transport. Conceptually, we can envisage at least three models of transport through the secretory pathway (Figure 11.1): First, a constant formation of new cisternae move cargo, *en masse*, forward, much like a cisternal *conveyor belt* (cisternal progression/maturation, [1]). Second, vesicles move cargo from one cisterna to another (vesicular transport, [2]). Third, stable or transient membrane-tubules connect cisternae enabling protein and lipid diffusion (for review—see [3,4]). Since the late 1950s, attempts (some truly ingenious) have been made to resolve which model is correct. In some cases, even using the same system, reagents, and methods, investigators

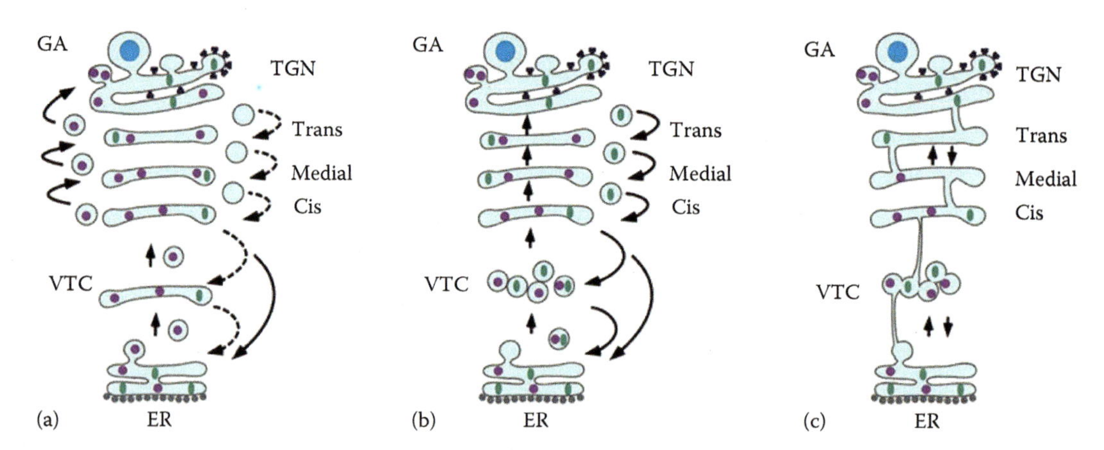

Figure 11.1 Three models to explain intracellular transport. (a) Cisternal maturation. Here, each cisterna carries newly synthesized material forward, whereas vesicles ferry resident proteins such as glycosylation enzymes in the opposite direction. (b) Vesicular transport, stable compartment. In this model, vesicles ferry newly synthesized material forward. Each cisterna represents a stable compartment and does not change over time. (c) The exocytic pathway is a stable but interlinked network. Preexisting as well as transient tubules, but not vesicles, are envisaged newly. (Abbreviations: GA: Golgi apparatus, TGN: trans-Golgi network, VTC: vesicular tubular cluster). (From Elsner, M. et al., *Mol. Membr. Biol.*, 20, 221–229, 2003.)

have reached opposing conclusions (for review—see [5]). More work is clearly needed to decode the inner workings of the Golgi apparatus.

Overview of membrane trafficking through the Golgi apparatus

Initial emphasis was put on interpreting function from morphology at the ultrastructural level with the aid of electron microscopy and cell fractionation, revealing membrane structures such as the transitional ER, vesicular carriers, vesicular tubular clusters, Golgi cisternae, Golgi stacks, and the Golgi ribbon. Morphological work based on transmission electron microscopy suggested early on that biosynthetic material move through the secretory pathway via membrane-bound carriers that leave the ER and gradually morph into Golgi cisternae at the *cis*-side of the Golgi stack. As this is an ongoing process, that is, new cisternae form continuously, this leads to the formation of the Golgi stack. At the *trans*-side, cisternae would then disassemble into transport carriers that continue to move cargo forward to the plasma membrane, to the endocytic pathway (e.g., lysosomal proteins), or to other compartments of the cell. Cisternal *assembly* and *disassembly* processes were envisaged to take place at the *cis*- and *trans*-Golgi networks, respectively, both membrane systems marked by a relatively high degree of fenestrae. Similar fenestrae were also observed between Golgi stacks linking each cisterna and stack laterally giving rise to a Golgi ribbon (the Golgi apparatus as originally observed through microscopy by Camillo Golgi in 1898) [6]. The envisaged cisternal maturation process explained how large protein structures (e.g., collagen and scale) could gradually assemble in the secretory pathway but rested on the assumption that the Golgi stack relied on a constant supply of new biosynthetic material. This assumption was proven wrong by Jamieson and Palade [7] demonstrating that protein synthesis could be dissociated from intracellular transport (i.e., they showed that inhibiting protein synthesis had no effect on the structure of the Golgi apparatus). This led to the view that the Golgi apparatus was not a transient structure. Rather, it was a stable organelle. Small vesicles were also seen in abundance between the ER and the Golgi interface and adjacent to the Golgi cisternae. These were proposed to serve as transport carriers transporting biosynthetic cargo from the ER to the Golgi apparatus (termed COPII vesicles) and subsequently, from cisternae to cisterna in a *cis* to *trans* direction (termed COPI vesicles). In this scenario, the Golgi stack was proposed as a stable structure in which enzymes (e.g., glycosylation enzymes and peptidases) would act on biosynthetic cargo in a stepwise and orderly fashion. Immunoelectron microscopy and biochemical fractionation experiments supported this in that enzymes (e.g., thiamine pyrophosphate and glycosylation enzymes involved in the maturation of complex N-linked carbohydrates) were shown to localize to distinct parts of the Golgi stack. Based on this, Rothman, Balch, and colleagues devised a clever *in vitro* complementation assay depicting vesicular transport of the G-protein of vesicular stomatitis virus via COPI vesicles, an assay that resulted in the identification of multiple key components in vesicular transport [8]. This array of components was complemented and paralleled by yeast genetics and together, provided most of the basic components of the secretory pathway (in yeast, termed sec mutants, for example, sec13/33 and sec23/24 of the COPII coat). With such an overwhelming amount of evidence, the cisternal maturation/progression model was abandoned and replaced by vesicular transport/stable compartment model leading to a long-standing paradigm that dominated the field throughout the 1970s and the 1980s (evident from most of the textbooks of that era. Eventually, this model collapsed, in part, as it failed to explain how macromolecular complexes, too large to fit into a (COPI) transport vesicle, could still make it through the stack. Coatomer (the 7 subunit coat of COPI vesicles) was also found to selectively bind to the K(X)KXX recycling motif present on ER-resident proteins and proteins cycling between the ER and the Golgi apparatus [9] and COPI vesicles, to mainly transport resident components of the Golgi apparatus including glycosylation enzymes of the trans-most part of the Golgi stack [10–12].

The third model, tubular connections could explain some of these discrepancies. Observations of tubular connections between and within the Golgi stack date back to the 1970s and 1980s [13,14]. Although a vast amount of data are available for the molecular mechanisms that participate in vesicle transport, (for review—see [15]), very little is known about the factors that regulate the biogenesis of tubular carriers. Few if any candidates emerged from the genetic and functional screens conducted in the 1980s and 1990s. Rather, they all point to a vesicular mechanism, at the very least, an essential and conserved (between yeast and mammal) mechanism with specific coat proteins, GTPases, and SNARE proteins to name a few. A technical challenge when studying membrane tubules is that these are very fragile and thus, difficult to fix in cells. Equally, fixation conditions might induce fusion of membranes resulting in the appearance of tubular connections. This makes them difficult to study by electron microscopy, light microscopy (in fixed cells), and biochemistry. As such, elucidating their existence and role(s) poses a challenge, one that perhaps, could be overcome by light microscopy of living cells. Here, caution is warranted as the Golgi apparatus, with its intricate membrane system, appears to have a high degree of plasticity. Overexpression of proteins (e.g., VSV-G) may lead to the formation of tubules (as may indeed many other *artificial* probes), not because these would highlight an existing transport process but rather, because they could very well induce artificial ones (e.g., tubules) by overwhelming the transport machinery (e.g., coat). Whether or not this could explain why overexpressed VSV-G can reach the plasma membrane in minutes (rather than the typical 10–20 minutes as deduced by the maturation of attached N-linked oligosaccharides) remains to be tested.

Visualization of membrane trafficking

For most part of the last century, research in the field of membrane trafficking has been synonymous with microscopy (Figure 11.2). Electron microscopy, in combination with immunolabeling and cytochemical techniques, remains powerful tools for determining the localization and structure of specific proteins and compartments. Before the end of the last century, this view was dramatically changed by the emergence of innovative genetic, molecular, and biochemical approaches that revolutionized and invigorated the field. However, such approaches have revealed little about the physical nature and the dynamics of these carriers in the context of living cells.

Genomics and proteomics studies carried the promise of the identification of the full inventory of vesicular trafficking machinery components and regulator. However, due to the dynamic nature of membrane compartments in cells, the most exciting advances in cell biology, which will continue to shape the future of this field, have been the development of Nobel Prize winning concepts of fluorescent protein tags and optical subdiffraction imaging methods [16–18], that together, allow single molecules to be monitored in living cells. As most proteins function in complexes, the future of this field also belongs to techniques such as fluorescence cross-correlation spectroscopy (FCCS) and förster resonance energy transfer (FRET) that enables identification and quantification of protein interactions *in situ*. Studies using these techniques should, in real time, provide a detailed picture of the molecular machines that mediate intracellular trafficking.

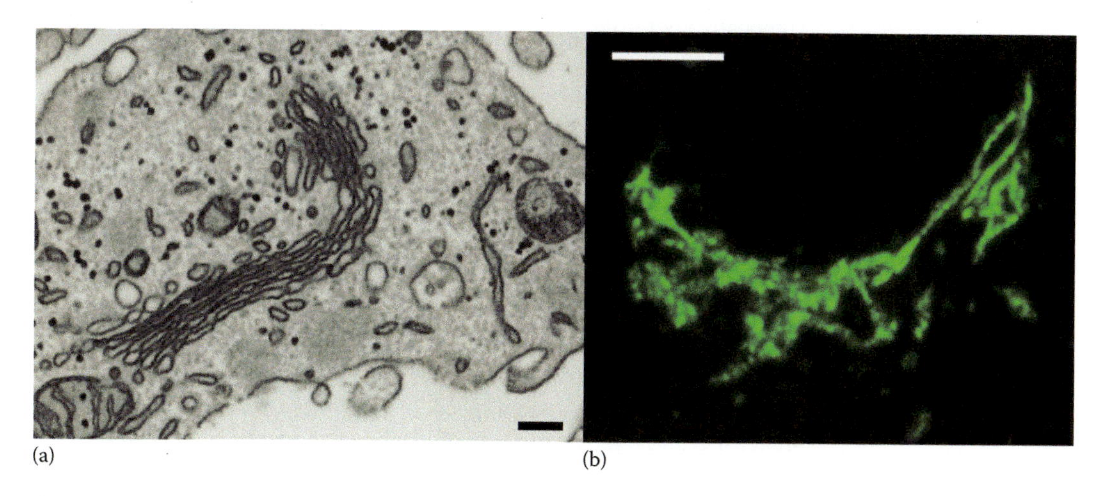

Figure 11.2 Golgi apparatus in HeLa cells. (a) Transmission electron micrograph, scale bar 200 nm and (b) laser scanning confocal microscopy image, scale bar 5 μm.

Imaging has become so central to today's cell biology discoveries that more than 70% of all high-impact bioscience publications rely on advanced microscopy techniques. Such high usage stimulates additional advances in technology development, pushing frontiers forward and creating imaging technologies central for future research. In this chapter, we focus on imaging technologies such as correlative microscopy and live-cell imaging that have already highlighted the plastic nature of the Golgi apparatus, an organelle that is sensitive to perturbation yet highly efficient in regaining both structure and function. Single-molecule microscopy and super-resolution microscopy are also discussed.

Tools for imaging membrane trafficking—types of microscopy

Mammalian Golgi architecture—electron microscopy and correlative light electron microscopy

The first ultra-structural images of the Golgi apparatus were published in the 1950s obtained with the then, newly developed electron microscope (EM) [19,20]. Since then, the structural complexity of the Golgi has been a topic of intense morphological research and one of the most photographed organelle in the cell. Driven by a collective desire to understand how changes in the dynamic structure of this organelle relate to its vital role(s), the comparatively high-resolution glimpses of the Golgi and other compartments of the secretory pathway offered through EM inspired development and application of some of our most informative molecular and genetic approaches. Despite this, relating basic molecular mechanisms to ultrastructure remains a challenge. In the many years of morphological research that followed, the visualization of the Golgi has gone hand-in-hand with the development of EM techniques. Rambourg and Clermont [13] were the first to investigate the Golgi apparatus in three dimensions (3D) using stereoscopy. Over the years, stereoscopy was applied to a variety of cells and greatly contributed to our current understanding of Golgi architecture. An alternative approach, to study 3D structure through serial sectioning, by which a series of adjacent thin sections are collected, enabled detailed reconstructions of the Golgi apparatus in 3D revealing stacks of Golgi cisternae interconnected laterally through highly fenestrated membranes [14]. In the 1990s, 3D-EM was boosted by the introduction of high-voltage, dual axis 3D electron tomography [21–24]. Sections were here photographed in tilt series at different angles, then reconstructed into 3D tomograms that enabled one to *look beyond* a given structure to determine how it relates to other cellular compartments. These EM tomography approaches afforded new insights into structure–function relationships of the Golgi apparatus and provoked a reevaluation of older paradigms. Multiple vesicles adjacent to cisternae were observed as well as cisternal ER membranes intercalating with the *trans*-most cisternae of the Golgi stack. This observation supported the early notion by the late Novikoff of a golgi-endoplasmic reticulum-lysosome (GERL) system, a close juxtapositioning of Golgi, ER, and lysosomal membranes [25].

Correlative light and electron microscopy

Another approach that holds great potential for Golgi research and that overcomes the limitation of EM in acquiring temporal information is to combine light and electron microscopy through correlative light electron microscopy (CLEM). Live-cell imaging of fluorescent proteins has revolutionized cell biology by the real-time visualization of dynamic events. However, live-cell imaging does not reveal membrane complexity and context. By CLEM, live cells are first viewed by light microscopy, fixed, and then prepared for direct EM or in combination with 3D electron tomography [26,27]. A further variation of CLEM was recently introduced to obtain ultrastructural information regarding dynamics of clathrin-coated vesicle budding at the plasma membrane. First, time-lapse light microscopy was used to characterize the temporal association of fluorescently tagged regulatory factors with newly forming endocytic vesicles. This information was then used to correlate the ultrastructure of endocytic intermediates with protein dynamics of the regulatory factors by cryoelectron tomography [28]. CLEM is a growing field that includes a large variety of strategies and one that reaches an impressive degree of spatial precision. The main advantage and strength of CLEM is that it provides ultrastructural context to molecules observed at the light microscopy (LM) level. No matter what the resolution obtained at the LM level, interpretation as to structure is limited without the ultrastructural context of the lipid bilayer, adjacent membrane structures, and other cellular structures (e.g., vesicles and tubules).

Efforts to better link fluorescence and electron microscopic techniques have been hindered for long times by methodical difficulties. Our group has developed a correlative photooxidation method that builds on earlier correlative studies that used concentrated light sources of particular wavelengths to generate an electron-dense reaction product by photooxidation, a highly localized precipitate specific to the structure of interest [29,30]. Our method, termed GRAB, for green fluorescent protein (GFP) recognition after bleaching, allows for the direct ultrastructural visualization of the GFP on illumination and uses oxygen radicals generated during the GFP bleaching process to photooxidize 3,3′-diaminobenzidine (DAB) into an electron-dense precipitate that can be visualized by routine transmission electron microscopy and electron tomography [31]. This ability, to bring together correlative labeling with time-resolved microscopy represents a powerful tool for future studies to dissect membrane traffic to, through, and from the Golgi.

As GFP and related variants are so commonly used, we focused on developing conditions in which the free radicals generated by GFP can

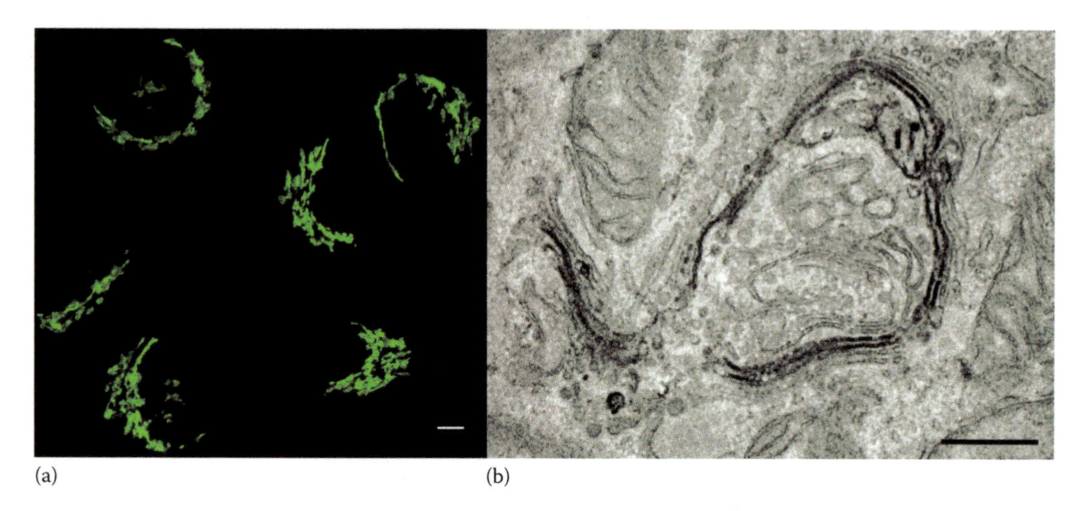

(a) (b)

Figure 11.3 Photooxidation of enhanced fluorescence protein polymerizes DAB to an electron-dense precipitate. (a) Fluorescence enzyme GalNAc-T2^{EGFP} stably expressed in HeLa cells and exclusively localized to the juxtanuclear region, scale bar 5 μm, and (b) after photooxidation and on embedding, the DAB staining can be identified by electron microscopy resembling the EGFP localization at the juxtanuclear Golgi stack, scale bar 1 μm.

be harnessed to precipitated DAB. We found, that the DAB precipitated appears to be locally distributed, indicating a close spatial relationship between the chromophore (GFP) and the final precipitate (Figure 11.3b). The goal was to establish conditions that produce a DAB precipitate of sufficient quality to permit high-resolution investigations by both 2D and 3D EM. Visualization in 3D is necessary to extract spatial and temporal information of GFP fusion proteins in highly convoluted membrane structures such as those involved on the Golgi architecture. As established Golgi markers, we used the human Golgi-resident glycosylation enzyme, N-acetylgalactosaminyltransferase-2 (GalNAc-T2) fused to green- or cyan-enhanced fluorescent protein (Figure 11.3a). The GRAB technique offers a method to quantitatively correlate fluorescence to protein distribution at the ultrastructural level: The density of the DAB product generated by this method appears proportional to the initial fluorescence [31].

Of interest was the notion that GalNAc-T2^{EGFP}-derived DAB precipitate could be found in vesicular structures adjacent to, or attached to, cisternal membranes (Figure 11.4a and b). Also, that the density of DAB precipitation was not uniform throughout the cisterna suggestive of intracisternal segregation (Figure 11.4c). This is in line with our *in vitro* studies finding Golgi-resident glycosylation enzymes in COPI vesicles [10,12,32]. In conclusion, the possibility of performing correlative and quantitative illumination-based electron microscopy with fluorescent proteins and their variants is and will be of great value for the community working with live cells as it offers a direct route from fluorescence to the ultrastructural level.

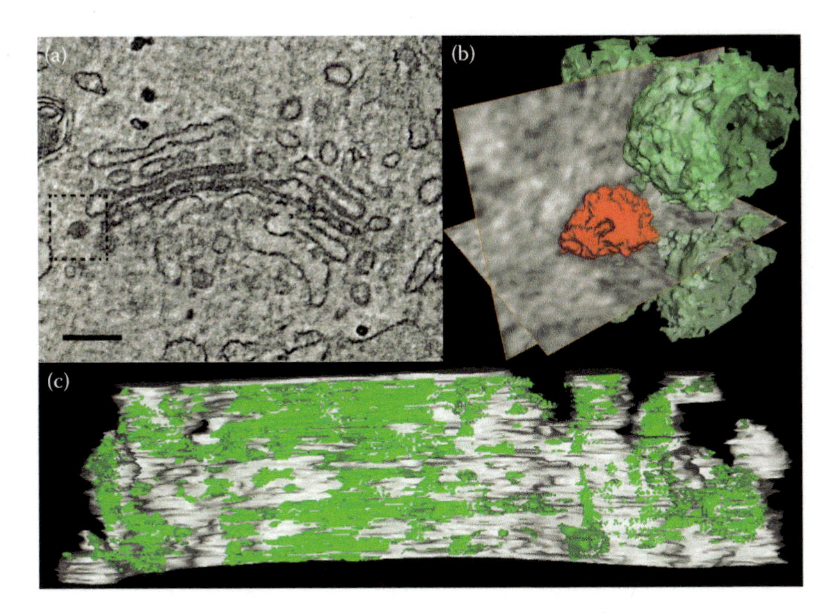

Figure 11.4 Photooxidation of GalNAc-T2^{EGFP} to generate an electron-dense DAB precipitate. In (a), GalNAc-T2^{EGFP}-derived DAB precipitate can be seen in two cisternae and in an adjacent vesicular-tubular structure. In (b), following a two-axis tilt, a series of recordings were used to reconstruct a 3D tomogram revealing a GalNAc-T2^{EGFP}-derived DAB-positive vesicle adjacent to the cisternal membrane. In (c), the GalNAc-T2^{EGFP}-derived DAB precipitate appears segregated within the cisterna. (From Grabenbauer, M. et al., *Nat. Methods.*, 2, 857–862, 2005.)

Importance of the Golgi intracellular transport components on lipid droplets biogenesis

Maintaining cellular lipid homeostasis is vital to the cell and the body and often involves transient or long-term storage in the form of cytoplasmic lipid droplets (CLDs). These are spherical structures that form throughout the cytoplasm, mainly in white adipocytes of white adipose tissue, and in foaming macrophages involved in the clearance of modified LDL and cholesteryl ester storage [33]. Under continued excessive energy intake or metabolic dysfunction, fat accumulation also takes place in peripheral tissue including liver, muscle, and the pancreas. Over time, this results in obesity and with it, the metabolic syndrome along with comorbidities such as cardiovascular disease, fatty liver disease, and type 2 diabetes, all with marked increase in mortality rates. CLDs have for a long time been viewed as static structures. Gradually, this view has been replaced by an increased appreciation that lipid droplets are highly dynamic and regulated organelles, having a distinct shape, composition, and cellular function.

In mammalian cells, three related ARF GTPase-activating protein termed ARFGAP1-3 have been extensively characterized in the context

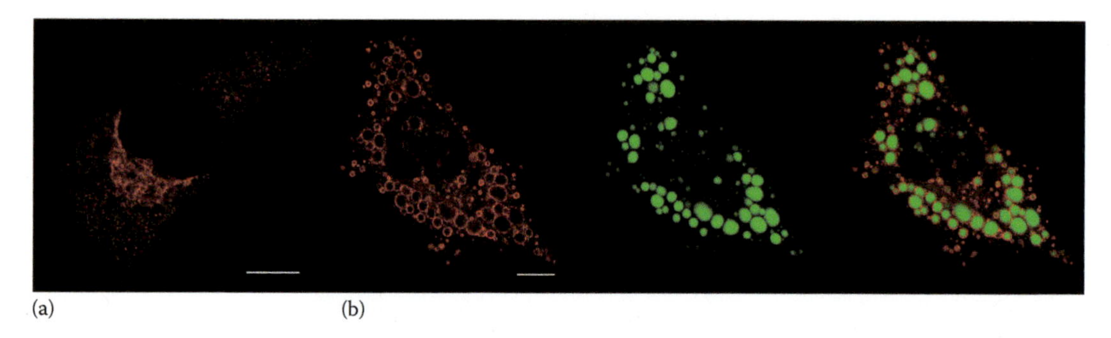

(a) (b)

Figure 11.5 ARFGAP1 relocates from the Golgi apparatus to CLDs on addition of oleate. HepG2 cells were fixed and processed for indirect immunofluorescence before (a) and 4 hours after addition of 0.5 mM oleate (b) and stained to reveal endogenous ARFGAP1 (red) and CLDs (green), by using a rabbit polyclonal to ARFGAP1 and BODIPY 493/503, respectively. Bars = 10 μm.

of COPI vesicle biogenesis including vesicle coat formation, dissociation of coat, cargo sorting, and membrane scission. Of the three, ARFGAP1 appears sensitive to both membrane lipid composition and membrane curvature and has also been proposed to act as a structural component alongside coatomer to form the COPI coat. The binding of ARFGAP1 to Golgi membranes appears to require diacylglycerol since addition of a phosphatidic acid phosphatase inhibitor, propranolol, ARFGAP1 is rapidly displaced from the Golgi to the cytosol [34]. Propranolol addition results in a decrease in the membrane buds, vesicles, and tubules though selective knockdown of ARFGAP1 suggesting a role at the stage of vesicle scission rather than bud formation [35]. Although, previous proteomics suggest the presence of ARFGAP1 on the surface of the lipid droplet [33], previous knockdown studies of ARFGAP1 showed no alteration on CLD formation and utilization [36]. We found that ARFGAP1 is present on CLD and that it appears important for their formation [37]. CLD association of ARFGAP1 was observed transiently in cultured cells on addition of oleate (Figure 11.5), and at steady state, in hepatocytes of human liver tissue. Overexpression of ARFGAP1 promotes CLD formation in both HeLa and HepG2 cells, whereas knockdown of ARFAGP1 impairs CLD formation in both cell lines.

As ARFGAP1 is historically linked to COPI vesicle function through its role as an activating protein for ARF1 in the recruitment of coatomer to Golgi membranes as well as in the scission of COPI vesicles, an involvement in COPI function at the level of CLD biogenesis and homeostasis is possible [38]. It is also conceivable that ARFGAP1 acts on ARF1 in the recruitment of phospholipid-modifying enzymes [39] or PLIN2 [40] to the surface of lipid droplets. ARFGAP1 is also known to interact with adaptor proteins for clathrin-coated vesicles as well as dardarin (LRRK2), a kinase with a GTPase domain. Variants of LRRK2 are associated with both Parkinson's and Chron's disease. The role(s) of ARFGAP1 on CLDs remains to be determined.

Protein diffusion—not as simple as it looks—fluorescence recovery after bleach, fluorescence loss in photobleaching, fluorescence correlation spectroscopy, fluorescence cross-correlation spectroscopy, and Förster resonance energy transfer

The dynamics of lipid and protein diffusion can be studied at great precision providing these can be labeled and/or expressed as fluorescent molecules. Diffusion is a main contributor to protein interaction as it influences on- and off-rates. Diffusion also plays a major role in signal transduction, membrane organization, and vesicle trafficking, the two latter as many enzymes (e.g., lipid-modifying enzymes) and all coat proteins diffuse through the cytosol to their site of membrane action. In addition, small GTPases, along with their exchange factors and activating proteins, shift between the cytoplasm and membrane depending on their activity. Whereas most small proteins (approximately 30 kDa or smaller) can diffuse *freely*, larger proteins and complexes experience restricted diffusion, a phenomenon imposed at the molecular, temporal, and spatial scale through molecular crowding (see below). This is often overlooked, yet causes major limitations in conclusions drawn when interpreting results based on simplified diffusion models. Of the many bleaching techniques that exist, fluorescence recovery after bleach (FRAP) and fluorescence loss in photobleaching (FLIP) are the most common, used to demonstrate mobility, interconnection, and recycling; and have been extensively applied in the Golgi field, showing that Golgi-resident enzymes move through the Golgi ribbon, presumably by diffusing laterally between adjacent and interconnected Golgi stacks [41] (for review—see [42]). In addition, the Golgi-resident enzymes can interchange their content even when distributed as separate stacks situated in close proximity to endoplasmic reticulum (ER) exit sites [43]. Such exchange is made possible through the constant recycling via the ER highlighting the latter as a recycling compartment for Golgi constituents. Indeed, about 7% of Golgi-resident proteins reside in the ER, at steady state. Golgi to ER recycling is both COPI-dependent (from the *cis*-side) and independent (from the *trans*-side) keeping Golgi membranes and their constituents in a constant flux.

FRAP and FLIP measure mobility (and diffusion) over relatively large distances and timescales. It is important here to take into account restricted diffusional mobility when comparing the behavior of different proteins, especially when determining the rate of membrane binding of cytosolic components. For example, FRAP experiments showed that the small GTPase, ARF1 (ADP ribosylation factor 1), had different binding kinetics than did coatomer, the major component of the COPI coat [44]. This was taken as evidence that GTP hydrolysis by ARF1 (through ARFGAP) is independent of coatomer dissociation challenging the long-standing model of ARF^{GTP}-dependent coatomer binding to

Golgi membranes. Rather, that study highlights the limitation of FRAP (and FLIP). Using fluorescent correlation spectroscopy (FCS) to monitor diffusion of singular molecules as they move through a confocal volume, coatomer was found to exhibit a diffusion-limited behavior, presumably because of its comparatively larger size (586 kDa compared to the 22 kDa of ARF1). When taking into account the significantly slower diffusion rate of coatomer, its binding kinetics to Golgi membranes is comparable to that of ARF1 and ARFGAP1 [45]. A possible explanation for why coatomer exhibits a diffusion-limited behavior is that it is restricted in its movement due to cytoplasmic crowding. Indeed, using FCS, we showed that membrane-bound as well as soluble proteins exhibit *restricted* diffusion, a phenomena termed subdiffusion, which is a form of anomalous diffusion where the size of the area or volume explored by a given molecule does not grow linearly with time [46,47]. This is evident when examining a simplified equation that approximates diffusion: $\Delta x \sim t^{\alpha}$ where Δx denotes the area (or volume) explored, t time, and α the degree of anomalous diffusion. When α is 1 (normal diffusion), diffusion is time-independent, that is, it does not matter at what time frame diffusion is measured, the rate of diffusion remains the same. When α is smaller (or larger) than 1, diffusion becomes time-dependent (Figure 11.6). It then matters a great deal at what timescale diffusion is measured, the smaller the timescale, the more accurate, which is why FCS is more advantageous than FRAP and FLIP as this approximates a temporal and spatial molecular scale. This greatly changes the scope whereby protein function can be understood. In layman's terms, proteins exhibiting subdiffusional behavior will hang around for a longer period of time. This translates into an *in vivo* reality where biological processes can take place even when *in vitro*-based estimates of affinities are in the millimolar range. In other words, with subdiffusion, the propensity for molecular self-organization is significantly enhanced. A major limitation therefore exists with protein–protein interaction studies based on pull-downs as these all rely on that protein's *stick together* through relatively *high* affinities. Affinities are consequently best estimated *in vivo*, not *in vitro*. In our opinion, FCCS, offers the best experimental avenue for such experiments.

Showing that proteins do interact, even transiently, is a mainstream activity in life sciences. Apart from antibody-based colocalization and proximity approaches using antibodies, two *in vivo*-based imaging techniques stand out when wanting to determine protein–protein interaction *in vivo*: FRET and FCCS. Both rely on the expression of fluorescent tags or the introduction of fluorescent probes (e.g., fluorescent antibodies or lipids). FRET, Förster resonance energy transfer, relies on close proximity and can be used to demonstrate minute changes in conformation between two interacting proteins or within a protein as well as showing that two proteins interact. An excited donor chromophore can instead of emission, transfer energy to an acceptor chromophore through dipole–dipole coupling. Though sensitive, FRET is limited to close proximity,

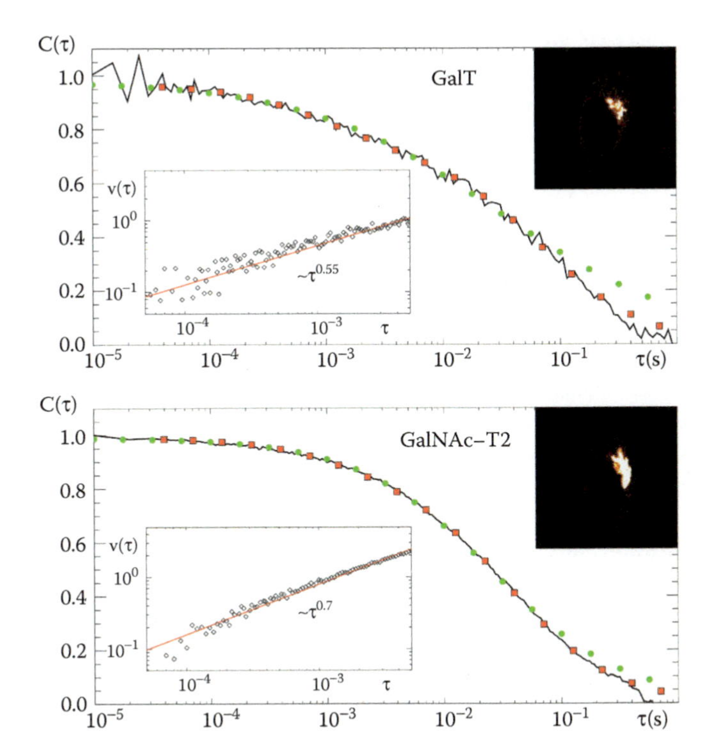

Figure 11.6 Anomalous diffusion of recycling Golgi enzymes in the ER, *in vivo*. HeLa cells expressing GalT(GFP) or GalNAc-T2(GFP) were analyzed by FCS. Insets (right top corners) show pronounced fluorescence over the juxtanuclear Golgi area and a small ER pool of fluorescence. Autocorrelations for both are displayed (black line-red squares) with close fits to a two-component diffusion model (green circles). Insets (lower left corners) show that the variance of increments $v(\tau) \sim \tau^{\alpha}$ follow a power law of 0.55 GalT(GFP) or 0.7 GalNAc-T2. All times τ are units in seconds. (From Weiss, M. et al., *Biophys. J.*, 84, 4043–4052, 2003.)

as the efficiency of radiation transfer is inversely proportional to 10^{-6}. Such limitation does not exist with FCCS, making this technique more suitable for determining *in vivo* affinities. A major limitation of FCS if wanting to determine diffusion rates of membrane components (proteins and lipids) is that the underlying topology of the membrane is rarely perfectly smooth. This introduces great variability and uncertainty when wanting to determine diffusion rates [46]. In contrast, FCCS monitors the movement of two different fluorescent molecules; the underlying uncertainty is canceled out when determining the affinity between the two components [48]. As such, FCCS can be readily applied to both membranes bound as well as soluble components. When estimating affinity it is nevertheless important to incorporate degree of subdiffusion (for review—see [49]). The main limitation of FCS and FCCS is that both techniques relies on correlation, that is, the detected fluorescent fluctuations have to be correlated to *a priori* constructed fitting models (e.g., one, two, three, or multiple component models). As such, there always exists an ambiguity or uncertainty with respect to the *trueness* of observed interactions, something that can be mitigated by control proteins known not to interact but introduced at similar concentration.

What next? Current and future prospects for Golgi visualization

Technologies such as live-cell imaging and correlative microscopy have highlighted the plastic and dynamic nature of the Golgi apparatus, one that is sensitive to perturbation yet highly efficient in regaining both structure and function. The highly dynamic membrane systems of Golgi apparatus are delicate and prone to fixation artifacts. Therefore, the understanding of Golgi morphology and its function should improve significantly with the development of better preparation methods for visualization. Improvements on high-pressure freeze fixation and sample preparation in terms of reproducibility and increased freezing quality in combination with the dynamics of *in vivo* light microscopy and probably in direct correlation to cryo-electron microscopy will help identifying the key players of Golgi function in the future. Most likely, this will not be solved using cellular cryo-electron microscopy as standalone technique, but rather in conjunction with various other technologies such as structural biology, biochemistry, light microscopy, and mass spectrometry-based proteomics.

Super-resolution microscopy approaches

Super-resolution light microscopy will further add to this picture. This technique may help to answer important outstanding questions, such as formation, fission, and fusion of transport vesicles that have a role in the early secretory pathway. There are three emerging approaches to super-resolution microscopy: (1) structured illumination microscopy (SIM), (2) single-molecule localization microscopy (SMLM), and (3) stimulated emission depletion (STED) microscopy; all three are commercialized for use in biological sciences. Most of these technologies use mathematical methods to extend resolution beyond the diffraction limit. SIM improves resolution to 120 nm in the *XY* direction and 250 nm in the *Z* direction [50], whereas SMLM techniques improve resolution to 20 nm in the *XY* direction and 50 nm in the *Z* dimension [18,51]. As a comparison, STED improves resolution to less than 50 nm in the *XY* direction and 80 in the *Z* dimension under optimal conditions [17]. These new exciting microscopy techniques greatly increase the resolving power of light microscopes, revealing new details of organelle structure, function, and dynamics.

Different super-resolution approaches to potentially resolve structures that are relevant for vesicular tubular clusters formation and intra-Golgi transport already exist. In STED the excitation light to resolve structural details is spatially confined, which works well for live-cell imaging and provides temporal resolutions that should allow the imaging of intra-Golgi transport carriers with a typical size between 60 and 80 nm. Thus far, STED has not provided any new conceptual insights regarding the intra-Golgi transport. One possible reason is that the complex requirements for nanoscopy pose real challenges for fluorophore design and labeling: the fluorophore must be bright, photostable, and live-cell compatible, that is, the labeling must yield a high fluorophore density that is benign to organelle function. As nanoscopes push the resolution to tens of nanometer, there is a critical need for high-density yet photostable probes to demark organelle boundaries and to study their dynamics. It is important to point out that commercial STED systems, although very powerful, do not yet provide sufficient resolution in the z-direction and thus do not allow one to address questions regarding vesicular tubular cluster formation and intra-Golgi transport mechanisms.

In the past few years, super-resolution microscopy using structured illumination microscopy (SIM) comes closer to fulfill the necessary requirements in order to visualize Golgi structure and intra-Golgi transport (Figures 11.7 and 11.8). Its primary limitation is that its resolution *in vivo* has been limited to ~100 nm for GFP, or only twice beyond the diffraction limit.

Recent SIM developments, by adopting a nonlinear SIM method and photo-switchable fluorescent probes or extending SIM resolution via high-numerical-aperture optics, showed that it is possible to image cellular structures, such as the nuclear pore and the actin cytoskeleton at

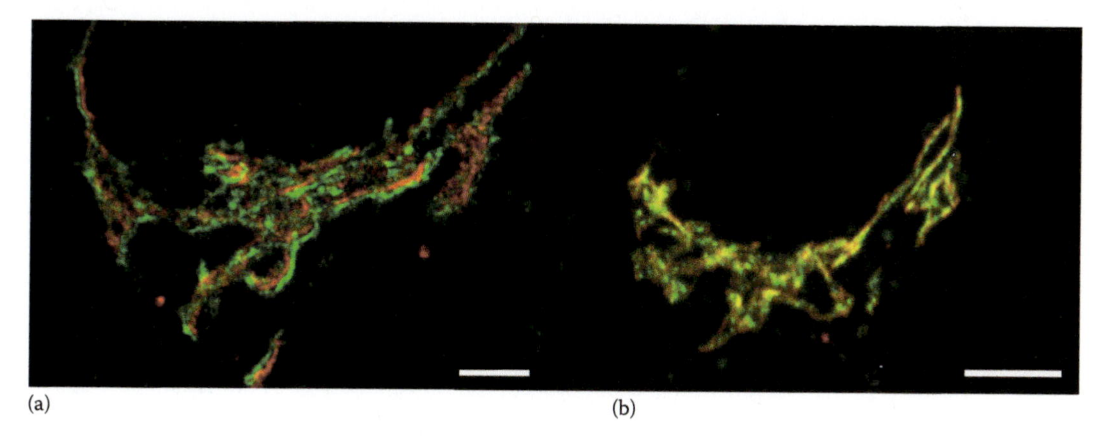

(a) (b)

Figure 11.7 HeLa cells fixed and processed for indirect immunofluorescence and stained to reveal endogenous gp27 (TMEM7) [52] (shown with pseudo color in red) at the *cis* Golgi cisternae, and exogenously expressed of VSV-G epitope tagged α-2,6-Sialyltransferase [53] (shown with pseudo color in green) at the *trans* Golgi cisternae, using an affinity purified rabbit polyclonal directed to the cytoplasmic domain of gp27 [52], and a VSV-G epitope-specific mouse mAb, P5D54 [54]; (a) Super-resolution structured illumination microscopy, scale bar 2 μm and (b) laser scanning confocal microscopy image, scale bar 5 μm.

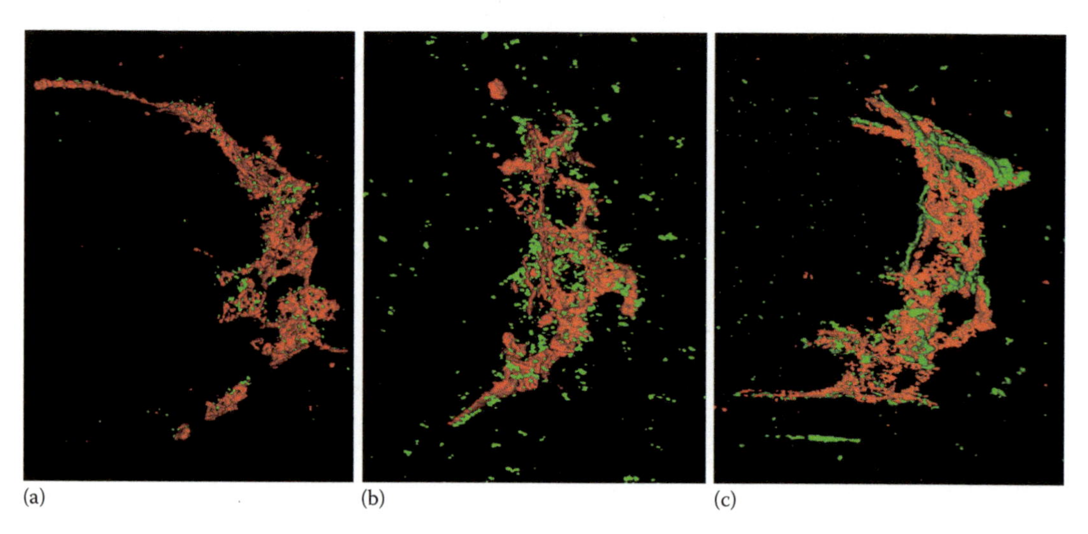

Figure 11.8 Super-resolution structured illumination microscopy of indirect immunofluorescence stain and surface-rendered 3D reconstruction of stably expressed VSV-G epitope-tagged α-2,6-Sialyltransferase in HeLa cells using an affinity purified rabbit polyclonal VSV-G epitope antibody [53] (shown with pseudo color in red). (a) Costaining with endogenous β-1,4-galactosyltransferase (shown with pseudo color in green) as revealed using a mouse monoclonal antibody. (From Hoffmeister, K.M. et al., *Science,* 301, 1531–1534, 2003), (b) costaining with endogenous coatomer (shown with pseudo color in green) as revealed using a mouse monoclonal antibody CM1A10. (From Palmer, D.J. et al., *J. Biol. Chem.,* 268, 12083–12089, 1993), and (c) with endogenous gp27 (TMEM7) was revealed using an affinity purified rabbit polyclonal directed to the cytoplasmic domain of gp27 [52]. Here, stably expressed VSV-G epitope-tagged α-2,6-Sialyltransferase revealed a VSV-G epitope-specific mouse mAb, P5D4 [54].

a resolution of 50 nm [57,58], and to achieve two-color 3D imaging of clathrin-coated vesicles with a spatial resolution of 120 nm [59]. Although, these exciting results indicate that SIM is a promising technique enabling multilabel live-cell imaging with common fluorophores, the spatial and temporal resolution of current commercial SIM systems are not yet sufficient to allow imaging of small structures such as COPII- or COPI-coated vesicles.

Another super-resolution approach uses stochastic single-molecule localization. In this case, photo-switchable or photo-activatable probes are used to generate thousands of images, in which single molecules can be localized with very high precision. The success of this technique depends on high molecular labeling density and low nonspecific sample background. The collected images are then used to reconstruct the density labeling of the marker of interest in the cell with a resolution down to 15 nm [60]. A major disadvantage of localization microscopy is that imaging of very fast events in the (sub)-second range, such as those important in intra-Golgi transport, is difficult because the number of single-molecule events that can be imaged within this time frame is often not sufficient to allow for a faithful reconstruction of the distribution of the molecule of interest in a living cell.

The combination of super-resolution microscopy with other methods, such as FRET [61], FCS, FCCS, or correlative life-cell imaging [62], might reveal details of intra-Golgi transport or the formation of transport carriers for large cargo that, to date, cannot be accessed owing to the resolution limits of conventional fluorescence time-lapse microscopy. In addition, it should be noted that, in contrast to EM, super-resolution microscopy methods do not provide information about the ultrastructural cellular context of the molecules analyzed. In order to achieve this, multilabeling of the samples to highlight cellular landmarks next to the structures of interest would be necessary, and this is still a challenge in super-resolution microscopy. Thus, the combination of conventional electron and cryo-electron microscopy with novel super-resolution light microscopic approaches of live cells might help to answer outstanding questions as to the function of the Golgi apparatus inside a living cell.

References

1. Grasse PP (1957) Ultrastructure, polarity and reproduction of Golgi apparatus. *C R Hebd Seances Acad Sci* 245: 1278–1281.
2. Jamieson JD, Palade GE (1967) Intracellular transport of secretory proteins in the pancreatic exocrine cell. I. Role of the peripheral elements of the Golgi complex. *J Cell Biol* 34: 577–596.
3. Elsner M, Hashimoto H, Nilsson T (2003) Cisternal maturation and vesicle transport: Join the band wagon! (Review). *Mol Membr Biol* 20: 221–229.
4. Rabouille C, Klumperman J (2005) Opinion: The maturing role of COPI vesicles in intra-Golgi transport. *Nat Rev Mol Cell Biol* 6: 812–817.
5. Rothman JE, Orci L (1990) Movement of proteins through the Golgi stack: A molecular dissection of vesicular transport. *FASEB J* 4: 1460–1468.
6. Golgi C (1989) On the structure of the nerve cells of the spinal ganglia. *J Microsc* 155: 9–14.
7. Jamieson JD, Palade GE (1968) Intracellular transport of secretory proteins in the pancreatic exocrine cell. III. Dissociation of intracellular transport from protein synthesis. *J Cell Biol* 39: 580–588.
8. Balch WE, Dunphy WG, Braell WA, Rothman JE (1984) Reconstitution of the transport of protein between successive compartments of the Golgi measured by the coupled incorporation of N-acetylglucosamine. *Cell* 39: 405–416.
9. Cosson P, Letourneur F (1994) Coatomer interaction with di-lysine endoplasmic reticulum retention motifs. *Science* 263: 1629–1631.
10. Lanoix J, Ouwendijk J, Lin CC, Stark A, Love HD, et al. (1999) GTP hydrolysis by arf-1 mediates sorting and concentration of Golgi resident enzymes into functional COP I vesicles. *EMBO J* 18: 4935–4948.
11. Love HD, Lin CC, Short CS, Ostermann J (1998) Isolation of functional Golgi-derived vesicles with a possible role in retrograde transport. *J Cell Biol* 140: 541–551.
12. Gilchrist A, Au CE, Hiding J, Bell AW, Fernandez-Rodriguez J, et al. (2006) Quantitative proteomics analysis of the secretory pathway. *Cell* 127: 1265–1281.
13. Rambourg A, Clermont Y, Marraud A (1974) Three-dimensional structure of the osmium-impregnated Golgi apparatus as seen in the high voltage electron microscope. *Am J Anat* 140: 27–45.

14. Rambourg A, Clermont Y (1990) Three-dimensional electron microscopy: Structure of the Golgi apparatus. *Eur J Cell Biol* 51: 189–200.
15. Lee MC, Miller EA, Goldberg J, Orci L, Schekman R (2004) Bi-directional protein transport between the ER and Golgi. *Annu Rev Cell Dev Biol* 20: 87–123.
16. Chalfie M, Tu Y, Euskirchen G, Ward WW, Prasher DC (1994) Green fluorescent protein as a marker for gene expression. *Science* 263: 802–805.
17. Hell SW, Wichmann J (1994) Breaking the diffraction resolution limit by stimulated emission: Stimulated-emission-depletion fluorescence microscopy. *Opt Lett* 19: 780–782.
18. Betzig E, Patterson GH, Sougrat R, Lindwasser OW, Olenych S, et al. (2006) Imaging intracellular fluorescent proteins at nanometer resolution. *Science* 313: 1642–1645.
19. Dalton AJ, Felix MD (1954) Cytologic and cytochemical characteristics of the Golgi substance of epithelial cells of the epididymis in situ, in homogenates and after isolation. *Am J Anat* 94: 171–207.
20. Farquhar MG, Rinehart JF (1954) Cytologic alterations in the anterior pituitary gland following thyroidectomy: An electron microscope study. *Endocrinology* 65: 857–876.
21. Ladinsky MS, Mastronarde DN, McIntosh JR, Howell KE, Staehelin LA (1999) Golgi structure in three dimensions: Functional insights from the normal rat kidney cell. *J Cell Biol* 144: 1135–1149.
22. Koster AJ, Klumperman J (2003) Electron microscopy in cell biology: Integrating structure and function. *Nat Rev Mol Cell Biol* 4(Suppl): SS6–SS10.
23. Marsh BJ (2007) Reconstructing mammalian membrane architecture by large area cellular tomography. *Methods Cell Biol* 79: 193–220.
24. Noske AB, Costin AJ, Morgan GP, Marsh BJ (2008) Expedited approaches to whole cell electron tomography and organelle mark-up in situ in high-pressure frozen pancreatic islets. *J Struct Biol* 161: 298–313.
25. Novikoff PM, Novikoff AB, Quintana N, Hauw JJ (1971) Golgi apparatus, GERL, and lysosomes of neurons in rat dorsal root ganglia, studied by thick section and thin section cytochemistry. *J Cell Biol* 50: 859–886.
26. Mironov AA, Polishchuk RS, Luini A (2000) Visualizing membrane traffic in vivo by combined video fluorescence and 3D electron microscopy. *Trends Cell Biol* 10: 349–353.
27. Mironov AA, Mironov AA, Jr., Beznoussenko GV, Trucco A, Lupetti P, et al. (2003) ER-to-Golgi carriers arise through direct en bloc protrusion and multistage maturation of specialized ER exit domains. *Dev Cell* 5: 583–594.
28. Kukulski W, Schorb M, Welsch S, Picco A, Kaksonen M, et al. (2012) Precise, correlated fluorescence microscopy and electron tomography of lowicryl sections using fluorescent fiducial markers. *Methods Cell Biol* 111: 235–257.
29. Capani F, Martone ME, Deerinck TJ, Ellisman MH (2001) Selective localization of high concentrations of F-actin in subpopulations of dendritic spines in rat central nervous system: A three-dimensional electron microscopic study. *J Comp Neurol* 435: 156–170.
30. Sharma DK, Brown JC, Choudhury A, Peterson TE, Holicky E, et al. (2004) Selective stimulation of caveolar endocytosis by glycosphingolipids and cholesterol. *Mol Biol Cell* 15: 3114–3122.
31. Grabenbauer M, Geerts WJ, Fernadez-Rodriguez J, Hoenger A, Koster AJ, et al. (2005) Correlative microscopy and electron tomography of GFP through photooxidation. *Nat Methods* 2: 857–862.
32. Lanoix J, Ouwendijk J, Stark A, Szafer E, Cassel D, et al. (2001) Sorting of Golgi resident proteins into different subpopulations of COPI vesicles: A role for ArfGAP1. *J Cell Biol* 155: 1199–1212.
33. Bartz R, Zehmer JK, Zhu M, Chen Y, Serrero G, et al. (2007) Dynamic activity of lipid droplets: Protein phosphorylation and GTP-mediated protein translocation. *J Proteome Res* 6: 3256–3265.

34. Fernandez-Ulibarri I, Vilella M, Lazaro-Dieguez F, Sarri E, Martinez SE, et al. (2007) Diacylglycerol is required for the formation of COPI vesicles in the Golgi-to-ER transport pathway. *Mol Biol Cell* 18: 3250–3263.
35. Asp L, Kartberg F, Fernandez-Rodriguez J, Smedh M, Elsner M, et al. (2009) Early stages of Golgi vesicle and tubule formation require diacylglycerol. *Mol Biol Cell* 20: 780–790.
36. Guo Y, Walther TC, Rao M, Stuurman N, Goshima G, et al. (2008) Functional genomic screen reveals genes involved in lipid-droplet formation and utilization. *Nature* 453: 657–661.
37. Gannon J, Fernandez-Rodriguez J, Alamri H, Feng SB, Kalantari F, et al. (2014) ARFGAP1 is dynamically associated with lipid droplets in hepatocytes. *PLoS One* 9: e111309.
38. Wilfling F, Thiam AR, Olarte MJ, Wang J, Beck R, et al. (2014) Arf1/COPI machinery acts directly on lipid droplets and enables their connection to the ER for protein targeting. *Elife* 3: e01607.
39. Godi A, Pertile P, Meyers R, Marra P, Di Tullio G, et al. (1999) ARF mediates recruitment of PtdIns-4-OH kinase-beta and stimulates synthesis of PtdIns(4,5) P2 on the Golgi complex. *Nat Cell Biol* 1: 280–287.
40. Nakamura N, Akashi T, Taneda T, Kogo H, Kikuchi A, et al. (2004) ADRP is dissociated from lipid droplets by ARF1-dependent mechanism. *Biochem Biophys Res Commun* 322: 957–965.
41. Cole NB, Smith CL, Sciaky N, Terasaki M, Edidin M, et al. (1996) Diffusional mobility of Golgi proteins in membranes of living cells. *Science* 273: 797–801.
42. Snapp EL (2013) Photobleaching methods to study Golgi complex dynamics in living cells. *Methods Cell Biol* 118: 195–216.
43. Storrie B, White J, Rottger S, Stelzer EH, Suganuma T, et al. (1998) Recycling of golgi-resident glycosyltransferases through the ER reveals a novel pathway and provides an explanation for nocodazole-induced Golgi scattering. *J Cell Biol* 143: 1505–1521.
44. Presley JF, Ward TH, Pfeifer AC, Siggia ED, Phair RD, et al. (2002) Dissection of COPI and Arf1 dynamics in vivo and role in Golgi membrane transport. *Nature* 417: 187–193.
45. Elsner M, Hashimoto H, Simpson JC, Cassel D, Nilsson T, et al. (2003) Spatiotemporal dynamics of the COPI vesicle machinery. *EMBO* Rep 4: 1000–1004.
46. Weiss M, Hashimoto H, Nilsson T (2003) Anomalous protein diffusion in living cells as seen by fluorescence correlation spectroscopy. *Biophys J* 84: 4043–4052.
47. Weiss M, Elsner M, Kartberg F, Nilsson T (2004) Anomalous subdiffusion is a measure for cytoplasmic crowding in living cells. *Biophys J* 87: 3518–3524.
48. Weiss M, Nilsson T (2004) In a mirror dimly: tracing the movements of molecules in living cells. *Trends Cell Biol* 14: 267–273.
49. Guigas G, Weiss M (2015) Effects of protein crowding on membrane systems. *Biochim Biophys Acta* 1858: 2441–2450.
50. Gustafsson MG (2000) Surpassing the lateral resolution limit by a factor of two using structured illumination microscopy. *J Microsc* 198: 82–87.
51. Rust MJ, Bates M, Zhuang X (2006) Sub-diffraction-limit imaging by stochastic optical reconstruction microscopy (STORM). *Nat Methods* 3: 793–795.
52. Fullekrug J, Suganuma T, Tang BL, Hong W, Storrie B, et al. (1999) Localization and recycling of gp27 (hp24gamma3): Complex formation with other p24 family members. *Mol Biol Cell* 10: 1939–1955.
53. Rabouille C, Hui N, Hunte F, Kieckbusch R, Berger EG, et al. (1995) Mapping the distribution of Golgi enzymes involved in the construction of complex oligosaccharides. *J Cell Sci* 108(Pt 4): 1617–1627.
54. Kreis TE (1986) Microinjected antibodies against the cytoplasmic domain of vesicular stomatitis virus glycoprotein block its transport to the cell surface. *EMBO J* 5: 931–941.

55. Hoffmeister KM, Josefsson EC, Isaac NA, Clausen H, Hartwig JH, et al. (2003) Glycosylation restores survival of chilled blood platelets. *Science* 301: 1531–1534.
56. Palmer DJ, Helms JB, Beckers CJ, Orci L, Rothman JE (1993) Binding of coatomer to Golgi membranes requires ADP-ribosylation factor. *J Biol Chem* 268: 12083–12089.
57. Rego EH, Shao L, Macklin JJ, Winoto L, Johansson GA, et al. (2012) Nonlinear structured-illumination microscopy with a photoswitchable protein reveals cellular structures at 50-nm resolution. *Proc Natl Acad Sci USA* 109: E135–E143.
58. Li D, Shao L, Chen BC, Zhang X, Zhang M, et al. (2015) ADVANCED IMAGING. Extended-resolution structured illumination imaging of endocytic and cytoskeletal dynamics. *Science* 349: aab3500.
59. Fiolka R, Shao L, Rego EH, Davidson MW, Gustafsson MG (2012) Time-lapse two-color 3D imaging of live cells with doubled resolution using structured illumination. *Proc Natl Acad Sci USA* 109: 5311–5315.
60. Patterson G, Davidson M, Manley S, Lippincott-Schwartz J (2010) Superresolution imaging using single-molecule localization. *Annu Rev Phys Chem* 61: 345–367.
61. Cho S, Jang J, Song C, Lee H, Ganesan P, et al. (2013) Simple super-resolution live-cell imaging based on diffusion-assisted Forster resonance energy transfer. *Sci Rep* 3: 1208.
62. Balint S, Verdeny Vilanova I, Sandoval Alvarez A, Lakadamyali M (2013) Correlative live-cell and superresolution microscopy reveals cargo transport dynamics at microtubule intersections. *Proc Natl Acad Sci USA* 110: 3375–3380.

Chapter 12 Illuminating the cycle of life

Anabela Ferro,
Patrícia Carneiro,
Maria Sofia Fernandes,
Tânia Mestre,
Ivan Sahumbaiev,
João M. Sanches, and
Raquel Seruca

Contents

Introduction

Cells are the fundamental units of life. The basic function of cells is to pass on to the next generation its genetic information. Through a timely ordered series of events, known as the cell cycle, cells are able to reproduce and generate two new genetically identical daughter cells [1,2]. Despite varying from organism to organism, the cell cycle fundamentally ensures that the genomic and cytoplasmic contents of cells are precisely duplicated and then accurately segregated in each cell-division cycle [3]. Paradoxically, cells maintain functionality by coordinating their growth with their division. However, cellular divisions cannot continue indelibly, as tissue homeostasis depends on the tight equilibrium between cell growth, cell proliferation, and cell death allowing a single-celled organism to remain unicellular and a human being to be generated from a single fertilized egg. Dysregulation of this stringent control is at the basis of many diseases, including cancer, ischemia/reperfusion injury, infection, and neurodegenerative disorders [4]. Malignant transformation is frequently associated with alterations in key regulators of the cell cycle and, indeed, a plethora of components from the cell-cycle network has emerged as prognostic markers and even as potential therapeutic targets [5,6]. However, the scenario is far more complex, given that a transformed cell has to overcome several control mechanisms, namely apoptosis and immune surveillance, and has to be sustained by neoangiogenesis [5]. Ultimately, it will be the comprehensive understanding of the multiple pathways regulating the switch between cell cycle and cell death that will unravel key molecular targets. A proficient control of the eukaryotic cell cycle will allow us, not only to tackle predominantly proliferative diseases, but also to prevail over the biological brakes undermining regenerative medicine.

The cell cycle in brief

In 1765, Abraham Trembley published the first drawings of cell division in protozoan and algae, following the development of the first microscopic lenses [7]. However, it was only in 1882 that the German anatomist Walther Flemming reported the first *real* eukaryotic cell division and coined the process *mitosis*. Throughout the following century, scientists have remained mesmerized with the complexity of events leading to cell division and essential for reproduction, now known as the cell cycle.

The cell cycle consists of three *gap* phases, G0, G1, and G2, which are interspersed between the DNA synthesis phase (S phase) and the mitosis phase (M phase). The G0/G1, S, and G2 phases are collectively known as interphase. The cell cycle is a coordinated network of genes and proteins, cyclically regulated by transcription, posttranslational modifications, as well as dynamic genetic and protein interactions [8,9]. Key regulatory molecules include the cyclin-dependent kinases (CDK), a family of serine/threonine kinases that are specifically activated at different phases of the cell cycle by cyclins. Cell-cycle progression is driven by the periodic oscillation of CDK/cyclin activities, which are in turn regulated by a number of mechanisms, including (1) cyclin synthesis; (2) activation of CDKs by CDK activating kinases (CAKs); (3) inhibition of CDKs by CDK inhibitors (CKIs); and (4) ubiquitin-mediated proteasomal degradation of cyclins [9]. Indeed, protein degradation plays a key role in driving cell-cycle transitions through two major E3 ubiquitin ligases, the Skp1–Cul1-F box protein (SCF) complex and the anaphase-promoting complex/cyclosome (APC/C), which ubiquitinate G1 cyclins from late G1 to early M phase and mitotic cyclins from anaphase till the end of G1 phase, respectively [9,10]. Further regulation is accomplished by a myriad of cell-cycle checkpoints that induce cell-cycle arrest on detection of defects and ensure the progression of the cycle in an orderly fashion while minimizing genomic instability [11,12]. Briefly, the G1/S checkpoint induces an arrest induced by DNA damage in a p53-dependent manner, whereas the S phase checkpoint delays initiation or elongation of DNA replication to minimize replications errors. The G2/M checkpoint restricts entry into mitosis minimizing chromosome missegregation, whereas the spindle assembly checkpoint (SAC) detects improper alignment of chromosomes on the mitotic spindle thus ensuring fidelity of chromosome segregation. Finally, postmitotic arrest prevents abnormal daughter cells from entering the next interphase [2,12]. A schematic representation of the cell cycle is displayed in Figure 12.1.

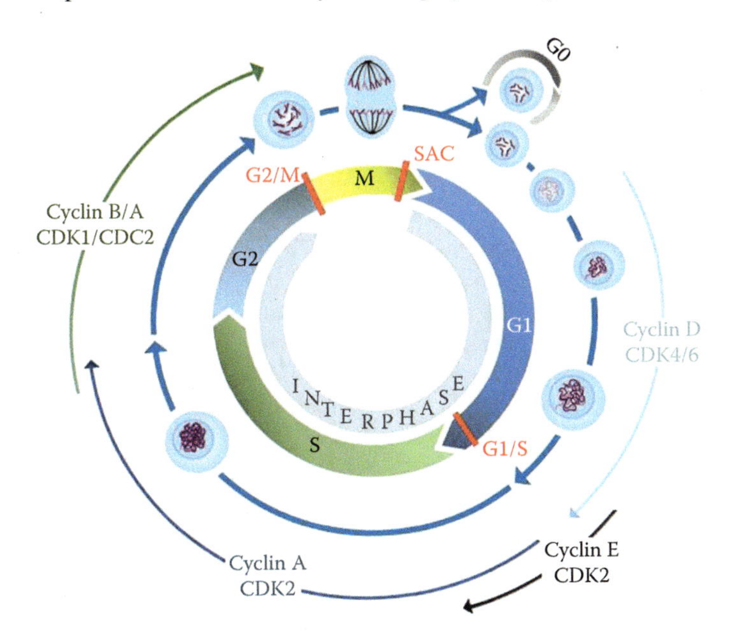

Figure 12.1 *Schematic representation of the cell cycle.* The eukaryotic cell cycle consists of three *gap* phases, G0, G1, and G2, the DNA synthesis phase (S), and the mitosis phase (M). The G0/G1, S, and G2 phases are collectively known as interphase. Cell-cycle progression is coordinated by the periodic oscillation of CDK/cyclin activities as depicted. Further regulation is accomplished by a myriad of cell-cycle checkpoints (in red) that induce cell-cycle arrest on detection of defects.

Herein we aim to provide a comprehensive overview of current methodologies enabling the assessment of cell-cycle progression in living cells, tissues, and whole animals. We hope to highlight traditional and state-of-the-art approaches and their inherent applications and limitations. The tools developed throughout the past decades have set the stage for unprecedented studies in the area of cell-cycle regulation, revealing its important therapeutic insights for a variety of diseases [8].

Experimental procedures

The DNA content and its variation during the cell-cycle phase is a specific signature of each biological sample. Briefly, during the cell cycle, nondividing eukaryotic cells remain quiescent by entering the G0 state; these are diploid cells, possessing two sets of chromosomes (2N). On reentering the cell cycle, these diploid cells start growing while progressing through Gl phase. In normal conditions, cells then proceed to an intermediate phase of synthesis (S phase) during which DNA and protein content is duplicated. On completion of this phase, cells are at G2 phase and are tetraploid (4N). At this point, mitosis (M phase) ensues and cells split in two new diploid daughter cells. Depending on their DNA content, cells may be categorized in three major classes of the cell cycle*:* (1) *prereplicative* for G0/G1 phase cells, (2) *replicative* for S phase cells, or (3) *postreplicative* and mitotic cells during G2/M phase [13]. As cells' cycle through each division, the surveillance of the fidelity of this process is fundamental, and the determination of DNA content during the cell cycle of individual or subpopulations of cells may highlight important biological cues on their physiological status [8]. The possibility to measure rapidly and accurately the DNA content of individual cells from large cell populations is harnessed to the evolution of cytometry-based methodologies and their applications in different cell types. In this chapter, we dissect the most widely used methods of DNA content analysis applied to cell-cycle analysis, as well as, the up-to-date cell-based fluorescent methodologies that are highly impacting on cell-cycle profiling.

DNA analyses have always been performed by profiling the DNA content of individual cells from cell preparations or tissue specimens. DNA quantification allows information concerning the distribution of cells from a cellular population across the different cell-cycle phases, the quantification of apoptosis by the estimation of the fractional DNA, and it establishes the DNA ploidy *status* of cells. The latter is particularly relevant especially for hematological neoplasms due to its predictive value [13]. DNA quantification methods critically rely on dyes that, due to their DNA-binding properties, are expected to reflect accurately the DNA content of a cell, in a given moment or throughout time [13]. The variation of DNA content of a cell population correlates with the cell-cycle *status*, which is typically represented as a DNA histogram.

It provides a snapshot of the distribution/frequency of cells in different cell-cycle phases, though, when obtained at different times, DNA histograms yield information on cell-cycle dynamics. However, cell-cycle analysis based solely on DNA quantification is rather insufficient. On the other hand, the combination of DNA content analysis with DNA synthesis estimation and multivariate immunofluorescence analysis unveils a far more realistic scenario on cell-cycle profiling, allowing the comprehensive characterization of the regulatory mechanisms that orchestrate cell-cycle progression, its components, and subcellular structures [7,8].

Several aspects must be taken into account when selecting the fluorescence molecules for analysis. The cell type used and its applications (live, fixed/permeabilized samples), the DNA-specific fluorophore chosen, and the spectral characteristics of each fluorescent molecule in multiplexed analysis are important aspects to be considered [13].

DNA staining of fixed cells

Often samples are to be stored indefinitely or transported, as is the case of clinical specimens that will be analyzed later or elsewhere. In these cases, it is preferable that samples are fixed to minimize cell autolysis [13]. Usually, in cell cycling analyses of fixed/permeabilized cells encompassing wide-field fluorescence microscopy or cytometry, the DNA is stained and multiplexed with specific antibodies for cell-cycle regulators with well-established spatiotemporal expression patterns, such as cyclins and/or cyclin-dependent kinases, or for histone modifications [7]. Cells may be fixed with cross-linking agents (formaldehyde, glutaraldehyde) or precipitating chemicals (ethanol, methanol, or acetone). The latter are preferable in cell-cycle analysis as the cross-linking of chromatin hinders DNA staining stoichiometry and diminishes DNA profiling accuracy [13]. Propidium iodide (PI) is the most widely used DNA dye on DNA analysis of fixed or permeabilized samples. It can be excited within the visible spectral range, which facilitates its application, although PI staining requires RNase pretreatment since it also binds to dsRNA. DRAQ7 and 7-aminoactinomycin D (7-AAD) are also commonly used, and like PI are far-red emitting fluorophores excitable within the visible range. Within the UV or violet spectral range, DAPI (4′,6-diamidino-2-phenylindole) is the most popular cell-impermeant dye used for DNA content applications, such as cell-cycle phase estimation [14]. Recently, Tembhare et al. reported a new method suitable for DNA content analysis in polychromatic cell-cycle studies of fixed samples, based on a novel, highly specific DNA-binding dye, the FxCycle™ Violet (Invitrogen) that is excitable within the violet range, sparing the remaining spectra for simultaneous surface markers immunophenotyping. This is highly discriminative for minimal residue disease determination, namely for identification of cancer stem cells or circulating tumor cells from its normal hematopoietic equivalents [15]. The inherent cellular differences concerning the rate of fluorophores uptake often requires optimization of staining protocols, or else, the resolution on DNA content analysis falls behind the expected [13]. Moreover, DNA staining discordances are commonly a consequence of cytoplasmic stainability, not only due to organelle's autofluorescence and mitochondrial DNA cross staining but also due to nuclear proteins that restrict dye access to DNA. All of these are known to diminish DNA content precision not only for cell-cycle phase estimation but also for DNA ploidy determination. The permeabilization of cells with lysing agents, such as detergents, hypotonic salt solutions, or enzymes, allow higher accessibility of fluorophores to nuclei, ultimately improving the stoichiometry and accuracy of DNA staining [13]. Identification of cell-cycle subphases is often required for the phenotypic characterization of eukaryotic cells.

There are several methods enabling the detection of newly synthesized DNA, allowing the identification of cells that have progressed to the S phase of the cell cycle and the detection of rapidly proliferating cells. The most common markers are the BrdU (5-bromo-2′-deoxyuridine, a nucleoside analog of thymidine), Ki-67, proliferating cell nuclear antigen (PCNA), and phospho-histone H3, which are used in antibody-dependent assays. Less commonly, *in situ* hybridization assays identifying cells actively moving through the cell cycle based on histone messenger RNA detection are also used. ^{3}H-thimidine approaches are phasing out, due to its radioactive nature [16].

The development of ethynyl-deoxyuridine (EdU)-based click chemistry reactions surpassed the classical BrDU antibody-dependent detection, particularly when S-phase labeled cells are meant to be sorted, recultured, and then used in subsequent assays [7].

Supravital cellular staining

Permeabilization or fixation-dependent cell-imaging procedures only provide a *glimpse* of highly dynamic cellular mechanisms. Often, proliferation studies, cell-cycle distribution assays, DNA ploidy assessment, estimation of apoptotic fractions, and concomitant subcellular characterizations on cell populations, tissue, organs, or animals are required to perform thorough live imaging. As such, live-cell analysis (supravital imaging) requires the use of cell-permeant fluorescence molecules that stoichiometrically stain the DNA. Commonly used dyes in supravital imaging are Hoechst 33342, which is excitable within the UV range, and the far-red emitting dihydroanthraquinone analog, DRAQ5 [17]. Between the two, a wide range of spectra is available, markedly extending their applications for DNA analysis in multiparametric assays. The introduction of Vybrant DyeCycle cell-permeant dyes expanded these possibilities for supravital imaging within the visible spectrum [18]. The latter shows relatively low cytotoxicity in live cell-cycle profiling, allowing the subsequent sorting and reculture of stained cells, with minimal cell loss [19]. In DNA analysis, it is important that dyes bind to DNA with high affinity while reflecting the DNA content of cells, without tampering with the live-cell data acquisition. However, cytotoxic assays have shown that DRAQ5 is associated with DNA damage, whereas Hoechst seems to have an impact on chromatin organization and histone–DNA interactions [20]. As such, care must be taken when choosing a supravital imaging dye.

Cell-based fluorescent reporters

When choosing the least cytotoxic available DNA dye and the most amenable for multiparametric analyses on live cells, several aspects must be taken into account. This is especially important if supravital staining is to be conjugated with fluorescence-activated cell sorting (FACS) in order to select cells in a specific cell-cycle phase among a heterogeneous sample, to further subculture them [20]. The development and application of GFP technology, or its fluorescent variants, revolutionized the biochemistry and cell biology fields. It provided the possibility to visualize and track any protein (ectopically expressed with a tag) in living cells, tissue, or animals [21,22]. Its application to cell-cycle analysis repositioned our knowledge on cell cycle, as well as its periodicity and dynamic nature, allowing visualization of the spatiotemporal expression and regulation of cell-cycle proteins [7]. The development of live-cell reporters comprising fluorescent tags and active portions of well-established spatial and temporal characteristics of regulators of the cell cycle, nuclear translocation, or protein degradation, provided the means to precisely discriminate each cell-cycle phase in living cells. The cell cycle is often targeted based on specific nuclear proteins' localization as these vary throughout the cell cycle. For this, several PCNA-based cell-cycle reporters have been designed. PCNA is an S-phase cell-cycle marker that is targeted to active DNA replicative *foci* [23,24]; moreover, these PCNA-based cell reporters estimate, with minimal perturbations, the overall cell-cycle progression and the length of each cell-cycle phase, in different cell types [25]. Monitoring G1/S transition during the cycle is also possible by assessing the differential cytoplasmic expression of human DNA helicase B fused with GFP in mammalian cells [26]. The combined variations of nuclear expression of fusion proteins DsRed1-DNA ligase I and GFP-Dnmt1 DNA methyltransferase in G1 and G2, respectively, allow the visual discrimination of all cell-cycle phases [23]. Visual discrimination of interphase/mitosis transitioning has also been possible due to the expression of a dual tool comprising a genetically encoded YFP fusion of a plasma membrane (PM)-targeting domain and a nuclear localization sequence (NLS), which combined, elegantly monitor the nuclear membrane collapse and restoration during mitosis [27]. Fluorescent ubiquitination (FUCCI)-based cell-cycle indicator) is an ingenious and powerful system based on cell-cycle transitions through differential accumulation and degradation protein patterns. It was developed to monitor live cell-cycle progression and to characterize

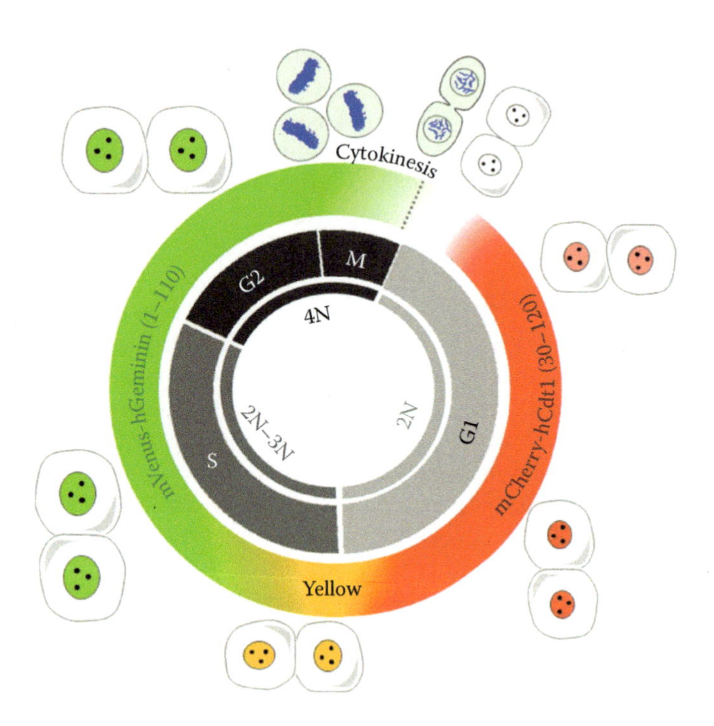

Figure 12.2 *The FUCCI technology.* FUCCI labels the cell nuclei in the G1 phase as red, whereas those in S, G2, and M phases are colored in green. Cells transitioning from G1 to S phase appear in yellow due to the coexpression of the FUCCI vectors. On cytokinesis, early G1 cell nuclei are colorless and gradually become red throughout G1 phase.

the spatiotemporal patterns of cell-cycle dynamics. This reporter system, and its upgraded variants, combines two oscillating proteins tagged to spectrally distinct proteins that mark specific cell-cycle stages. The encoded fluorescent proteins tagging Cdt1 cyclically accumulate in G1, whereas geminin counterparts accumulate in S/G2/M phases and are degraded in G1. This cyclic expression/degradation profile enables a visual report on cell-cycle progression and spatiotemporal dynamics of single cells from cultured cells, brain slices, or even transgenic models constitutively expressing FUCCI (Figure 12.2). Indeed, this powerful tool has been generalized to more complex systems, namely in the generation of whole transgenic animals and shows a remarkable applicability in several fields, namely development, regeneration, cell-cycle–dependent diseases (as cancer), as well as in drug screens [28–36]. Even with its unprecedented relevance in cell-cycle dynamics, FUCCI presents some drawbacks, such as the impossibility to infer on the biochemistry of the central-core cell-cycle enzymes themselves. In FUCCI, the early G1 and G0 cells are colorless hampering the visual discrimination and tracking of these cells. The same difficulty is observed for S, G2, and mitotic cells as they all are labeled with green fluorescence [37]. Moreover, FUCCI-based systems are dependent on genetically encoded proteins, to which a putative toxicity is associated with and may ultimately impact on the cell's expression behavior. Thus, the ability to extract high-content data from unperturbed complex cell populations while assessing single cells has obvious advantages. One of the most recent examples is the label-free cell-cycle analysis method [38], which enables the quantification of DNA content and mitotic cell-cycle phases of different eukaryotic cell through flow-cytometry imaging. This new bioimaging approach explores morphological parameters such as size, shape, and granularity of fixed or live cells, extracted from brightfield and darkfield images. With a supervised machine-learning framework, this system infers the cell-cycle phase of single cells. These label-free morphometric measurements using nonfluorescent channels, such as bright and darkfield, maximize the potential of simpler cytometers, as it releases the available fluorescent channels for biological measurements other than cell-cycle analyses. In the long run label-free cell-cycle profiling potentially reduces the concomitant confounding effects of polychromatic analyses [38].

Fifty shades of cytometry

Classical flow cytometry Since the seminal studies headed by Leland Hartwell, Paul Nurse, and Tim Hunt in yeast and sea urchin, on the regulation of the cell cycle, huge advances have taken place on how we look at cells [39]. Nowadays, the challenge is to surpass the qualitative assessment of the biological processes and obtain information-rich measurements by high-throughput approaches at the single-cell resolution [38]. Flow cytometry is unique in its ability to rapidly assess millions of cells, and simultaneously dissect multiple cellular signatures of individual cells [40]. In 2015 classical flow cytometry reached seniority. Eighty years have gone by since Moldavan presented the first optical device used to detect air particles flowing in a capillary tube. It was the first attempt and the dawn of the flow cytometry principle [41]. The 1960s were a great decade for flow cytometry, with two seminal works reported independently: (1) the principle of hydrodynamics focusing as the base of detection and (2) counting of cells in a fluidic medium were indeed major breakthroughs for the development of the flow cytometer as we know it today [42,43]. Moreover, the first cell sorter was developed [44,45], and for the first time, fluorescence was integrated on flow cytometric analyses [46,47]. The first works on cell-cycle analysis using flow cytometry were then reported [48,49] and Milstein, with the advent of organic dyes (fluorescein and rhodamine) in flow cytometry and monoclonal antibodies, supported the development of the two-color FACS [47,50,51]. Ever since, the progressive development of flow cytometers enabled the development and application of organic and inorganic dyes themselves on increasing medical applications. Further, the comprehensive understanding of the absorption/transfer energy principle that characterizes fluorescence molecules [52] headed the development and application of tandem conjugated dyes in flow cytometry [53]. Later on, the development of small organic dyes (Alexa dye series), which are the improved versions of fluorescein and rhodamine molecules, embossed a superior resolution on flow cytometric analyses, enabling the usage of fluorophores virtually on every wavelength of the optical spectrum [54]. As technology advanced, a greater flexibility in fluorescence dyes was required, prompting the development and application of quantum dots to biology, suppressing the void on available and useful violet-excitable fluorophores; these are highly tunable and photostable, inorganic semiconductor nanocrystals that enable a very discrete, *atomic-like* resolution on fluorescence analyses [55–57]. These fluorescence-based multilabeling advances along with Roederer's and Hazenber's work on applying them to flow cytometry boosted massively the quality and the multidisciplinary applications of this technique, turning it into an increasingly polychromatic tool [58]. Altogether, the five-color analyzer evolved up to the currently available state-of-the-art 18-color-based flow cytometers [59]. The chemical properties of violet-excitable compounds (e.g., organic polymers and brilliant violet dyes) have also improved and are now significantly more stable and brighter than quantum dots [60]. With applications on immunology, cell biology, cancer research, regeneration, and therapy, flow cytometry revolutionized the way we perceive and address the challenges that come across humankind. It has been the dominant technology to dissect and correlate multicharacteristics of single cells, within complex cell populations, enabling high rate analyses, as well as live-cell sorting reliant on specific properties of cells [61–63]. The feasibility of this technique has been recently expanded to the possibility of measuring specific, discrete nucleic acid sequences. The low abundance of nucleic acids sequences on high fluorescent background signals has rendered the detection of nucleic sequences otherwise impossible by conventional flow cytometry. Soh et al. developed the *branched DNA technique*, which comprises target-specific probes and sequentially hybridized amplification reagents that massively amplify the fluorescence signals above the background noise content. This novel technique allows multiplexing the quantification of native and unperturbed mRNA species with the immunophenotyping of single cells through flow cytometry. The branched DNA technique opens new avenues for the correlation of small quantities of mRNA with various biological measurements, at the single-cell level [64]. Yet, the development of flow cytometry as the powerful technique that we know today, enabling *all kind of flavors* analyses (18-colors plus two physical parameters) arose, not only, from the technical milestones in laser and fluorophore technology, but also from the engineering advances in the optical components, such as light collectors and detectors, as well as from constant advances in signal processing (electronic) components. Concomitantly, software improvements allow collecting and processing the ever-growing raw data generated from such complex multicolored analyses. The development of software correcting the fluorescence spillover that often occurs between the myriad of dyes used in polychromatic analyses (fluorescence compensation) was particularly important [63,65]. Flow cytometry is instrumental to our ability to deeply evaluate cells. The comprehensive knowledge of the biology of cells, their underlying mechanisms, and intercellular cross talk, will enable us to fully perceive how cellular *function* interplays with the outcome *phenotype.*

In parallel to flow cytometry, other cytometric technologies have been developed enabling high-throughput imaging as well as molecular and mass spectrometric analyses [63]. Whereas in flow cytometry the sample preparation is, *per se*, critical for the optimal analysis of single cell in a homogeneous stream, for other studies it may be *the* major disadvantage. To obtain fluidic preparations suitable for analysis by flow cytometry, samples have to be disintegrated and the natural architecture of the tissue sample or cell preparations is lost, hampering studies based on cell–cell interactions, for instance. Moreover, flow cytometry fails to provide real images of analyzed cells. The concomitant use of flow and imaging cytometry in laser-scanning microscopy suppressed this drawback. Laser-scanning cytometry merges the advantages of the highly discriminative, polychromatic, and quantitative power of flow cytometry, while preserving the natural architecture of heterogeneous cellular or tissue samples (cell structure resolution). Given that, cells or tissue samples are on slides and are not discarded as is the case in flow cytometry, laser-scanning cytometry allows multivariate, photometric, and morphometric analysis, allowing repetition and subsequent assessment of other cell events [66]. Depending on the stains used, laser-scanning cytometry enables cell's immunophenotyping, as well as collecting information on cell functions, protein expression, DNA content measurements, and cell-cycle staging, and/or subnuclear cytogenetic profiling of each cell (object) analyzed. Coupling the scanning cytometer with a microscope equipped with a CO_2 chamber enables the study of intercellular and intraorganelle trafficking and signal transduction *in vivo.* Assessment of differences in cell proliferation or apoptotic index of cells growing on slides, or multiwells, live and/or fixed has been possible with laser-scanning cytometry [67]. Another powerful and popular flow cytometric platform, enabling real-time cell imaging, is imaging flow cytometry. This technology conveys the beauty of the polychromatic, multivariate, and quantitative properties of flow and laser-scanning cytometry, while generating a collection of images of each individual cell in a flow stream. Resembling laser-scanning cytometry, it allows spectrophotometric and morphometric analysis of individual cells. However, the ability to perform quantitative measurements (cell shape, size, texture, and probe location) endows imaging cytometry a differential capacity on profiling heterogeneous samples. The possibility to interrogate, individually, thousands of cells per second came to circumvent the relative low-throughputness of laser-scanning cytometry. The imaging cytometers (of which Amnis Imagestream cytometer is the most popular) and the associated image analysis software excelled other available technology. Current versions of imaging cytometers allow the detection of up to

12 fluorophore-tagged molecules, including quantum dots and the brilliant violet [63,68]. Numerous biological applications exploring the high-throughput analysis of imaging cytometry have been reported, such as cell–cell interactions, cell signaling, protein–protein interactions, colocalizations, internalization events, or even analysis of cellular morphological changes on different stimuli or treatments. Cell-cycle profiling of tumor samples has been possible with imaging flow cytometry by scoring normal or abnormal proliferation, through the quantification of DNA content of cells or tissue samples [69,70]. Imaging flow cytometry is well suited for these types of studies, as it allows the identification and classification of cancer cells, apoptosis (fragmented nuclei), and the evaluation of the effects of immune- or chemotherapeutic treatments through internalization and intracellular trafficking assays. Through image flow cytometry, cell's immunophenotyping correlated with quantitative morphometric measurements enable the identification of rare subpopulations of potentially metastatic cells, including circulating tumor cells [71,72]. Imaging cytometry has been also used in the study of the role of extracellular vesicles, namely exosomes, on the occurrence and progression of cancers [70].

Acoustic-based flow cytometry Despite the development witnessed, flow cytometry is still being progressively improved. In fact, in recent years, classical hydrodynamic focusing of flow cytometry has been challenged with the introduction of acoustic focusing. Acoustic-based flow cytometry allows (even) higher flow rate analysis, with higher accuracy and precision analysis, at the single-cell level. Enabling the analysis of higher volumetric flow rates than classical flow cytometry, through multiparalleled acoustic flow streams of focused cells, this technology impacted massively in oncology [73]. Identification of very small and rare cells in blood samples, as is the case of circulating tumor cells in cancer detection, or during treatment follow-ups, stem cells inventory (on chemotherapy treatments), minimal residual derived-disease cells (for leukemia assessments), and natural killer cells (in immunodeficiency diseases), requires the analysis of higher number of events for accuracy purposes. More so, acoustic-based flow cytometers offer higher analyses rates (ten times faster than classical cytometry), and cell sorting without loss of high-throughput and quality data. The possibility to perform polychromatic analysis (as it accommodates the most common fluorophores and fluorescent proteins available) further boosts its applicability [73,74]. The higher sensitivity rates obtained with acoustic flow cytometry is one of the key characteristics of this technology. Any application requiring cell-cycle phasing, estimation of apoptotic or proliferative indexes of cell subpopulation, therapy follow-ups, and so on benefits tremendously from the minute analysis variability of acoustic-based cytometry [75].

The most striking technology in the cytometry field is cytometry by time-of-flight (CyTOF) [76]. The underlying principle combines flow cytometry methodology and mass spectrometry analysis. The method is based on the use of antibodies tagged with rare, nonradioactive isotopes of heavy metals (e.g., lanthanides), circumventing the commonly observed limitations associated with spectral overlap of fluorophores. The ingenuity of the method lays on the simple fact that these rare isotopes are not, ordinarily, present in the cell, and thus allow analyses with superior analytical resolutions and virtually no background. This is particularly important for low abundance proteins that are normally lost in background noise. CyTOF raised the dimensionality of cell analyses to unprecedented levels, as it is now possible to analyze up to 40 unique parameters (theoretically, there are up to 100 usable isotopes) on individual cells. The possibility of high throughput analysis allows mass cytometry and cell barcoding, which is particularly relevant in large-scale drug screens and immunology research. It allows the analysis of multiple samples simultaneously, while minimizing intersample variability due to technical limitations, as sample handling and instrument variability, as well as it reduces the time of analysis and reagent consumption [77]. Despite its tremendous potential, one major drawback of mass spectrometry is its incompatibility with live-cell sorting, as the biological samples need to be vaporized. Nevertheless, mass cytometry has initiated a postfluorescence era for cytometry of fixed samples, setting new paradigms for the study of cell cycle [40].

Bioimaging software

Bioimaging techniques are accelerating our ability to *see* cells and monitor multiple cellular mechanisms, either basally or in response to stimuli, homeostatically, or in pathology. This ability to uncover biological systems is, however, highly dependent on the concomitant evolution of image analysis tools that yield quantitative features from biologically relevant features. This is particularly relevant for models based on high-throughput and high-content environment analyses that generate huge amounts of data. Increasing numbers of machine learning frameworks, derived either from supervised or unsupervised-manner, have been supporting the translation of qualitative biological features into highly reliable, biologically sound quantifications that would be difficult to achieve otherwise [78]. Several cell-cycle–based algorithms constitute useful tools for visualizing or tracking cell's cycling through the different phases, either from digital images of fixed, permeabilized cells or from live-cell imaging. CellTrak is a fluorescent-based cell-cycle indicator supporting higher spatiotemporal resolutions and quantifications on cell-cycle profiling than those pioneered by FUCCI. This is a semiautomated quantitative framework supporting cell-cycle phasing from time-lapse experiments, based on dynamic information retrieved from cell-cycle–dependent expression variations in living cells [79]. Another model-based approach has been recently reported by Roukos et al. which measures the DNA content of cells stained with a DNA dye, accurately determining all cell-cycle stages of individual cells imaged by fluorescence microscopy [3]. The framework is based on algorithms that, in an unsupervised manner, enable an accurate quantification of integrated nuclear intensity of DAPI-stained nuclei. The possibility to combine this imaging analysis method with other DNA-binding dyes or immunohistochemistry broadens its applicability [3]. In our unpublished work, we developed a new and semiautomated algorithm for cell-cycle phase determination based on DAPI-stained fluorescence images from widefield microscopy [80]. This clustering-based analysis tool is a two-stage classifier, using both the intensity of DAPI stain and nuclear size features. Nuclear area and DAPI-staining are dependent on cell's DNA content, which are known to be cell-cycle phase-dependent, and thus, highly discriminative of cell-cycle phase. With this work, we propose a novel imaging method to sensitively classify the cell-cycle phase of cells from 1D-cell populations. Importantly, the results obtained are consistent with the typical distribution of cells in all cell-cycle phases of a histogram, the G1, S, and G2/M phases (Figure 12.3). Our image analysis method is independent of any stably expressing cell-cycle reporters, such as FUCCI, or the expression of cell-cycle markers, such as cyclins, eliminating the putative fluorescence confounding the effects occurring when multiplexing cell-cycle analyses with antibody-based methods. It is a rapid, inexpensive, and disseminated method for nuclei staining that may be applied to the accurate estimation of cell-cycle status of all individual cells within a population, while preserving its architectural nature. This DAPI-stained computer-based image analysis frees-up other fluorescence channels available, allowing quantitative measurements of multicolor analyses combined with cell-cycle staging, in virtually all biological specimens. By using low cytotoxic, cell-permeant fluorescent dyes, as discussed earlier in this chapter, this new cell-cycle classifier surpasses the classical *broken cell* methods [8], empowering *in vivo* applications.

Bioimaging tools aim at tackling different biologically relevant functions, namely cell motility, mitotic index (proliferation), apoptosis, and cell-cycle progression, by extracting information at the single-cell level concerning the four cell-cycle phases [81]. This is particularly relevant when studying the effects of inhibitor drugs designed to block the replication of cancer cells. In fact, most clinically available drugs target S, G2, and mitotic cells but are ineffective in quiescent cancer cells, a therapeutic limitation that must be urgently tackled. Either automatically or semiautomatically, the algorithmic segmentation, tracking, and profiling of cell-cycle phases can assist on the definition of such specific therapeutic options, given that it has been proposed that cell-cycle staging of cancer cells can determine their drug responsiveness [82]. Based on the expression pattern of specific, endogenous genes or proteins, genetically encoded fluorescent proteins or morphological nuclei alterations occurring throughout the cell cycle, new labeling tools, and appropriate image analysis algorithms will be crucial to support the mathematical formulation of the cell-cycle *status* of cells, based on data generated from microscopy analyses (widefield or time-lapse confocal) or cytometry-based procedures.

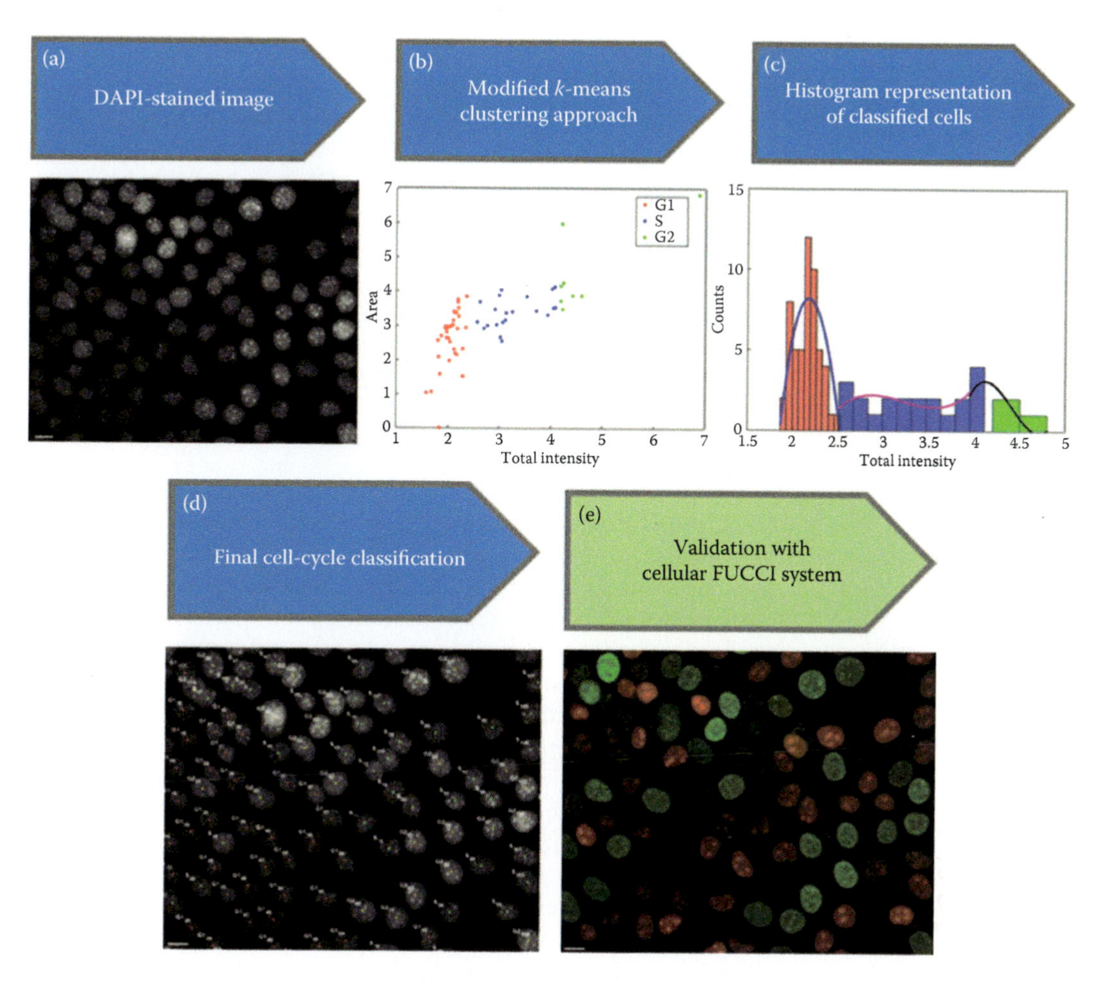

Figure 12.3 *Pipeline analysis of cell-cycle profiling based on DAPI-stained images.* Our bioimaging method is based on DAPI-stained images (unpublished data): (a) DAPI-plane images are acquired and then fed onto the bioimaging analysis tool, (b) cells are classified and distributed into three cell-cycle stages based on a modified *k*-means clustering strategy: *G1*, *S*, and *G2*, (c) a typical histogram is obtained from the distribution of cells within the three major phases, (d) the overall cell-cycle classification of cells is depicted in the original DAPI-stained fluorescence image, and (e) determination of the accuracy of our bioimaging method on its ability to profile the cell-cycle phases was determined with commercially available cellular FUCCI system, which enables a simple and fast cell-cycle readout. Bars in images represent 10 μm.

Applications

The importance of cell-cycle regulation to medicine was highlighted in 2001 when Leland H. Hartwell, R. Timothy Hunt, and Paul M. Nurse were jointly awarded with the Nobel Prize for physiology and medicine. Indeed, a myriad of diseases display alterations in cell-cycle regulatory networks stirring considerable research toward the identification of targets for therapeutic intervention and tumor prognosis biomarkers. Cell-cycle research has been a challenging issue, in particular, monitoring cell-cycle progression in living cells. To tackle the complexity of cell-cycle spatiotemporal dynamics, multidisciplinary approaches involving advanced imaging technologies combined with mathematical modeling have been developed [36]. Importantly, fluorescent tools now available function as live-cell sensors allowing subpopulations of living cells to be tracked and quantified throughout the cell cycle, while maintaining cell and tissue integrity. Further, it is now possible to perform high-throughput screening of compounds aimed at modulating the cell cycle [8]. Here we will focus on discussing a number of applications of cell-cycle assessment from biological processes to diagnostic and therapeutic interventions.

Stem cell research

Stem cells have a prominent capacity to self-renew and to differentiate into multiple lineages of cells, awarding stem cell research with the potential to yield promising therapies for a plethora of conditions, including neurodegenerative disorders, diabetes, and heart disease. Furthermore, stem cell therapy endorses tissue engineering as an auspicious treatment for replacing damaged tissues through the combination of scaffolds, cells, and biologically active molecules [83]. Emergent evidence suggests that cell-cycle mechanisms may be linked to the capacity of embryonic stem (ES) cells to differentiate into a desired lineage [83,84]. Several studies have demonstrated the usefulness of cell-cycle assessment, namely through FUCCI reporters, in mouse and human pluripotent cell biology [85]. Recently, using this fluorescence cell reporter, Pauklin and Vallier reported that the capacity of differentiation of human pluripotent stem cells varies during the progression of their cell cycle. Further, the authors demonstrated that cell-cycle manipulation can direct differentiation of human pluripotent stem cells (hPSCs) bypassing the need of exogenous growth factors [86]. Likewise, a study by Singh and coworkers took advantage of the FUCCI system and verified that the expression of developmental regulators is higher in G1 than in other phases of the cell cycle, thus leading cells to be more responsive to differentiation signals [87].

Further, in the field of regenerative medicine, Roccio et al. have recently reported the use of FUCCI and stem cell reporters to identify regulators of cell-cycle reactivation of progenitor cells within the auditory epithelium [88].

One of the top fields of stem cell research concerns cancer stem cells (CSCs), which represent the small subset of tumor cells that have the potential to proliferate indefinitely and generate all lineages of the tumor bulk. Initially described in hematopoietic malignancies, CSCs have subsequently been identified in a variety of solid tumors, where they are responsible for local tumor recurrence, distant metastasis, and resistance to conventional anticancer therapies [89–91]. Cell-cycle evaluation is essential in the pursuit for effective therapies targeting both CSCs and bulk tumor cells, and in fact, it has been proposed that effective strategies may encompass cell-cycle modulation [90,91]. Atashpour and colleagues have reported that quercetin in combination with doxorubicin chemotherapy is an effective strategy for treatment of both CSCs and bulk tumor cells in colorectal cancer, by inducing a G2/M arrest and enhanced cytotoxicity [10]. In a similar study, Gong et al. demonstrated that inhibition of phosphatidylinositol 3-kinase (PI3K) signaling induced a G0/G1 cell-cycle arrest and apoptosis in osteosarcoma CSCs [92]. Through a label retention assay, using BrdU, Filippi-Chiela, and coworkers investigated the role of resveratrol for targeting glioblastoma (GBM) CSCs and found that this natural compound induced autophagy and S-G2/M cell-cycle arrest, suggesting that it may be an effective antitumor agent [93].

Neurodegenerative disorders

The mechanisms triggering the onset of human neurodegenerative diseases remain unknown despite the enormous progress throughout the past two decades. Although terminally differentiated neurons of normal brain are incapable of cell division, aberrant cell-cycle activation has been proposed to be a key step in neuronal death in several degenerative diseases, including Alzheimer's disease, Parkinson's disease, and amyotrophic lateral sclerosis (ALS). Several studies have reported elevated expression of cell-cycle proteins and regulators such as cyclin D1, CDK4, hyperphosphorylated pRb, and E2F-1 in the brain of patients [94].

Through immunocytochemistry, many groups have reported the presence of cell-cycle proteins, including activators and inhibitors, in autopsy specimens from Alzheimer's disease patients [94,95]. More strikingly, Yang et al. reported actual DNA replication in at-risk neurons in the Alzheimer's disease brain through fluorescence *in situ* hybridization (FISH) and the role of cell-cycle reentry in the early stages of the disease [96].

Cardiovascular disease

Cardiovascular disease is the leading global cause of morbidity and mortality. The role of cell-cycle control in the maintenance of cardiac cell population is indubitable, and as such, much attention has been paid to the development of assays to assess the spatial and temporal patterns of mammalian cardiomyocyte cell cycle. Mammalian cardiomyocytes loose the capacity to proliferate soon after birth, as they withdraw from the cell cycle and differentiate, thus limiting cardiac repair following myocardial injury. Hashimoto et al. have recently established a time-lapse imaging tool using the FUCCI system to evaluate cardiomyocyte cell-cycle dynamics *ex vivo*, which revealed an elongated cardiomyocyte cell-cycle phase during development [97]. In a subsequent study, the same group aimed to analyze cardiomyocyte movement in the developing murine heart using time-lapse imaging of FUCCI expressing cardiomyocytes. The authors reported an upregulation of representative cell-cycle regulatory genes in G1/S, S, G2/M, and M phase transitions, despite that only a small number of genes related to motility were upregulated, and in fact, were not able to correlate dynamic movement with proliferating cardiomyocytes [98].

Following several studies demonstrating that cyclins and other cell-cycle regulators can induce myocardial regeneration and enhance cardiac function, the group of Shigetaka Kitajima proposed that manipulation of positive and negative cell-cycle regulators could be an option for ischemic heart failure treatment [99]. By modulating cell-cycle drivers (cyclin D1/CDK4) and a cell-cycle brake (Skp2), the authors were able to induce a steady proliferation of adult cardiomyocytes *in situ, which was verified by* immunohistochemically using antibodies against Ki67, phosphorylated histone H3 (H3P), and Aurora B [99].

Interestingly, zebrafish adult heart retains regeneration capacity by proliferation in a subset of existing cardiomyocytes [100]. Choi et al. have recently developed a FUCCI-based screen to identify modulators of cardiomyocyte proliferation in zebrafish, where Hedgehog, insulin-like growth factor, or transforming growth factor β signaling pathways were found to play an important role [101].

Undoubtedly, cardiovascular pathology arises due to an imbalance between pro- and antiproliferative molecules, and as such, a thorough understanding of the cell-cycle circuitries and inherent cyclin–CDK–CKI interactions will be essential for the rationale design of efficient therapies.

Metabolic disorders

Metabolic disorders associated to elevated body weight and obesity has reached epidemic proportions in industrialized countries, instigating research to identify novel biomarkers and potential molecular targets for efficient drug development. Metabolic disorders are a set of pathophysiological conditions displaying energy and redox imbalance due to disruption of normal metabolic processes. Several studies have provided evidence that this is closely associated with ROS-mediated oxidative stress signals that induce adipogenesis when terminally differentiated preadipocytes reenter the cell cycle and proliferate [102].

Another striking example of how cell-cycle assessment can be invaluable in the study of metabolic diseases is diabetes, which results from either β-cell loss or decreased β-cell growth. Indeed, the cell-cycle machinery has been implicated in the pathogenesis of diabetes through dysregulation of pancreatic β-cell proliferation, and it has been extensively proposed that modulation of the cell-cycle molecular circuitry could be a potential therapeutic route [103]. A key regulator of β-cell expansion is cyclin D2 shown to stimulate cell-cycle progression from G1 to S phase. Interestingly, it has been proposed that cyclin D2 expression in pancreatic cells is regulated by cannabinoids, and that in fact, activation of the cannabinoid 1 receptor (CB1R) leads to a cell-cycle arrest at G1 phase by decreasing cyclin D2 levels [103]. Recently, through BrdU analysis flow cytometry, Hobson et al. demonstrated that upregulation of aurora kinase A was sufficient to induce β-cell proliferation, highlighting the importance of the cell cycle in therapeutic interventions for diabetes [104].

Bacterial and viral infections

The importance of the cell cycle in pathogenic infections has been widely reported over the past decades.

Indeed, several bacterial pathogens have been shown to modulate the host cell cycle through the action of the effectors cyclomodulins to sustain infection [105]. For instance, pathogenic *Escherichia coli* block host cells at G2/M phase transition, whereas it was demonstrated that *Neisseria gonorrhoeae* inhibit cell proliferation by leading to a G1 arrest [106]. Interestingly, *Helicobacter pylori*, which is awarded a key role in the etiology of gastric cancer, has been reported to modulate the cell-cycle augmenting gastric epithelial cells proliferation [107]. In a recent study, Cabannes group carried out an extensive analysis of the host cell-cycle progression in response to infection by *Listeria monocytogenes* (Lm). Through flow cytometry and time-lapse imaging, the authors demonstrated that Lm infection leads to host DNA damage and activates checkpoint responses generating an advantageous dissemination niche [106]. In a similar study, using time-lapse-imaging of FUCCI-expressing cells, Morikawa and coworkers reported that the bacterial cycle inhibiting factor (CIF) interferes with the ubiquitination of host cell-cycle regulators, such as cyclin D1 and p27, causing a G1 cell-cycle arrest [108].

Likewise, regulation of the cell cycle on viral infections has been addressed for decades with the aim to identify potential antiviral therapeutic targets. Indeed, viruses target critical cell-cycle regulators and employ several strategies to modulate cell proliferation in order to replicate more efficiently [109]. The most virulent human pathogen causing influenza in humans, influenza A virus (IAV), was reported to induce a G0/G1 arrest and to affect the expression profile of key cell-cycle regulators. Through flow cytometry analysis of the cell cycle, he and colleagues demonstrated that the levels of hyperphosphorylated retinoblastoma protein (Rb) and cyclins E and D1 were decreased, whereas there was a significant increase of p21, consistent with the G0/G1 arrest profile. Furthermore, these alterations were associated with increased viral protein expression and viral progeny production, awarding biological relevance to the interference in the host cell cycle [110].

Hepatitis C virus (HCV) is a small enveloped virus that is one of the major causes of chronic hepatitis, cirrhosis, and hepatocellular carcinoma (HCC). HCV infection has been reported to both promote and inhibit the host cell-cycle progression, highlighting the importance of cell-cycle analysis. Several studies have demonstrated that HCV proteins can modulate cell-cycle regulatory genes affecting the G1/S checkpoint [109]. Sarfraz et al. reported that chronic HCV infection leads to an arrest in cell cycle associated to a reduced expression of S-phase cyclin and an increased expression of the G1 inhibitor p21 [110,111].

Similarly, the human herpesvirus 6 (HHV-6), an important immunosuppressive and immunomodulatory virus, is known to modulate the cell cycle of host cells. Specifically, through flow cytometry analysis, HHV-6A infection was shown to induce G2/M arrest in infected T cells [112].

Importantly, the cytotoxity associated with human immunodeficiency virus type 1 (HIV-1) infection has been demonstrated to be associated with effects on cell proliferation [113]. Indeed, over two decades ago, HIV vpr, a virion-associated protein, was shown to inhibit cell proliferation by causing a G2/M phase arrest in HIV-1 infected cells [53]. Many studies have since corroborated the importance of this arrest in both optimizing the conditions for viral replication and in hampering replication of the host immune cells [113].

Aside from being beneficial for the pathogen's viability, disruption of the host cell cycle ultimately contributes to several associated pathologies including cell transformation and cancer progression, thus highlighting the importance of assessing the cell-cycle circuitries on pathogenic infections.

Cancer research

The importance of cell-cycle assessment in cancer research is indubitable and immeasurable. Sustained proliferative signaling and evasion of growth suppressors are hallmarks of cancer intimately associated to defects in cell-cycle regulation [114]. Tumor samples often display cells with abnormal cell division and proliferation, and alterations in the expression and activity of cyclins and cyclin-dependent kinases are well documented in cancer, thus constituting biomarkers of proliferation and attractive pharmacological targets for development of anticancer therapeutics [115].

Importantly, two prototypical tumor suppressors encode the Rb (retinoblastoma) and p53 proteins, which are key regulators within two key complementary circuitries governing the decisions of cells to proliferate or perish by programmed cell death. The Rb protein (pRb), responsible for a major G1 checkpoint, blocking S-phase entry and cell growth, is universally disrupted in human cancer [116]. Inactivation of p53 plays a key role in a broad spectrum of human cancers and it is widely accepted that tumor formation can be inhibited by p53-mediated cell-cycle arrest, apoptosis, and senescence [117].

Tumor heterogeneity is a major issue in cancer therapy as it leads to chemosensitive and chemoresistant subpopulations within the tumor. Miwa et al. have addressed cell cycle and the fate of single cancer cells in a tumor using the FUCCI imaging system and found heterogeneous responses to UVB irradiation, including cell-cycle arrest, escape from the arrest, mitosis, and apoptosis in individual cells [118]. Similarly, in a previous study, the same group demonstrated that on treatment of cancer cells with chemotherapeutic drugs, the chemoresistant subpopulation of cells could be readily identified by an arrest in S/G2/M [118].

Cell-cycle analysis is also of utmost importance in studies concerning cancer cell invasion. Indeed, to target invading cancer cells one must determine the cell-cycle phase, given that most chemotherapeutic agents target S/G2/M cells [118]. In a recent study, Yano and coworkers demonstrated by real-time FUCCI imaging that cancer cells in G0/G1 can migrate further and faster than those at S/G2/M phase of the cell cycle. Furthermore, the authors observed that chemotherapy had little effect on this subpopulation of highly migratory cancer cells, which can explain why chemotherapy is so inefficient in preventing metastasis [118].

More recently, Sommariva et al. reported a meta-analysis assessing the value of the cell-cycle progression (CCP) score in prostate cancer treatment and found that the CCP score is an important tool to elucidate the aggressive potential of prostate cancer in an individual and its use improves prognosis [119].

Understanding cell-cycle progression and regulation in cancer will allow a better delineation of pharmacological solutions with the highest impact on cancer cells, without jeopardizing the growth of normal cells, and ultimately, minimizing the side effects of chemotherapy treatments [115].

Hematological disorders

Hematologic malignancies account for 9% of all cancers and represent a major cause of death in children. Hematopoiesis is initiated by relatively quiescent multipotent hematopoietic stem cells (HSC), which on cell-cycle entry will give rise to differentiating progenitor populations. Disturbance of cell-cycle regulation is known to play a crucial role in the development of hematological-derived cancers, where normal hematopoiesis is disrupted by uncontrolled growth [120]. Aurora kinases are important for mitosis and have been found to be overexpressed in a plethora of human tumors, namely hematologic malignancies [121]. Cell-cycle assessment has been instrumental in the search for Aurora kinase inhibitors as anticancer drugs [122,123]. Polo-like kinases (Plks) are a family of serine/threonine kinases that play important functions throughout the cell cycle also regarded as potential therapeutical targets in hematologic malignancies [124,125]. Renner and coworkers have highlighted Plk1 inhibition, both pharmacologically and by small interfering RNA, for the treatment of acute myeloid leukemia (AML) [126]. More recently, Münch et al. have shown that Plk1 inhibition in the bone marrow of AML patients results in mitotic arrest and subsequent apoptosis using live-cell imaging experiments [127].

Cell cycle and dysfunctional cell adhesion

Our group focuses in understanding the role of adhesion molecules in cancer. A specific interest is the impact of a pivotal adhesion molecule in epithelia (E-cadherin) and its mediated signaling underlying cancer. In hereditary diffuse gastric cancer E-cadherin, mutations are causative events and in contrast to many other cancer type these carcinomas to do not show a high proliferation rate [128]. However, little is known concerning how E-cadherin alterations lead to cell-cycle deregulation or how the dynamics of cell cycle impact on E-cadherin expression and function. We recently found that eukaryotic cells expressing mutant E-cadherin moieties comprising a juxtamembrane mutation show varying protein expression profiles along the cell-cycle phases in contrast to wild-type E-cadherin expressing cells (Ferro et al., unpublished data). The distinctive E-cadherin expression profile observed in E-cadherin mutant cells in G1/G0 versus G2 phase may be possibly related to the activation of ERAD as a compensatory cellular mechanism. Although preliminary, these results open a new possibility of identifying molecular cofactors that can interfere with the profile of expression and function of mutant proteins, namely dysfunctional E-cadherin in cancer. Therapeutically, their identification is crucial to target such factors and circumvent the expression of unfolded proteins through the exploitation of cell-cycle proteins acting in G1/G0 phase. Interestingly, the growth suppressor properties of normal E-cadherin, in many carcinoma cells, are due to an increase in the expression and activity levels of $p27^{Kip1}$ that lead to cell–cell contact inhibition and quiescence, with cell-cycle arrest at G1 phase [129]. In addition, $p21^{Cip1}$ and $p27^{Kip1}$ CDK inhibitors, which control G1/S phase transition, are often downregulated in gastric cancer (GC) [130].

A thorough evaluation of the cell-cycle dynamics in the context of E-cadherin dysfunction may unveil a mechanism underlying the prosurvival and prooncogenic characteristics of E-cadherin-deficient cells associated to cancer.

Limitations

Despite the tremendous possibilities awarded by microscopy and other cytometry methods applied to the study of cell-cycle dynamics, each has its own set of limitations. Allowing the preservation of the cell's natural architecture, microscopy is the gold standard technology to study unperturbed systems of *ex vivo*, *in vivo*, and processed samples. Despite the advantages of traditional fluorescence microscopy, the improvement of fluorescent dyes and probes, and the massive development of resolution-enhanced microscopes, the high throughputness of microscopy still falls behind the needs to address many biological questions [131,132]. As such, multiparametric analyses in thousands of cells, testing several cellular parameters simultaneously, are not easily manageable by microscopy. On the other hand, flow cytometry technology has underwent an amazing evolution throughout the past decades, boosting its high-throughput properties, and it is now possible to routinely perform polychromatic analyses of cell samples tagged and emitting eight or more colors. The multidimensional improvement/evolution witnessed on every aspect of cytometry, such as available reagents, light sources, detectors, and analysis software, eased-up the analysis of data retrieved from high-content experiments. Nowadays it is possible to quantitatively analyze millions of cells in few minutes, while focusing on each cell individually. However, the fundamental principle of classical flow cytometry and imaging cytometry, where one must use single cell-enriched samples to be analyzed in a hydrodynamic flow stream, does not provide information on unperturbed cellular systems. Samples have to be disintegrated, hence cellular responses and 3D information derived from the physiological tissue microenvironment is lost. Laser-scanning cytometry suppressed this limitation, while empowering cytometry with imaging, quantification, and the possibility to continuously reanalyze individual cells in tissues *in situ*. The development of acoustical focusing next-generation cytometers and the combination of mass spectrometry with cytometry (CyTOF) set the high-throughput and high-content analysis to unprecedented levels. Acoustic-based cytometry is particularly fascinating for rare event analysis, such as circulating tumor cells, minimal residual disease, and stem cell identification, given that it allows faster analyses of higher sample volumes. Mass cytometry allows next-generation multiparametric analysis with minimal background noise derived from signal overlap or endogenous cellular content. The possibility to use rare metal-tagged antibodies ensures a scalable multiparameterization of analysis thus enabling the study of functionally complex and heterogeneous biological systems, in a single celled manner. The major drawback of mass cytometry is its incompatibility with live-cell sorting. The need for more validated metal-tagged antibodies and the improvement of DNA-binding stoichiometry is yet to be tackled. These will be the next-generation cytometric technologies; however, their current pricing, both due to the machines it selves and required reagents, is a serious bottleneck issue for their availability to academic and clinical settings.

Future

In recent years, we have witnessed a strong trend toward the development of technologies devoted to the analysis of single cells. Cytometric methodologies, in all their dimensionality, are the most ranked nowadays, allowing the identification and characterization of individual cells, or of specific subcellular populations, derived from highly complex and heterogeneous samples. Undoubtedly, the analytical cellular systems available nowadays, despite the limitations associated with each method, have reached a discriminative and resolution power comparable to that of molecular analyses. The incessant improvement and development of highly sensitive and accurate methods proficient in the detection of minimal residual disease or allowing higher resolutions in the identification of rare disease cell subpopulations within a normal biological background will certainly be the rational approach in many pathologies. Multiplexed analyses, encompassing different analytical parameters, yielding high-throughput derived content, along with the development of bioimaging software enabling the concomitant integrative analysis of the data generated, will lead to in-depth knowledge of the disease and its progression, ultimately achieving patient-centered research and precision medicine.

Acknowledgments

This work was financed by FEDER—Fundo Europeu de Desenvolvimento Regional funds through the COMPETE 2020-Operacional Programme for Competitiveness and Internationalisation (POCI), Portugal 2020 and the Portuguese funding agency FCT-Fundação para a Ciência e a Tecnologia-Ministério da Ciência, Tecnologia e Inovação in the framework of the project "Institute for Research and Innovation in Health Sciences" (POCI-01-0145-FEDER-007274), PTDC/BIM-ONC/0171/2012, PTDC/BIMONC/0281/2014, PTDC/BBB-IMG/0283/2014, SFRH/BPD/97295/2013 (AF), NORTE-01-0145-FEDER-000029 (PC) and SFRH/BPD/63716/2009 (MSF). We also thank the American Association of Patients with Hereditary Gastric Cancer *No Stomach for Cancer* for funding the project "Today's present, tomorrow's future on the study of germline E-cadherin missense mutations."

References

1. Nurse P (2000) A long twentieth century of the cell cycle and beyond. *Cell* 100: 71–78.
2. Vermeulen K, Van Bockstaele DR, Berneman ZN (2003) The cell cycle: A review of regulation, deregulation and therapeutic targets in cancer. *Cell Prolif* 36: 131–149.
3. Roukos V, Pegoraro G, Voss TC, Misteli T (2015) Cell cycle staging of individual cells by fluorescence microscopy. *Nat Protoc* 10: 334–348.
4. Zhivotovsky B, Orrenius S (2010) Cell cycle and cell death in disease: Past, present and future. *J Intern Med* 268: 395–409.
5. Tessema M, Lehmann U, Kreipe H (2004) Cell cycle and no end. *Virchows Arch* 444: 313–323.
6. Perez de Castro I, de Carcer G, Malumbres M (2007) A census of mitotic cancer genes: New insights into tumor cell biology and cancer therapy. *Carcinogenesis* 28: 899–912.
7. Kurzawa L, Morris MC (2010) Cell-cycle markers and biosensors. *Chembiochem* 11: 1037–1047.
8. Henderson L, Bortone DS, Lim C, Zambon AC (2013) Classic "broken cell" techniques and newer live cell methods for cell cycle assessment. *Am J Physiol Cell Physiol* 304: C927–C938.
9. Santo L, Siu KT, Raje N (2015) Targeting cyclin-dependent kinases and cell cycle progression in human cancers. *Semin Oncol* 42: 788–800.

10. Atashpour S, Fouladdel S, Movahhed TK, Barzegar E, Ghahremani MH et al. (2015) Quercetin induces cell cycle arrest and apoptosis in CD133(+) cancer stem cells of human colorectal HT29 cancer cell line and enhances anticancer effects of doxorubicin. *Iran J Basic MedSci* 18: 635–643.
11. Malumbres M, Barbacid M (2009) Cell cycle, CDKs and cancer: A changing paradigm. *Nat Rev Cancer* 9: 153–166.
12. Dominguez-Brauer C, Thu KL, Mason JM, Blaser H, Bray MR et al. (2015) Targeting mitosis in cancer: Emerging strategies. *Mol Cell* 60: 524–536.
13. Darzynkiewicz Z (2011) Critical aspects in analysis of cellular DNA content. *Curr Protoc Cytom* Chapter 7: Unit 7.2.
14. Kapuscinski J (1995) DAPI: A DNA-specific fluorescent probe. *Biotech Histochem* 70: 220–233.
15. Tembhare P, Badrinath Y, Ghogale S, Patkar N, Dhole N, et al. (2016) A novel and easy FxCycle violet based flow cytometric method for simultaneous assessment of DNA ploidy and six-color immunophenotyping. *Cytometry A* 89(3): 281–291.
16. Wood CE, Hukkanen RR, Sura R, Jacobson-Kram D, Nolte T et al. (2015) Scientific and Regulatory Policy Committee (SRPC) review: Interpretation and use of cell proliferation data in cancer risk assessment. *Toxicol Pathol* 43: 760–775.
17. Martin RM, Leonhardt H, Cardoso MC (2005) DNA labeling in living cells. *Cytometry A* 67: 45–52.
18. Telford WG, Bradford J, Godfrey W, Robey RW, Bates SE (2007) Side population analysis using a violet-excited cell-permeable DNA binding dye. *Stem Cells* 25: 1029–1036.
19. Pastrana E, Cheng LC, Doetsch F (2009) Simultaneous prospective purification of adult subventricular zone neural stem cells and their progeny. *Proc Natl Acad Sci USA* 106: 6387–6392.
20. Zhao H, Traganos F, Dobrucki J, Wlodkowic D, Darzynkiewicz Z (2009) Induction of DNA damage response by the supravital probes of nucleic acids. *Cytometry A* 75: 510–519.
21. Shimomura O, Johnson FH, Saiga Y (1962) Extraction, purification and properties of aequorin, a bioluminescent protein from the luminous hydromedusan, Aequorea. *J Cell Comp Physiol* 59: 223–239.
22. Tsien RY (1998) The green fluorescent protein. *Annu Rev Biochem* 67: 509–544.
23. Easwaran HP, Leonhardt H, Cardoso MC (2005) Cell cycle markers for live cell analyses. *Cell Cycle* 4: 453–455.
24. Schonenberger F, Deutzmann A, Ferrando-May E, Merhof D (2015) Discrimination of cell cycle phases in PCNA-immunolabeled cells. *BMC Bioinformatics* 16: 180.
25. Hahn AT, Jones JT, Meyer T (2009) Quantitative analysis of cell cycle phase durations and PC12 differentiation using fluorescent biosensors. *Cell Cycle* 8: 1044–1052.
26. Gu J, Xia X, Yan P, Liu H, Podust VN et al. (2004) Cell cycle-dependent regulation of a human DNA helicase that localizes in DNA damage foci. *Mol Biol Cell* 15: 3320–3332.
27. Jones JT, Myers JW, Ferrell JE, Meyer T (2004) Probing the precision of the mitotic clock with a live-cell fluorescent biosensor. *Nat Biotechnol* 22: 306–312.
28. Sakaue-Sawano A, Kurokawa H, Morimura T, Hanyu A, Hama H et al. (2008) Visualizing spatiotemporal dynamics of multicellular cell-cycle progression. *Cell* 132: 487–498.
29. Mort RL, Ford MJ, Sakaue-Sawano A, Lindstrom NO, Casadio A et al. (2014) Fucci2a: A bicistronic cell cycle reporter that allows cre mediated tissue specific expression in mice. *Cell Cycle* 13: 2681–2696.

30. Sugiyama M, Sakaue-Sawano A, Iimura T, Fukami K, Kitaguchi T et al. (2009) Illuminating cell-cycle progression in the developing zebrafish embryo. *Proc Natl Acad Sci USA* 106: 20812–20817.
31. Fukuhara S, Zhang J, Yuge S, Ando K, Wakayama Y et al. (2014) Visualizing the cell-cycle progression of endothelial cells in zebrafish. *Dev Biol* 393: 10–23.
32. Bouldin CM, Snelson CD, Farr GH, 3rd, Kimelman D (2014) Restricted expression of cdc25a in the tailbud is essential for formation of the zebrafish posterior body. *Genes Dev* 28: 384–395.
33. Ogura Y, Sakaue-Sawano A, Nakagawa M, Satoh N, Miyawaki A et al. (2011) Coordination of mitosis and morphogenesis: Role of a prolonged G2 phase during chordate neurulation. *Development* 138: 577–587.
34. Zielke N, Korzelius J, van Straaten M, Bender K, Schuhknecht GF et al. (2014) Fly-FUCCI: A versatile tool for studying cell proliferation in complex tissues. *Cell Rep* 7: 588–598.
35. Zielke N, Edgar BA (2015) FUCCI sensors: Powerful new tools for analysis of cell proliferation. *Wiley Interdiscip Rev Dev Biol* 4: 469–487.
36. Saitou T, Imamura T (2016) Quantitative imaging with Fucci and mathematics to uncover temporal dynamics of cell cycle progression. *Dev Growth Differ* 58: 6–15.
37. Bertero A, Vallier L (2015) Fucci2a mouse upgrades live cell cycle imaging. *Cell Cycle* 14: 948–949.
38. Blasi T, Hennig H, Summers HD, Theis FJ, Cerveira J et al. (2016) Label-free cell cycle analysis for high-throughput imaging flow cytometry. *Nat Commun* 7: 10256.
39. Enders GH (2010) *Cell Cycle Deregulation in Cancer*. Springer-Verlag, New York.
40. Bendall SC, Nolan GP, Roederer M, Chattopadhyay PK (2012) A deep profiler's guide to cytometry. *Trends Immunol* 33: 323–332.
41. Moldavan A (1934) Photo-electric technique for the counting of microscopical cells. *Science* 80: 188–189.
42. Coulter WH (1953) Means for counting particles suspended in a fluid. Patent #2,656,508 Application.
43. Crosland-Taylor PJ (1953) A device for counting small particles suspended in a fluid through a tube. *Nature* 171: 37–38.
44. Kamentsky LA, Melamed MR, Derman H (1965) Spectrophotometer: New instrument for ultrarapid cell analysis. *Science* 150: 630–631.
45. Fulwyler MJ (1965) Electronic separation of biological cells by volume. *Science* 150: 910–911.
46. Van Dilla MA, Fulwyler MJ, Boone IU (1967) Volume distribution and separation of normal human leucocytes. *Proc Soc Exp Biol Med* 125: 367–370.
47. Bonner WA, Hulett HR, Sweet RG, Herzenberg LA (1972) Fluorescence activated cell sorting. *Rev Sci Instrum* 43: 404–409.
48. Dean PN, Jett JH (1974) Mathematical analysis of DNA distributions derived from flow microfluorometry. *J Cell Biol* 60: 523–527.
49. Crissman HA, Tobey RA (1974) Cell-cycle analysis in 20 minutes. *Science* 184: 1297–1298.
50. Hulett HR, Bonner WA, Barrett J, Herzenberg LA (1969) Cell sorting: Automated separation of mammalian cells as a function of intracellular fluorescence. *Science* 166: 747–749.
51. Pearson T, Galfre G, Ziegler A, Milstein C (1977) A myeloma hybrid producing antibody specific for an allotypic determinant on "IgD-like" molecules of the mouse. *Eur J Immunol* 7: 684–690.
52. Hardy RR, Hayakawa K, Parks DR, Herzenberg LA, Herzenberg LA (1984) Murine B cell differentiation lineages. *J Exp Med* 159: 1169–1188.

53. Elson LH, Nutman TB, Metcalfe DD, Prussin C (1995) Flow cytometric analysis for cytokine production identifies T helper 1, T helper 2, and T helper 0 cells within the human CD4+CD27-lymphocyte subpopulation. *J Immunol* 154: 4294–4301.
54. Berlier JE, Rothe A, Buller G, Bradford J, Gray DR et al. (2003) Quantitative comparison of long-wavelength Alexa Fluor dyes to Cy dyes: Fluorescence of the dyes and their bioconjugates. *J Histochem Cytochem* 51: 1699–1712.
55. Ekimov AI, Onushchenko AA (1981) Quantum size effect in three-dimensional microscopic semiconductor crystals. *ZhETF Pis ma Redaktsiiu* 34: 363.
56. Brus LE (1984) Electron–electron and electron-hole interactions in small semiconductor crystallites: The size dependence of the lowest excited electronic state. *J Chem Phys* 80: 4403.
57. Michalet X, Pinaud FF, Bentolila LA, Tsay JM, Doose S et al. (2005) Quantum dots for live cells, in vivo imaging, and diagnostics. *Science* 307: 538–544.
58. Roederer M, Bigos M, Nozaki T, Stovel RT, Parks DR et al. (1995) Heterogeneous calcium flux in peripheral T cell subsets revealed by five-color flow cytometry using log-ratio circuitry. *Cytometry* 21: 187–196.
59. Perfetto SP, Chattopadhyay PK, Roederer M (2004) Seventeen-colour flow cytometry: Unravelling the immune system. *Nat Rev Immunol* 4: 648–655.
60. Chattopadhyay PK, Gaylord B, Palmer A, Jiang N, Raven MA et al. (2012) Brilliant violet fluorophores: A new class of ultrabright fluorescent compounds for immunofluorescence experiments. *Cytometry A* 81: 456–466.
61. De Rosa SC, Herzenberg LA, Herzenberg LA, Roederer M (2001) 11-color, 13-parameter flow cytometry: Identification of human naive T cells by phenotype, function, and T-cell receptor diversity. *Nat Med* 7: 245–248.
62. Chattopadhyay PK, Price DA, Harper TF, Betts MR, Yu J et al. (2006) Quantum dot semiconductor nanocrystals for immunophenotyping by polychromatic flow cytometry. *Nat Med* 12: 972–977.
63. Chattopadhyay PK, Roederer M (2015) A mine is a terrible thing to waste: High content, single cell technologies for comprehensive immune analysis. *Am J Transplant* 15: 1155–1161.
64. Soh KT, Tario JD, Jr., Colligan S, Maguire O, Pan D et al. (2016) Simultaneous, single-cell measurement of messenger RNA, cell surface proteins, and intracellular proteins. *Curr Protoc Cytom* 75: 7.45.41–47.45.33.
65. Mair F, Hartmann FJ, Mrdjen D, Tosevski V, Krieg C et al. (2016) The end of gating? An introduction to automated analysis of high dimensional cytometry data. *Eur J Immunol* 46: 34–43.
66. Darzynkiewicz Z, Bedner E, Li X, Gorczyca W, Melamed MR (1999) Laser-scanning cytometry: A new instrumentation with many applications. *Exp Cell Res* 249: 1–12.
67. Grabarek J, Darzynkiewicz Z (2002) Versatility of analytical capabilities of laser scanning cytometry (LSC). *Clin Appl Immunol Rev* 2: 75–92.
68. Zuba-Surma EK, Kucia M, Abdel-Latif A, Lillard JW, Jr., Ratajczak MZ (2007) The ImageStream System: A key step to a new era in imaging. *Folia Histochem Cytobiol* 45: 279–290.
69. Thaunat O, Granja AG, Barral P, Filby A, Montaner B et al. (2012) Asymmetric segregation of polarized antigen on B cell division shapes presentation capacity. *Science* 335: 475–479.
70. Reyes EE, Kunovac SK, Duggan R, Kregel S, Vander Griend DJ (2013) Growth kinetics of CD133-positive prostate cancer cells. *Prostate* 73: 724–733.
71. Dent BM, Ogle LF, O'Donnell RL, Hayes N, Malik U et al. (2016) High-resolution imaging for the detection and characterisation of circulating tumour cells from patients with oesophageal, hepatocellular, thyroid and ovarian cancers. *Int J Cancer* 138: 206–216.

72. Ryszawy D, Sarna M, Rak M, Szpak K, Kedracka-Krok S et al. (2014) Functional links between Snail-1 and Cx43 account for the recruitment of Cx43-positive cells into the invasive front of prostate cancer. *Carcinogenesis* 35: 1920–1930.
73. Piyasena ME, Austin Suthanthiraraj PP, Applegate RW, Jr., Goumas AM, Woods TA et al. (2012) Multinode acoustic focusing for parallel flow cytometry. *Anal Chem* 84: 1831–1839.
74. Jakobsson O, Antfolk M, Laurell T (2014) Continuous flow two-dimensional acoustic orientation of nonspherical cells. *Anal Chem* 86: 6111–6114.
75. Pasternak MM, Strohm EM, Berndl ES, Kolios MC (2015) Properties of cells through life and death—An acoustic microscopy investigation. *Cell Cycle* 14: 2891–2898.
76. Ornatsky O, Bandura D, Baranov V, Nitz M, Winnik MA et al. (2010) Highly multiparametric analysis by mass cytometry. *J Immunol Methods* 361: 1–20.
77. Behbehani GK, Thom C, Zunder ER, Finck R, Gaudilliere B et al. (2014) Transient partial permeabilization with saponin enables cellular barcoding prior to surface marker staining. *Cytometry A* 85: 1011–1019.
78. Shamir L, Delaney JD, Orlov N, Eckley DM, Goldberg IG (2010) Pattern recognition software and techniques for biological image analysis. *PLoS Comput Biol* 6: e1000974.
79. Ridenour DA, McKinney MC, Bailey CM, Kulesa PM (2012) CycleTrak: A novel system for the semi-automated analysis of cell cycle dynamics. *Dev Biol* 365: 189–195.
80. Ferro A, Sahumbaiev I, Mestre T, Sanches JM, Seruca R (Unpublished data) Cell cycle profiling of heterogeneous cell populations based on fluorescence imaging.
81. Padfield D, Rittscher J, Thomas N, Roysam B (2009) Spatio-temporal cell cycle phase analysis using level sets and fast marching methods. *Med Image Anal* 13: 143–155.
82. Yano S, Zhang Y, Miwa S, Tome Y, Hiroshima Y et al. (2014) Spatial-temporal FUCCI imaging of each cell in a tumor demonstrates locational dependence of cell cycle dynamics and chemoresponsiveness. *Cell Cycle* 13: 2110–2119.
83. El-Badawy A, El-Badri N (2016) The cell cycle as a brake for beta-cell regeneration from embryonic stem cells. *Stem Cell Res Ther* 7: 9.
84. Boward B, Wu T, Dalton S (2016) Control of cell fate through cell cycle and pluripotency networks. *Stem Cells* 34(6): 1427–1436.
85. Singh AM, Trost R, Boward B, Dalton S (2015) Utilizing FUCCI reporters to understand pluripotent stem cell biology. *Methods* 101: 4–10.
86. Pauklin S, Vallier L (2013) The cell-cycle state of stem cells determines cell fate propensity. *Cell* 155: 135–147.
87. Singh AM, Chappell J, Trost R, Lin L, Wang T et al. (2013) Cell-cycle control of developmentally regulated transcription factors accounts for heterogeneity in human pluripotent cells. *Stem Cell Reports* 1: 532–544.
88. Roccio M, Hahnewald S, Perny M, Senn P (2015) Cell cycle reactivation of cochlear progenitor cells in neonatal FUCCI mice by a GSK3 small molecule inhibitor. *Sci Rep* 5: 17886.
89. Korennykh AV, Egea PF, Korostelev AA, Finer-Moore J, Stroud RM et al. (2011) Cofactor-mediated conformational control in the bifunctional kinase/RNase Ire1. *BMC Biol* 9: 48.
90. Bandhavkar S (2016) Cancer stem cells: A metastasizing menace! *Cancer Med* 5(4):649–655.
91. Doherty MR, Smigiel JM, Junk DJ, Jackson MW (2016) Cancer stem cell plasticity drives therapeutic resistance. *Cancers (Basel)* 8(1): 8.
92. Gong C, Liao H, Wang J, Lin Y, Qi J et al. (2012) LY294002 induces G0/G1 cell cycle arrest and apoptosis of cancer stem-like cells from human osteosarcoma via down-regulation of PI3K activity. *Asian Pac J Cancer Prev* 13: 3103–3107.

93. Filippi-Chiela EC, Villodre ES, Zamin LL, Lenz G (2011) Autophagy interplay with apoptosis and cell cycle regulation in the growth inhibiting effect of resveratrol in glioma cells. *PLoS One* 6: e20849.
94. Folch J, Junyent F, Verdaguer E, Auladell C, Pizarro JG et al. (2012) Role of cell cycle re-entry in neurons: A common apoptotic mechanism of neuronal cell death. *Neurotox Res* 22: 195–207.
95. Herrup K, Neve R, Ackerman SL, Copani A (2004) Divide and die: Cell cycle events as triggers of nerve cell death. *J Neurosci* 24: 9232–9239.
96. Yang Y, Geldmacher DS, Herrup K (2001) DNA replication precedes neuronal cell death in Alzheimer's disease. *J Neurosci* 21: 2661–2668.
97. Hashimoto H, Yuasa S, Tabata H, Tohyama S, Hayashiji N et al. (2014) Time-lapse imaging of cell cycle dynamics during development in living cardiomyocyte. *J Mol Cell Cardiol* 72: 241–249.
98. Hashimoto H, Yuasa S, Tabata H, Tohyama S, Seki T et al. (2015) Analysis of cardiomyocyte movement in the developing murine heart. *Biochem Biophys Res Commun* 464: 1000–1007.
99. Tamamori-Adachi M, Takagi H, Hashimoto K, Goto K, Hidaka T et al. (2008) Cardiomyocyte proliferation and protection against post-myocardial infarction heart failure by cyclin D1 and Skp2 ubiquitin ligase. *Cardiovasc Res* 80: 181–190.
100. Poss KD, Wilson LG, Keating MT (2002) Hear regeneration in zebrafish. *Science* 298: 2188–2190.
101. Choi WY, Gemberling M, Wang J, Holdway JE, Shen MC et al. (2013) In vivo monitoring of cardiomyocyte proliferation to identify chemical modifiers of heart regeneration. *Development* 140: 660–666.
102. Rani V, Deep G, Singh RK, Palle K, Yadav UC (2016) Oxidative stress and metabolic disorders: Pathogenesis and therapeutic strategies. *Life Sci* 148: 183–193.
103. Kim J, Lee KJ, Kim JS, Rho JG, Shin JJ et al. (2016) Cannabinoids regulate Bcl-2 and Cyclin D2 expression in pancreatic beta cells. *PLoS One* 11: e0150981.
104. Hobson A, Draney C, Stratford A, Becker TC, Lu D et al. (2015) Aurora Kinase A is critical for the Nkx6.1 mediated beta-cell proliferation pathway. *Islets* 7: e1027854.
105. Nougayrede JP, Taieb F, De Rycke J, Oswald E (2005) Cyclomodulins: Bacterial effectors that modulate the eukaryotic cell cycle. *Trends Microbiol* 13: 103–110.
106. Leitao E, Costa AC, Brito C, Costa L, Pombinho R et al. (2014) Listeria monocytogenes induces host DNA damage and delays the host cell cycle to promote infection. *Cell Cycle* 13: 928–940.
107. Peek RM, Jr., Blaser MJ, Mays DJ, Forsyth MH, Cover TL et al. (1999) Helicobacter pylori strain-specific genotypes and modulation of the gastric epithelial cell cycle. *Cancer Res* 59: 6124–6131.
108. Morikawa H, Kim M, Mimuro H, Punginelli C, Koyama T et al. (2010) The bacterial effector Cif interferes with SCF ubiquitin ligase function by inhibiting deneddylation of Cullin1. *Biochem Biophys Res Commun* 401: 268–274.
109. Bagga S, Bouchard MJ (2014) Cell cycle regulation during viral infection. *Methods Mol Biol* 1170: 165–227.
110. He Y, Xu K, Keiner B, Zhou J, Czudai V, et al. (2010) Influenza A virus replication induces cell cycle arrest in G0/G1 phase. J Virol 84: 12832–12840.
111. Sarfraz S, Hamid S, Siddiqui A, Hussain S, Pervez S et al. (2008) Altered expression of cell cycle and apoptotic proteins in chronic hepatitis C virus infection. *BMC Microbiol* 8: 133.
112. Behrman S, Acosta-Alvear D, Walter P (2011) A CHOP-regulated microRNA controls rhodopsin expression. *J Cell Biol* 192: 919–927.

113. Zhao RY, Elder RT (2005) Viral infections and cell cycle G2/M regulation. *Cell Res* 15: 143–149.
114. Hanahan D, Weinberg RA (2011) Hallmarks of cancer: The next generation. *Cell* 144: 646–674.
115. Peyressatre M, Prevel C, Pellerano M, Morris MC (2015) Targeting cyclin-dependent kinases in human cancers: From small molecules to Peptide inhibitors. *Cancers (Basel)* 7: 179–237.
116. Burkhart DL, Sage J (2008) Cellular mechanisms of tumour suppression by the retinoblastoma gene. *Nat Rev Cancer* 8: 671–682.
117. Vogelaar IP, Figueiredo J, van Rooij IA, Simoes-Correia J, van der Post RS et al. (2013) Identification of germline mutations in the cancer predisposing gene CDH1 in patients with orofacial clefts. *Hum Mol Genet* 22: 919–926.
118. Miwa S, Yano S, Kimura H, Yamamoto M, Toneri M et al. (2015) Cell-cycle fate-monitoring distinguishes individual chemosensitive and chemoresistant cancer cells in drug-treated heterogeneous populations demonstrated by real-time FUCCI imaging. *Cell Cycle* 14: 621–629.
119. Sommariva S, Tarricone R, Lazzeri M, Ricciardi W, Montorsi F (2016) Prognostic value of the cell cycle progression score in patients with prostate cancer: A systematic review and meta-analysis. *Eur Urol* 69: 107–115.
120. Aleem E, Arceci RJ (2015) Targeting cell cycle regulators in hematologic malignancies. *Front Cell Dev Biol* 3: 16.
121. Bavetsias V, Linardopoulos S (2015) Aurora kinase inhibitors: Current status and outlook. *Front Oncol* 5: 278.
122. Cullinane C, Waldeck KL, Binns D, Bogatyreva E, Bradley DP et al. (2014) Preclinical FLT-PET and FDG-PET imaging of tumor response to the multi-targeted Aurora B kinase inhibitor, TAK-901. *Nucl Med Biol* 41: 148–154.
123. Goldenson B, Crispino JD (2015) The aurora kinases in cell cycle and leukemia. *Oncogene* 34: 537–545.
124. Palmisiano ND, Kasner MT (2015) Polo-like kinase and its inhibitors: Ready for the match to start? *Am J Hematol* 90: 1071–1076.
125. Talati C, Griffiths EA, Wetzler M, Wang ES (2016) Polo-like kinase inhibitors in hematologic malignancies. *Crit Rev Oncol Hematol* 98: 200–210.
126. Renner AG, Dos Santos C, Recher C, Bailly C, Creancier L et al. (2009) Polo-like kinase 1 is overexpressed in acute myeloid leukemia and its inhibition preferentially targets the proliferation of leukemic cells. *Blood* 114: 659–662.
127. Munch C, Dragoi D, Frey AV, Thurig K, Lubbert M et al. (2015) Therapeutic polo-like kinase 1 inhibition results in mitotic arrest and subsequent cell death of blasts in the bone marrow of AML patients and has similar effects in non-neoplastic cell lines. *Leuk Res* 39: 462–470.
128. Pinheiro H, Oliveira C, Seruca R, Carneiro F (2014) Hereditary diffuse gastric cancer—Pathophysiology and clinical management. *Best Pract Res Clin Gastroenterol* 28: 1055–1068.
129. St Croix B, Sheehan C, Rak JW, Florenes VA, Slingerland JM et al. (1998) E-Cadherin-dependent growth suppression is mediated by the cyclin-dependent kinase inhibitor p27(KIP1). *J Cell Biol* 142: 557–571.
130. Gamboa-Dominguez A, Seidl S, Reyes-Gutierrez E, Hermannstadter C, Quintanilla-Martinez L et al. (2007) Prognostic significance of p21WAF1/CIP1, p27Kip1, p53 and E-cadherin expression in gastric cancer. *J Clin Pathol* 60: 756–761.
131. Toomre D, Bewersdorf J (2010) A new wave of cellular imaging. *Annu Rev Cell Dev Biol* 26: 285–314.
132. van de Linde S, Heilemann M, Sauer M (2012) Live-cell super-resolution imaging with synthetic fluorophores. *Annu Rev Phys Chem* 63: 519–540.

Chapter 13 Methods for the visualization of circadian rhythms: From molecules to organisms

Rukeia El-Athman,
Jeannine Mazuch,
Luise Fuhr, Mónica Abreu,
Nikolai Genov, and
Angela Relógio

Contents

Introduction

A variety of physiological and behavioral processes undergo time-dependent daily oscillations controlled by an endogenous timing system known as the circadian clock. The term *circadian* rhythm, derived from the Latin *circa diem* (about a day), refers to the internal biological rhythms of about 24 hours. External time cues, so-called zeitgebers, for example, light, synchronize the entire organism to the geophysical time and allow the anticipation and adaption to external daytime-dependent rhythms [1]. Circadian rhythms are evolutionarily conserved and have been reported for both prokaryotic and eukaryotic organisms. Their complexity ranges from simple clocks encoded by three different genes in unicellular cyanobacteria to intricate systems of multiple oscillators in mammals [2].

The mammalian circadian system is hierarchically organized into three major components: input signaling pathways, a central pacemaker in the suprachiasmatic nucleus (SCN) of the ventral hypothalamus, and numerous output signaling pathways to peripheral oscillators resulting in the rhythmic control of physiology and behavior. The phase of the pacemaker can be directly entrained to the solar day–night cycle via light signals perceived by photosensitive retinal ganglion cells. These are processed via neuronal and humoral signals and subsequently propagated to the oscillators in peripheral organs, which are present in virtually all cells [1]. Cellular oscillators are driven by regulatory transcriptional and translational networks whose molecular constituents interact via positive and negative feedback loops [3], as depicted in the lower panel of Figure 13.1. Together they generate robust endogenous rhythms with a period of approximately 24 hours, detectable in ~10% of all mammalian genes [2]. These so-called clock-controlled genes (CCGs) are involved in the regulation of a variety of cellular and biological processes, including metabolism, bone formation, the timing of the cell-division cycle, sleep–awake cycles, memory consolidation, and immunity [4]. Deregulations of the circadian clock have been reported to be associated with several pathological phenotypes ranging from sleep disorders to susceptibility to cancer [5]. Its biological and medical relevance motivated major developments in the circadian field during the past decades and led to the characterization of the function and dynamics of various circadian clock components.

This chapter provides an overview of established experimental methodologies and computational models in this field. Starting at the molecular level, experimental procedures used for the *in vitro* quantification of mRNAs and proteins are introduced together with the corresponding analysis methods and applications. An outline of *in vivo* methodologies for studying the circadian clock in various model organisms is provided. Finally, we discuss some of the emerging technologies and recent developments for the quantification of clock-associated components, which may lead to new insights into this remarkable time-generating system.

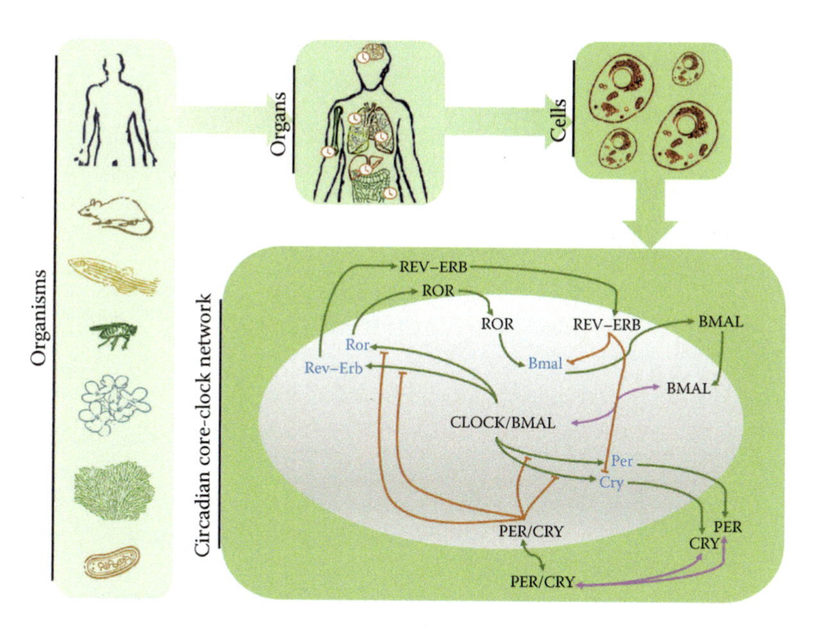

Figure 13.1 Circadian rhythms and the core-clock network in mammals. Circadian rhythms are present in many different organisms. In mammals, a main pacemaker is located in the SCN of the hypothalamus (formed by two clusters of coupled neurons) whose oscillations are propagated to peripheral clocks in all organs. At the cellular level, elements of the core clock form an intricate network. The heterodimer complex CLOCK/BMAL regulates the transcriptional activation of the target gene families *Per, Cry, Ror,* and *Rev–Erb*. The translated proteins of the PER and CRY families translocate into the nucleus where they inhibit the CLOCK/BMAL-mediated transcription (PC loop). The newly translated proteins of the REV–ERB and ROR families fine-tune *Bmal* expression by acting as inhibitors and activators, respectively (RBR loop). (From Fuhr, L. et al., *Comput. Struct. Biotechnol. J.*, 13, 417–426, 2015.)

Experimental procedures and applications

In vitro methods

Visualization and quantification of gene expression The ability to quantify oscillations in gene expression at the cellular level is one of the most powerful tools for the investigation of circadian rhythms. Different methods can be used to measure the expression of a gene as illustrated in Figure 13.2. These include real-time bioluminescence, real-time quantitative reverse transcription polymerase chain reaction (RT-qPCR), northern blots, RNase protection assays, RNA *in situ* hybridization, RNA expression microarrays (REMs), and RNA sequencing (RNAseq). The combination of bioluminescence data with high-throughput genomic approaches offers complementary information on circadian function *in vitro* and allows for further investigation using *in vivo* models.

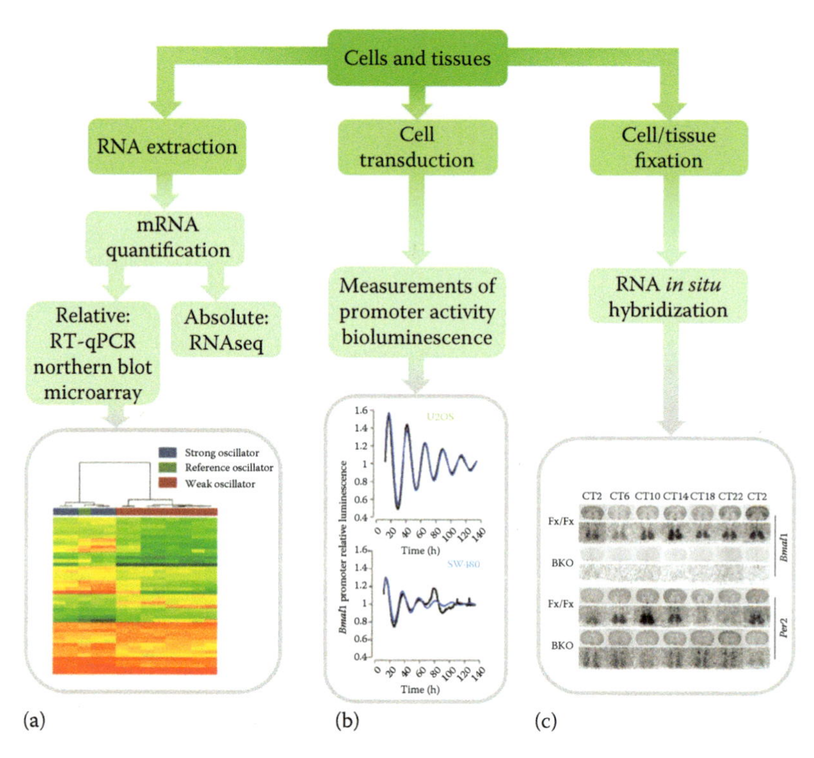

Figure 13.2 Experimental workflows for the visualization and quantification of circadian rhythms at the RNA level. Each of the branches of the diagram contains a corresponding example of visualization of the respective type of data: (a) A heatmap generated from mRNA microarray data for different cell lines was used to cluster the different experimental conditions (columns) as well as the expression of genes (rows). The color code denotes the intensity of gene expression (From Relógio, A. et al., *PLoS Genet.*, 10, e1004338, 2014.). (b) depicted is the circadian bioluminescence data for the activity of the Bmal1:luc promoter analyzed during several days in two cell lines, previously synchronized (From Relógio, A. et al., *PLoS Genet.*, 10, e1004338, 2014.). and (c) shown are *Bmal1* and *Per2* levels in coronal brain section and SCN of Fx/Fx and BKO mice from an *in situ* hybridization. (From Izumo, M. et al., *Elife*, 3, e04617, 2015.)

Real-time bioluminescence Real-time bioluminescence recording approaches based on luciferase reporter systems allow for the evaluation and characterization of sustained molecular rhythms [7–10]. One of the remarkable properties of this technique is the possibility to record the promoter activity of the gene of interest continuously during several days.

The lentivirus-mediated transduction permits a stable integration of the reporter system into the host genome [11]. The reporter system usually consists of an expression cassette containing the luciferase gene under the control of a clock-gene promoter of interest. A subsequent enzymatic reaction between the expressed luciferase and exogenously added substrate (luciferin) leads to the release of photons. The amount of photons correlates with the activity of the clock gene promoter and can be detected in real time with photomultipliers. Several devices are commercially available to record bioluminescence data. The LumiCycle photomultiplier system (Actimetrics) is a commonly used tool that allows for the analysis of 32 or 96 samples in parallel. Circadian parameters such as period, phase, amplitude, and damping rate can be extracted using the software provided by the manufacturer [7].

Real-time bioluminescence recording approaches are useful to evaluate the circadian phenotype on genetic perturbations (e.g., overexpression or RNA-mediated knockdown by shRNA and siRNA technologies) and knockout by genome editing using TALEN and CRISPR technologies. The effect of pharmacological compounds on the circadian clock function is also commonly tested using this method, providing new targets for *in vivo* validation and exploration [12–14].

Relative gene expression quantification by RT-qPCR This technique provides information about the relative expression level of a gene at a specific time point. To investigate whether core-clock genes or other genes of interest are rhythmic at the mRNA level, cells need to be harvested in regular time intervals to quantify their relative expression over one or more circadian cycles. RT-qPCR is frequently used to validate high-throughput data resultant from transcriptome-wide studies such as microarrays or RNA seq. RT-qPCR data can be analyzed with R, a freely available software environment for statistical computing. A common method to analyze RT-qPCR data is the $2^{-\Delta\Delta C_t}$ method that can be conducted with the R package *ddCt* [15].

Northern blot This technique is used to study gene expression levels by evaluating the amount of RNA (or mRNA) present in a sample. The technique involves RNA extraction and size-separation of RNA molecules by electrophoresis, transfer of the RNA to a membrane, and detection of the specific RNAs by hybridization with a labeled complementary probe [16]. Northern blot (NB) was used by Zhao and colleagues to observe circadian oscillations of the mRNA of the mammalian circadian clock gene *Cipc* in multiple mouse tissues [17].

In comparison with RT-qPCR, this technique has the advantage of a higher specificity that comes, however, at the cost of a lower sensitivity.

RNase protection assays These assays are used to measure RNA levels based on the selective digestion of single-stranded RNA by RNase, whereas the double-stranded RNAs are protected. When comparing this technique to NB and RT-qPCR, two main advantages can be pointed out: (1) An RNase protection assay (RPA) is more sensitive, allowing for the detection of rare or lowly expressed RNAs and (2) it gives information concerning possible mutations. The main steps in RPA include the preparation and purification of the probe (usually antisense riboprobes), probe hybridization with the RNA from the sample, and digestion with RNase and polyacrylamide gel electrophoresis (PAGE) to visualize the results [18]. RPA was used in immortalized rat fibroblasts cultured for more than 25 years to show that its clock machinery was still producing oscillations [19].

RNA in situ hybridization This methodology allows for the spatiotemporal detection of several transcripts in small tissues, tissue slices, and cells. It can be performed using fluorescent or radioactive labeling systems [20]. Using *RNA* in situ *hybridization* (RNA ISH), Watanabe and colleagues evaluated the expression of *Per2* in the SCN of teleost fish and found it to be rhythmic [21].

A variation of ISH is the fluorescence *in situ* hybridization (FISH) method. By using complementary probes loaded with fluorescent elements that bind to specific nucleotide sequences, the presence or absence of such nucleotide segments can be visualized and measured. This can be applied to both DNA and RNA to evaluate the expression of a particular gene under predefined conditions or for the detection of mutations or deletions in chromosomes. Since the fluorescence light emitted is rather weak, the application of very sensitive digital cameras is required. Usually, this includes a subsequent manual data analysis; however, automated approaches are also being developed [22,23]. An R package that can be used in the context of FISH data analysis is *FISHalyserR* [24].

Microarray technology This technology is a very powerful tool for the large-scale analysis of gene expression. The most commonly used type of arrays are provided by the company Affymetrix [25] and allow for genome-wide screening in a single sample. Microarrays are glass-coated chips with a matrix type distribution of oligonucleotides of known sequences, named probes. The hybridization of the target sample (fragmented and labeled) to its complementary probe allows for a relative quantification of gene expression via measurement of optical intensity of the hybridized sample fragments. A pipeline of open-source tools provides the means for the data analysis, mainly based on R [26]. Graphical user interface (GUI) tools such as the RStudio IDE allow for better development of individual scripts [27]. The Bioconductor/Biobase [28] software repository provides an increasing number of packages that automate many steps related to computational biology and bioinformatics. For exon arrays, the R package *oligo* provides all necessary tools for reading and normalizing the raw data [29]. The most common normalization technique is the robust multiarray average (RMA) [30] that is also part of the *oligo* package. The quality control of the arrays can be performed with the R package *arrayQualityMetrics* [31]. Subsequent analysis can be done with the R package *limma* [32] as it provides the tools for the creation of linear models that are fitted to the data of the arrays given an appropriate design matrix. The design matrix represents a numerical vector or a matrix which delivers information on the experimental design. The genes that show the biggest difference between the elements specified in the matrix can be determined and further investigated.

Microarrays are also an appropriate tool for time-course experiments. To observe gene expression over a certain time period, samples are taken at distinct time points and arrays are prepared for each individual time point. The resulting information can be used to determine the circadian characteristics of the genes, for example, using the R packages *rain* [33] or *MetaCycle* [34].

A current standard in the visualization of microarray data are heatmaps, which allow for the clustering of genes according to their expression level. The utilization of such techniques is of major importance to understand the circadian network at the transcriptional level and time-series microarrays are frequently used to find oscillating genes on a genomic scale [35]. Kamphuis and colleagues focused solely on this technique to evaluate the circadian gene expression of clock-controlled genes in the rat retina, concluding that despite the expression of all clock genes in this tissue, only a few of them showed a circadian expression [36].

RNA Sequencing RNAseq is a more recent approach for genome-wide gene expression studies that attempts to overcome some of the weaknesses of the microarray technology and which profits from fast development and substantial price decrease. The major advantage of RNAseq is the unbiased approach toward transcript detection [37]. While microarrays are limited by the existing probe libraries, RNAseq is able to detect previously unknown elements and offers a broader range of signal which means that the detection and quantification of elements with very low or very high expression is improved [38].

Visualization and quantification of proteins Different methods are used to measure and visualize the behavior of selected clock proteins. These include western blotting, spectroscopic, and microscopic methods. This section focuses on microscopic methods for live-cell imaging including fluorescence and bioluminescence microscopy.

Live-cell microscopy This methodology aims at determining the localization and the quantification of proteins using fluorescent or bioluminescent markers. This technique allows to measure clock protein oscillations in single cells and gives insights into protein dynamics and spatial and temporal distribution of clock proteins over time. In addition, it enables the measurement of different events in one sample at the same time, for example, the circadian phase, protein localization, and cell division.

Fluorescent fusion proteins or fluorescent dyes are used for time-lapse microscopy for which the need for special incubating chambers for sample cultivation, minimization of cytotoxicity, and focus drift has to be

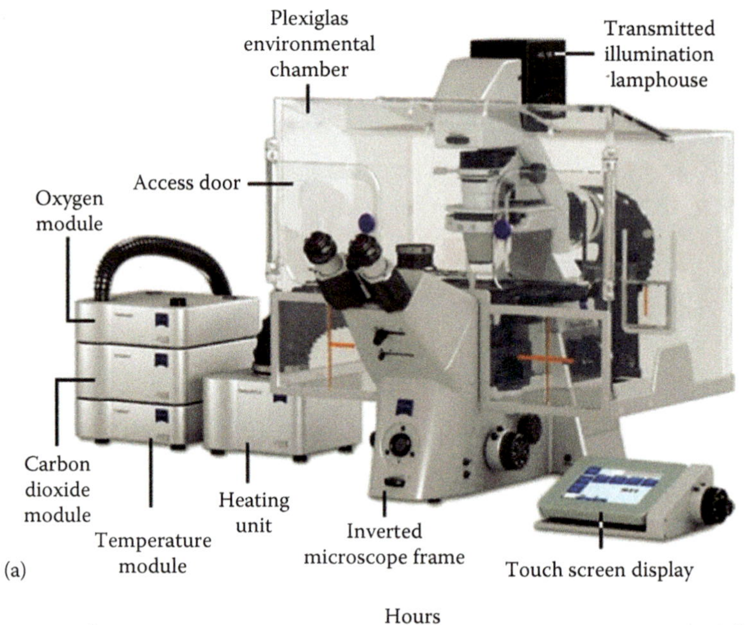

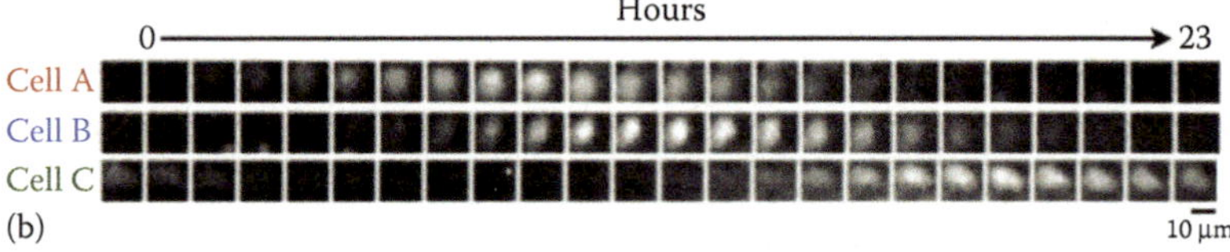

Figure 13.3 Live-cell microscopy to monitor the spatial and temporal distribution of proteins. (a) Principal setup of a live-cell imaging microscope. A general live-cell imaging microscopy setup consists of an inverted microscope, a CCD camera system, a host computer, a plexiglas chamber for sample cultivation, CO_2, and temperature control. Independent from the illumination method for image acquisition (such as brightfield, phase contrast, fluorescence, and bioluminescence), these components are necessary for live-cell imaging. To ensure high resolution, the system should be positioned on a vibration isolation platform. (Courtesy of Carl Zeiss Microscopy GmbH, Jena, Thuringia.). (b) perl-d2EGFP expression in optically identified single neurons of Perl::d2EGFP transgenic mice. Displayed is an hourly montage of images of three individual SCN neurons. (From Enoki, R. et al., *J. Neurosci. Meth.*, 207, 72–79.)

taken into account [39]. The basic setup of a microscope for live-cell imaging is shown in Figure 13.3a.

Fluorescence microscopy This technique is used to image the localization and to quantify the abundances of fluorescent fusion proteins or fluorescent-labeled proteins over time. The sample is illuminated with light of a specific wavelength that is absorbed by the fluorophores, which leads to the emission of light of a longer wavelength [40].

Basic components of a fluorescence microscope are a light source, an excitation filter, a dichroic mirror, and an emission filter. Based on the characteristics of the fluorophore, brightness and rate of photobleaching, different filters and mirrors are chosen [41]. The basic differentiation is made between widefield microscopy that uses full-field illumination and confocal microscopy techniques, which remove out-of-focus fluorescence [41].

Fluorescent tags can be added either to the N-terminus or to the C-terminus of the protein of interest. Tags inserted into an internal loop exist as well. A flexible linker is often introduced between the protein of interest and the fluorescent protein to avoid steric hindrance and to ensure the proper folding of each protein domain [39].

A common problem of fluorescence microscopy is photobleaching and phototoxicity as a consequence of sample illumination. Photobleaching can be minimized by choosing the lowest possible level of illumination and by eliminating oxygen from the media [41]. Phototoxicity can be prevented by using low illumination as compared to fluorescence microscopy in fixed samples, as the measurements take place over a longer time period [42].

Various studies used fluorescence microscopy to study clock proteins in different organisms. In cyanobacteria, the clock proteins KaiA and KaiC were fused to yellow fluorescent protein (YFP) or green fluorescent protein (GFP) for *in vivo* studies [39]. Enoki and colleagues described an alternative method of single-cell resolution fluorescence imaging of circadian rhythms, which aims to overcome the problems of single-point laser scanning. Their imaging system consists of a multipoint scanning Nipkow disk confocal unit, a high-sensitivity EM–CCD camera, and an inverted microscope with an autofocus function. SCN slices of Per1::d2EGFP mice were cultured on collagen-coated glass dishes and thousands of images were acquired over three days of measurements without considerable phototoxicity, bleaching, or focus drift. Furthermore, individual neurons could be detected within the SCN slices [43] (Figure 13.3b).

Bioluminescence microscopy This microscope technique detects light that is produced via a chemical reaction of the enzyme luciferase and its substrate luciferin. As the luciferin/luciferase reaction requires ATP, only physiologically active cells will produce light which increases the specificity of the method. An advantage of bioluminescence microscopy is therefore that it does not need excitation light and consequently does not induce phototoxicity and photobleaching [43].

Furthermore, specially designed filter sets to separate excitation and emission light are not needed, nor is autofluorescence an issue, leading to significantly lower background levels [42]. On the other hand, low absolute signal levels may constitute a problem and the spatial and temporal resolution is rather poor as compared to fluorescence imaging [43]. However, due to the low background, bioluminescence microscopy can still be used to observe small changes of signal intensity [42].

Enoki and colleagues used bioluminescence microscopy to acquire images of SCN slices of Per1::Luc mice. Bioluminescence images of SCN slices were obtained with a Luminoview 200 imaging system from Olympus with an EM–CCD camera [43].

Resonance energy transfer Resonance energy transfer (RET) is a phenomenon of quantum mechanics that can be used to investigate protein–protein interactions, based on the genetic labeling with fluorescence and bioluminescence proteins. It allows the visualization of protein interactions in real time in living cells [44]. First described by Theodor Förster, the principle of RET consists of an energy transfer from a donor chromophore in an excited state to an acceptor molecule [44]. Since the energy transfer can only take place in a limited distance range, RET can be used for distance calculation. Proteins of interest are fused to donor and acceptor fluorophores attached either on the C- or N-terminus or internally. A fluorophore can serve as a fluorescence/Förster resonance energy transfer (FRET) donor if its emission spectrum overlaps the excitation spectrum of another fluorophore functioning as the acceptor. The donor fluorophore is excited with specific monochromatic light and transfers its excited state energy to an acceptor fluorophore, which normally emits fluorescence of a different wavelength [45].

Ma and colleagues used FRET to measure circadian clock–protein interactions of the cyanobacterial clock proteins KaiB and KaiC [46]. The authors used a FRET-based assay by tagging Cerulean (an improved cyan fluorescent protein) and Venus (an improved YFP) to KaiC and KaiB, respectively. The resulting FRET signal was used to measure the degree of interaction between both proteins. In a study by Meyer and colleagues, FRET was used to monitor the interaction of period (PER) and timeless (TIM), two clock proteins, in living Drosophila cells [47]. The results gained from the FRET analysis suggested that after prolonged association of PER and TIM in the cytoplasm they dissociate and enter the nucleus independently.

Among the most widely used techniques for FRET quantification are donor dequenching, enhanced acceptor emissions, and the comparison of donor–acceptor emission ratios [48]. In the donor dequenching approach, FRET is measured as the increase in the donor emission after the acceptor is destroyed by photobleaching with the result that energy transfer can no longer occur between donor and acceptor. For enhanced acceptor emission, FRET is measured as the increase in acceptor intensity by calculating the ratio of the acceptor emissions in the absence and the presence of the donor. The comparison of the donor–acceptor emission peak ratio before and after FRET is a simple and fast method for FRET detection that does not take into account possible contamination factors such as cross-talk and bleed-through [48]. Cross talk describes the problem of overlapping excitation spectra of donor and acceptor leading to a direct excitation of the acceptor by the donor excitation light. Bleed-through is due to overlapping emission spectra and refers to the donor fluorescence emission that is detected in the range of acceptor fluorescence.

While continuously imaging PER–CFP and TIM–YFP in live single cells, Meyer and colleagues calculated FRET in subcellular compartments on a pixel-by-pixel basis and by subsequent averaging over the whole cell [47]. FRET measurements were regarded as a function of time after PER and TIM induction in order to create temporal profiles of FRET level changes. For each image, the profiles were compared to the contemporaneous nuclear accumulation profiles of PER by calculating the ratio of mean pixel value in the nucleus to mean pixel value in the whole cell.

Bioluminescence resonance energy transfer Bioluminescence resonance energy transfer (BRET) combines bioluminescence and fluorescence, in contrast to FRET [45]. In BRET, the donor is an enzyme (e.g., luciferase) that catalyzes a substrate (e.g., luciferin) that becomes bioluminescent. As a result, there is a reduction in the donor emission and a consequent fluorescence increase of the acceptor. The RET between a light-emitting luciferase and an acceptor fluorophore is a phenomenon that also occurs in nature in some marine animals such as *Aequorea victoria* and *Renilla reniformis* [44].

When analyzing protein–protein interactions with BRET, a luciferase is fused to one candidate protein and a GFP mutant is fused to the second candidate protein. Interactions between both candidate proteins bring the luciferase and GFP in close proximity, so that RET occurs and the bioluminescent emission changes [45]. BRET efficiency is defined as the ratio of GFP emission to that of luciferase emission [49].

Some problems of FRET can be avoided in BRET techniques, because of the combination of bioluminescence and fluorescence. Photobleaching and autofluorescence, as well as instantaneous excitation of the donor and acceptor fluorophore do not occur [45]. BRET was used to demonstrate that the cyanobacterial clock protein KaiB interacts to form homodimers [45]. In their study on the interaction of the cyanobacterial clock proteins KaiA and KaiB, Xu and colleagues measured the BRET luminescence intensity that passed the interference filters of 480 and 530 nm [45]. The 530/480 nm luminescence ratio was calculated for each strain enabling them to screen for colonies expressing an above-background ratio whose DNA sequence could then be further characterized.

Fluorescence recovery after photobleaching Fluorescence recovery after photobleaching (FRAP) is a method to measure the recovery of fluorescence signals after photobleaching in a specific subcellular region. The speed of fluorescence recovery in that region is dependent on the diffusion level or transportation rate of the fluorophores. Hence, the technique can be used to measure diffusion, import or export of the target molecule in a cellular region-specific manner [50]. The initial analysis steps for FRAP datasets include the quantification of fluorescence intensity in the regions-of-interest (ROIs) of the raw images, removing noise, and normalizing the background in order to reduce systematic bias and artifacts. Free software solutions such as easyFRAP [51] aim to facilitate the qualitative and quantitative analysis of FRAP datasets.

Öllinger and colleagues used live-cell fluorescence microscopy to analyze nuclear shuttling kinetics of PER2 in U2OS cells transduced with lentiviral particles containing vectors harboring PER2-Venus or PER2–Dendra2 fusion proteins. To measure nuclear import dynamics, cells expressing PER2-Venus were bleached and the exchange of bleached molecules with fluorescent molecules was measured [52].

Data analysis for bioluminescence and fluorescence microscopy The analysis of bioluminescence and fluorescence data is complex and often requires specialized mathematical and statistical methods [53]. Time-lapse movies or rather sequences of successive images can be recorded with highly sensitive digital cameras. In the course of a circadian cycle, cells may go through high and low fluorescence intensity with the lowest expression level very close to the background intensity, which may lead to over- or underexposure of images [54]. In addition, individual cells exhibit a significant variability in fluorescence intensity that further complicates the accurate image quantification [55]. This variability is termed noise and is usually defined as the ratio between the standard deviation and the mean expression level of a cell population [56]. The freely available tool Circadian Gene Express (CGE) detects and tracks moving cells and measures their fluorescent/bioluminescent protein expression level in a semiautomatic way [54]. In addition to the oscillation profile, the software also evaluates cell nucleus size, cell motility, and cell division time. Similarly, the Cell Automated Segmentation and Tracking Platform (CAST) is a tool for the automatic and robust detection of position and size of cells or nuclei that was developed as a set of custom MATLAB functions accessible through a graphical interface [57].

There are several methods to estimate the period length and signal amplitude of circadian oscillations. To calculate the peak-to-peak distance between two sequential pairs of peaks, Lande-Diner and colleagues identified peaks after smoothing the trace using an in-built MATLAB function [58]. The overall period distribution was estimated by pooling the distances from all cells. The single-cell traces were Fourier transformed and averaged across cells with the resulting peak taken as frequency. The signal amplitude was measured at the positions of the peaks identified for determining the peak-to-peak distance.

In silico visualization of circadian rhythms

The expression of genes and proteins involved in or controlled by the circadian clock can also be predicted *in silico* with the help of computational models (Figure 13.4).

Gene regulatory network models Gene regulatory network models describe the behavior of molecular elements with mathematical methods. They are comprised of biological entities such as genes and proteins whose value (e.g., concentration level or phosphorylation status) at a specific time point is represented by a local state. According to the representation of state values and time, computational models for regulatory network analysis can be roughly divided into three classes: (1) logical models that describe both time and/or concentration levels qualitatively and require the discretization of experimental data (e.g., Boolean regulatory networks [59]), (2) continuous models that use real-valued measurements over a continuous timescale (e.g., models using ordinary differential equations (ODEs) [59]), and (3) single-molecule level models that take the stochastic nature of gene expression into account (e.g., Bayesian networks [60]). The so-called hybrid models combine both discrete and continuous aspects of modeling in order to describe components in more than one category [61].

Circadian computational models Circadian computational models are commonly characterized by six properties: (1) they should display rhythmic behavior, (2) their period should be of circadian nature (~24 hours) in addition to being, (3) temperature-compensated, (4) able to entrain to external zeitgebers, (5) their amplitude must be high enough to generate outputs, and (6) should not damp in constant conditions [61]. Due to their dynamic behavior, circadian oscillators particularly lend themselves to mathematical modeling using ODEs [4,61]. Oscillatory changes in gene expression can be obtained by time-series measurements *in vitro* using, for example, RNA microarrays or luciferase and fluorescence assays. Parameters such as degradation rates of RNAs and proteins can be obtained by literature search, whereas missing parameters can be estimated using global optimization algorithms [4]. The quality of the model is assessed by its ability to simulate overall circadian behavior, as well as the expression levels of specific genes and proteins as determined by experimental data. Initial approaches of circadian modeling focused mostly on *Neurospora crassa* [62] and *Drosophila melanogaster* [63]. An early model for the intracellular mammalian clock was developed using differential equations with mass-action kinetics in order to clarify the effect of changes

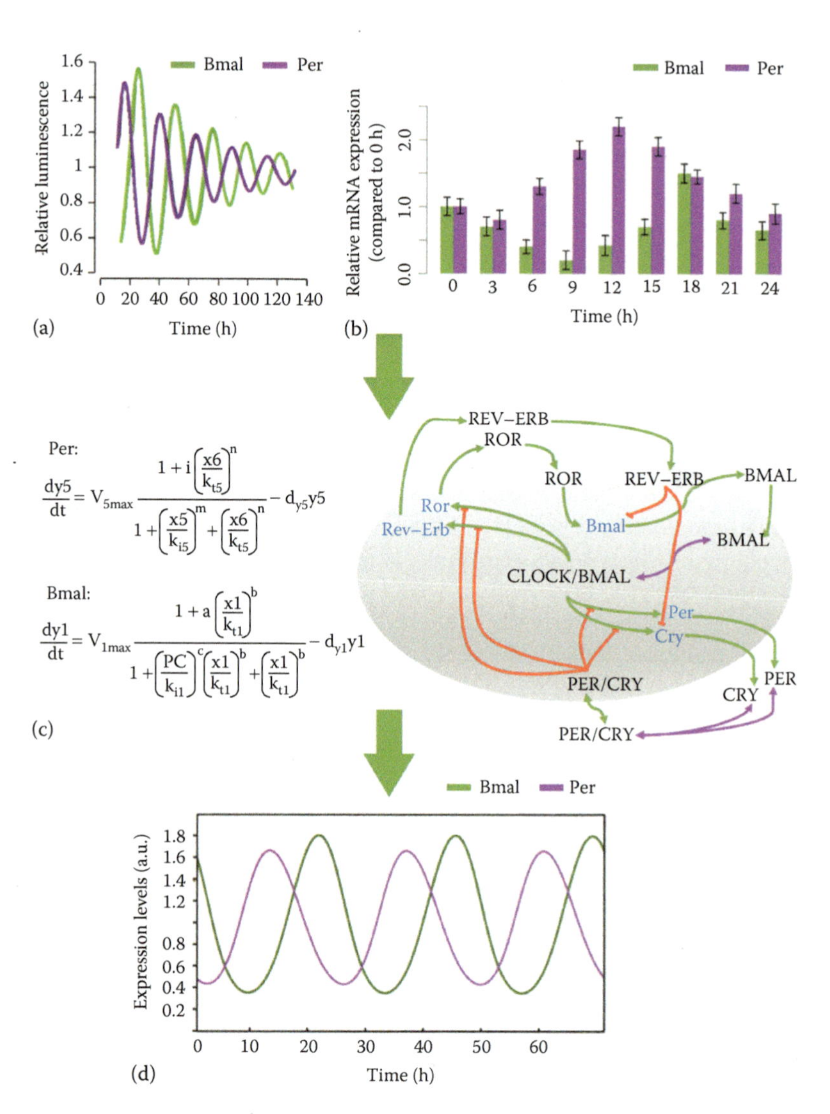

Figure 13.4 Quantitative computational modeling of the mammalian circadian clock. Experimental observations such as (a) bioluminescence data, (b) microarray gene expression data (exemplary data representation) can be used for the parameterization of (c) a quantitative gene regulatory network model of the mammalian core-clock. (d) the *in silico* expression profiles generated by the model represent detailed and testable predictions concerning the behavior of clock-associated components. (From Relógio, A. et al., *PLoS Comput. Biol.*, 7, e1002309, 2011.)

in kinase activity on the clock [64]. It could be verified experimentally that a mutation in casein kinase-I increases the phosphorylation and degradation of the circadian regulators PER1 and PER2 confirming the predictions made by the model [65]. A more recent model for the mammalian core clock was used to investigate how the mRNA degradation rate of the core-clock gene *Per* affects the overall period of the system [3]. This model elucidated the role of the ROR/Bmal/REV–ERB loop as an independent oscillator. The same model was employed in a recent study that examined the potential impact of the Ras oncogene on the circadian clock [66].

In vivo methodologies

Visualization and quantification of circadian rhythms in living organisms In the following, established methods to visualize circadian rhythms *in vivo* are described that allow to investigate the molecular clockwork or clock-driven processes in living animals. To this end, strictly controlled conditions of the environment are required, such as temperature of air or water, food availability, and defined light/dark regimes.

Circadian rhythms in unicellular organisms It was only in the twentieth century that daily rhythms in protozoa, such as *Chlamydomonas reinhardtii*, were observed in cell division, in response to UV irradiation, and in their light and gravity dependent activity (phototaxis, gravitaxis) [67]. The circadian clock of prokaryotes (cyanobacteria) has been well- studied as well. Daily rhythms in photosynthesis can be analyzed by measuring the soluble amount of oxygen using a polarographic dissolved oxygen meter. Robust circadian rhythms are also present in the fungus *Neurospora crassa*. The production of asexual spores, so-called conidiae, is subjected to daily changes. Clock-controlled conidiation rhythms can be analyzed by densitometry [68].

Circadian rhythms in multicellular animals One well-studied organism in the field of chronobiology is the zebrafish (*Danio rerio*) whose swimming activity pattern can be analyzed using a video image analysis system. Locomotor activity of larvae placed in a plate and illuminated with infrared light can be recorded with a CCD camera [69]. In addition, infrared photocells, attached to the aquarium wall, can be used to study locomotor activity patterns. When the fish crosses an infrared beam, activity is recorded and images are generated, using the software provided by the manufacturer (El temps, Barcelona) [70]. *Drosophila melanogaster* is also a commonly studied circadian model organism. Following their entrainment in defined light/dark conditions, the flies are transferred to activity tubes. As for the zebrafish, locomotor behavior of individual flies can be recorded when the fly crosses an infrared beam by using activity movement systems, such as *Drosophila activity monitors* (DAM) (TriKinetics). Activity events can be analyzed with corresponding software, such as the Fly Activity Analysis Suite (Faas) [71]. Similar techniques are used to analyze circadian behavior in rodents. Single individuals are housed in running wheel cages placed in light-tight chambers under constant conditions. By opening/closing of a switch in the wheel, rotation is recorded and registered in a connected computer, for example, using the ClockLab Aquisition System (Actimetrics). Activity patterns of individual animals are visualized in actograms [72]. Wheel running analysis is a common tool to investigate circadian behavior (free-running period, phase angle of entrainment, and amplitude) of different animal strains, mutants, and animals treated with pharmacological drugs, kept under jetlag conditions or suffering from different diseases [73].

In vivo *imaging of reporter proteins* *In vivo* imaging of reporter proteins is a common tool to characterize the circadian clock itself, as well as clock-controlled biological processes or drug treatment effects in real time.

In vivo *imaging of bacterial LUX reporter proteins* The Lux operon contains five genes, luxA, B, C, D and E. Bioreporters contain the luxA and luxB genes that generate light following the administration of the reporter's substrate decanal. By using constructs harboring bacterial *luxA* and *B* gene sequences fused downstream of various promoter sequences, circadian bioluminescence rhythms in strains from cyanobacteria can be detected [74]. Stable integration of luxA, B constructs into the genome of cyanobacteria allows for studying properties of the molecular clock and global circadian gene expression by bioluminescence screening [75,76].

In vivo *imaging of Firefly reporter proteins* Reporter constructs which express luciferase under the control of a clock gene promoter are frequently used for circadian measurements *in vivo.* Firefly luciferase (Fluc) from *Photinus pyralis* is a frequently used bioluminescence proteins to study *in vivo* rhythms of single cells or multicellular organisms. Briefly, Fluc catalyzes the oxidation of luciferin to oxyluciferin in the presence of ATP and Mg^{2+}. Following administration of the reporter's substrate luciferin, resulting photons can be detected with ultralight sensitive devices, such as photomultiplier tubes (Hamamatsu, Actimetrics) or CCD camera devices (IVIS Imaging Systems, Xenogen, Alameda, CA). The data quantification and the background correction are usually done with corresponding software (LumiCycle Analysis software, Actimetrics, Wilmette, IL; Chronostar) [77]. An attractive model system for *in vivo* imaging of clock-driven rhythms is the zebrafish due to its transparency and its ability to be directly entrained by external light. By subcloning promoter sequences of the *Per2* gene, a key regulator in light entrainment of the fish, upstream of a fluorescence reporter gene, clock gene-driven rhythms of the reporter gene can be visualized in zebrafish embryos by fluorescence microscopy [78]. Constructs in which E-box sequences are fused to the luciferase reporter gene and controlled by a minimal promoter, allow for the monitoring of rhythmic CLOCK/BMAL transcriptional activity. Using this tool, maturation of the circadian clock during developmental processes as well as core-clock activity in the adult brain was analyzed successfully [79]. The generation of a transgenic zebrafish line expressing firefly luciferase under the control of the *Per3* promoter, a strongly oscillating core-clock gene in the zebrafish, provides a remarkable tools to investigate circadian rhythms *in vivo* [80]. Larvae can be placed in white 96 well plates containing D-luciferin. Bioluminescence can be recorded by a Topcount multiplate scintillation counter (Perkin–Elmer) and analyzed using appropriate software (Import and Analysis macro by Steve Kay, Scripps Institute). Transgenic rats expressing firefly luciferase under the control of the murine *Per1* promoter allow for monitoring rhythmic *Per1* promoter activation in peripheral tissues *ex vivo* [81]. To study circadian bioluminescence rhythms of peripheral clocks *in vivo*, beetle D-luciferin can be injected directly into the rat via a sterile injection port or constantly released by a mini-osmotic pump implanted into the peritoneal cavity [82].

To elucidate whether clocks in other brain regions are operating independently from the SCN clock, bioluminescence imaging of the olfactory bulb in SCN intact and lesioned Per1:luc rats was performed in a daytime-dependent manner. By placing a sterile port and a glass window into the skull of the rat, imaging can be done at different time points around circadian cycles following the injection of D-luciferin via the port. Signals from reporter rats can be imaged with an ultrasensitive CCD camera device using *In Vivo* Imaging Systems (IVIS, Alameda, CA) and provided software (Living Image, Xenogen; Igor, WaveMetrics, Portland, OR) [81]. Commonly used for studying *in vivo* rhythms is the PER2::LUCIFERASE knock-in mouse that expresses a PER2::LUC fusion protein [83]. Injection of D-luciferin into anesthetized PER2::LUC reporter mice enables the imaging of circadian rhythms in peripheral tissues [84–86]. Bioluminescence is recorded by IVIS and analyzed by Living Image software (IVIS Kinetics; Caliper Life Sciences, Perkin Elmer), as illustrated in Figure 13.5.

Recently, a method for real-time recordings of liver-specific gene expression in moving mice was developed. A RT biolumicorder (Lesa technology) consists of a cylindrical cage with photoreflecting walls. A reflecting cone on top of the cage channels photons to a photomultiplier tube, whereas another cone in the centre of the cage projects photons to the reflecting walls. Following luciferin injection via miniosmotic pumps, *in vivo* imaging of reporter gene expression and activity of adenoviral transduced Bmal1:luc and Rev-Erbα:luc SKH1 and PER2::LUC mice can be recorded via RT-Biolumicorder data acquisition software (Instrument Control based on the LabView platform) [87].

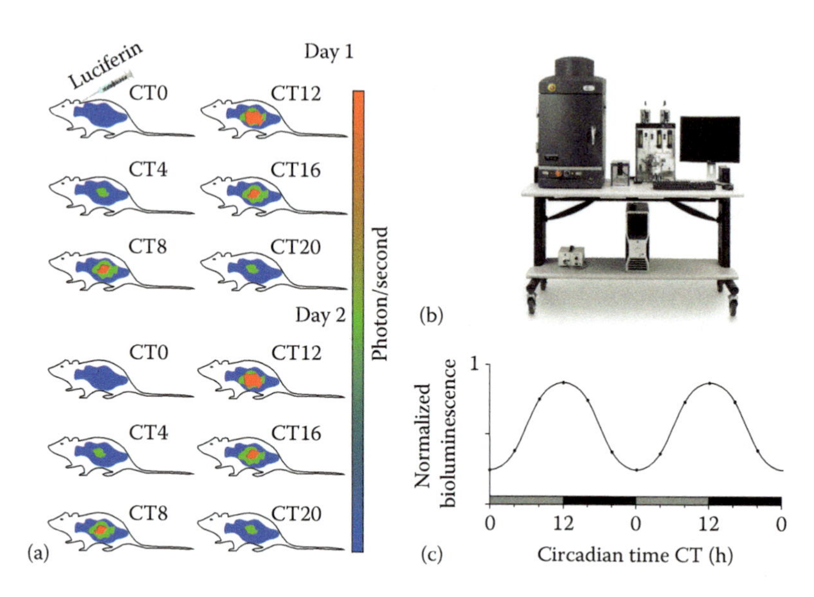

Figure 13.5 Schematic representation of *in vivo* imaging in mice with the IVIS Lumina System (Xenogen). (a) Mice expressing clock-gene controlled luciferase enzymes are monitored at different circadian time points (CT) during two circadian cycles (day 1, day 2) following the administration of luciferin. Light emission corresponds to the location and intensity of luciferase expression, (b) bioluminescence signals from anesthetized individuals are recorded by a cooled charged-coupled camera device (CCD, IVIS® Lumina System, courtesy of Xenogen, Alameda, CA) and stored in a connected computer. (c) the resulting data are normalized and plotted in reference to the circadian time point of bioluminescence measurements.

Bmal1 is an essential core-clock gene and shows an antiphasic expression pattern compared to genes of the *Per* family. By using Bmal1–Eluc transgenic mice expressing luciferase under the control of the *Bmal1* promoter, circadian rhythms were detected in peripheral tissues [88]. Among *in vivo* measurements of clock protein-driven luciferase, expression analysis of firing rates of the SCN is a useful method to study the interaction of molecular, electrical, and behavioral circadian rhythms in living animals [89]. The combination of optogenetic manipulation in the firing rate from neurons of the SCN with bioluminescence recordings *in vivo* allows to study the impact of electrophysiology on circadian behavior [90]. An overview concerning the experimental procedure for *in vivo* bioluminescence imaging is provided in Figure 13.6.

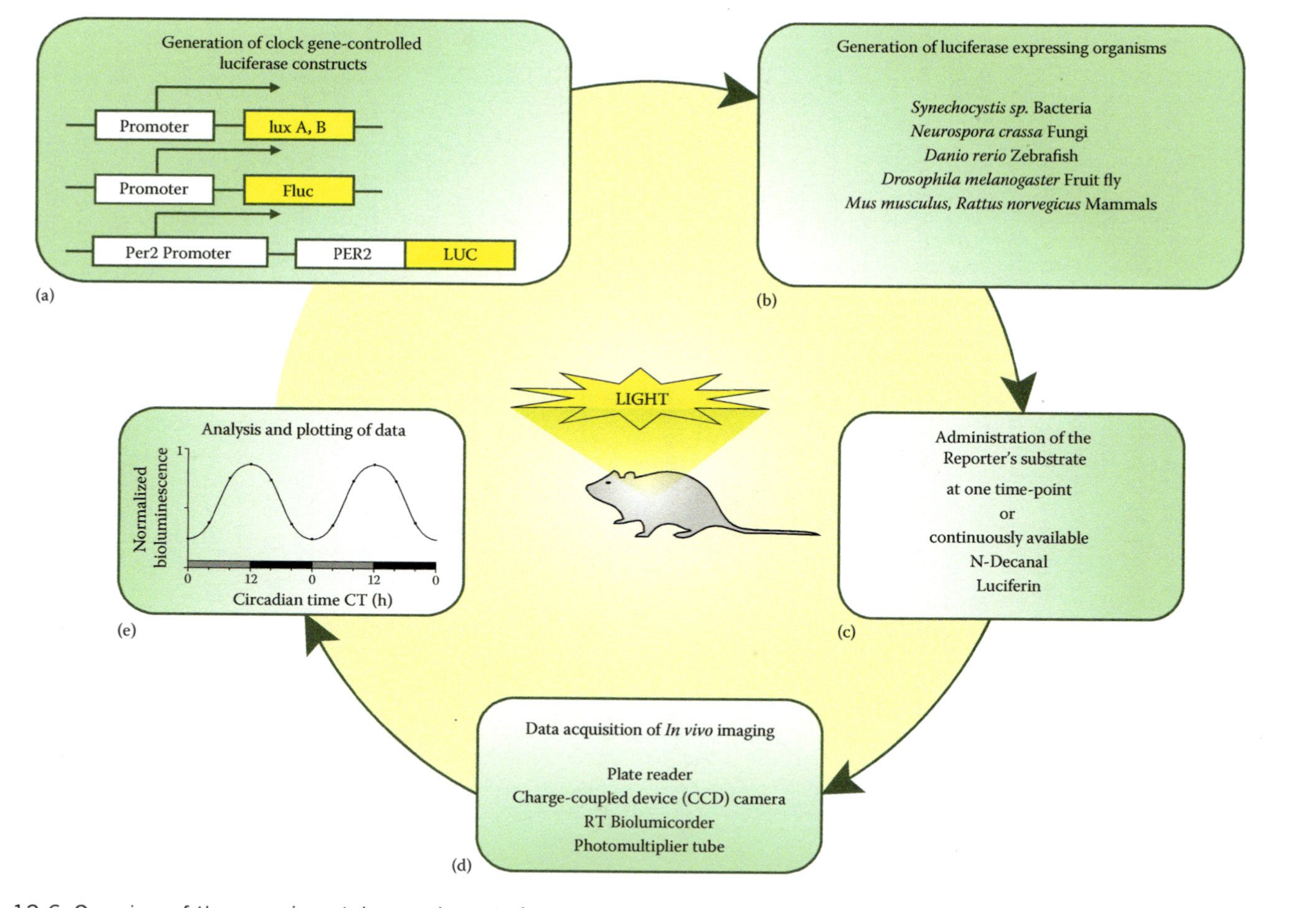

Figure 13.6 Overview of the experimental procedures to image the circadian clock *in vivo* via bioluminescence signals: (a) Several clock gene promoter-driven luciferase reporter constructs were established to generate transgenic or knock-in models (b). (c) Following the administration of the reporter's substrate, the resulting photon emission can be detected by ultralight-sensitive devices (d). The data are stored in a connected computer and analyzed by appropriate software. (e) Normalized bioluminescence is plotted and allows for studying properties of peripheral circadian clocks and clock-controlled biological processes *in vivo*.

Limitations and future perspectives

The imaging and quantification of circadian oscillations and clock-controlled components faces a number of challenges some of which are common to other fields of biological research, whereas others are limited to the study of circadian rhythms.

At the cellular level, cell populations tend to lose their synchronization after some time, resulting in a marked damping of oscillations. Although synchronization can be achieved by changing the medium, or by the administration of chemicals, the loss of synchronization of individual cells leads to a dampened circadian rhythm over time. Using devices for performing luminometry on self-luminous tissue, it is not possible to record spatial information at single-cell resolution. Another limitation of circadian research *in vitro* lies in the inherent need for time-series measurements, which require multiple arrays and potential replicates. Despite the rapidly declining cost for high-throughput data generation in the past years, microarray and RNAseq experiments remain cost-intensive, which often limits the number of biological replicates and corresponding controls. For the theoretical analysis of high-throughput datasets, users are not constrained to rely solely on commercially licensed software due to the existence of freely available open-source pipelines. However, most of these pipelines require at least basic programming skills and an understanding of the underlying algorithms, thus posing a challenge to many scientists lacking computational knowledge. In some cases, the discontinued development of analysis pipelines due to the experimental shift from RNA microarrays to RNAseq results in outdated databases and software that is no longer supported or developed.

A common problem of time-lapse microscopy of living cells is due to the focus drift of the microscope that occurs when imaging cells over a long time period. The toxicity for the cells also increases with time necessitating a carefully prepared environment for live-cell imaging. Other common problems of fluorescence microscopy include biological and (bio-) chemical issues such as autofluorescence, as well as phototoxicity and photobleaching as a consequence of sample illumination. In comparison to fluorescence microscopy, bioluminescence microscopy yields a lower signal leading to a lower spatial and temporal resolution. When using RET-based analysis methods such as FRET or BRET, only proteins in close proximity can be detected. Furthermore, fusing the protein of interest with a fluorescent protein tag may alter the original protein–protein interaction or function that is meant to be detected. Though there have been promising developments in recent years, the analysis of fluorescent and bioluminescent single-cell time-series data remains a challenge due to cell movement and changes in fluorescence/bioluminescence intensity levels. Many pipelines require manual intervention or have been developed to deal with specific experimental setups. Clearly, further efforts are needed to robustly and automatically analyze diverse data sets originating, for example, from different tissue types.

Although mathematical models of the circadian clock can help in understanding the underlying molecular mechanisms of the core-clock and predict possible outcomes of corresponding experiments, they do not represent the exact biological environment. Models are limited by the abstractions and assumptions that have to be made in order to reduce a highly complex biological system to a gene regulatory network model whose size is appropriate for computational analysis. Differential equation models for circadian rhythms require long spans of experimental data on kinetic parameters, which are difficult to obtain and therefore often not available. Another problem of the parameter estimation process is the heterogeneous origins of the experimental data, which fail to incorporate interindividual or tissue-specific differences. However, introducing too many parameters may increase the risk of overfitting the model to the underlying experimental noise resulting in a decrease of the model's predictive performance in addition to a higher computational expense. If several different models are able to simulate the same circadian phenotype, the simplest model is to be preferred.

The analysis of circadian processes *in vivo* is influenced by the necessity of adding the exogenous reporter's substrate to the medium or injecting it into the animal. In addition, the bioluminescence reaction of luciferin is dependent on ATP, Mg^{2+}, and oxygen in the culturing medium. The degradation of luciferin may impair reporter signals as well. *In vivo* bioluminescence recordings can also be influenced by body temperature or blood flow requiring a parallel control of these parameters. Only a few knock-in animal models for studying circadian clock gene-driven luciferase expression are available, which makes it difficult to study clock gene-dependent biological processes. *In vivo* imaging of clock-controlled luciferase knock-in mice requires anesthesia, a procedure that is very sophisticated and time consuming and that might affect biological processes. Furthermore, it has been observed that anesthesia by ketamine has an influence on circadian rhythms in physiology and behavior [91].

In the past decades, the study of circadian rhythms has largely benefitted from ongoing developments in the fields of biological imaging and quantification, yet the detailed mechanisms driving the molecular clock circuitry remain unclear. New experimental and computational methodologies may help to further elucidate the role of the circadian clock and clock-controlled genes and pathways in health and disease.

The methods for the measurement and analysis of *in vitro* gene and protein expression discussed in this chapter focus mainly on features and information derived from a population of cells. A foreseeable development that is going to influence the coming years of scientific research is the increase of applications and techniques addressing the single-cell level. This will allow for a much more in-depth analysis of circadian behavior of tissues and will have a significant impact on our understanding of tissue development, cancer emergence and progression, and the influence of an individual cell on the cell population, the microenvironment, and vice versa. Instead of looking at averaged results from cell populations, the individual measurements derived from single cells provide an opportunity to reconstruct possible spatial and temporal harmonics and to adjust theoretical models and clinical application accordingly. Fluorescence and bioluminescence live-cell microscopy is a powerful tool to study the spatial and temporal protein distribution in different sample types such as cell lines, mouse SCN slices, bacteria, or primary cells. An essential prerequirement is the generation of fluorescence or bioluminescence tags of the protein of interest. Improvement of techniques for the generation of these fusion proteins could lead to an increased use of live-cell microscopy. The development of incubating chambers allows the study of live samples over increased time periods, which is especially interesting for the long-term study of clock proteins. The advance of single-cell measurements highlights the importance of specialized analysis methods for time-series data in the form of automated software solutions for background correction and normalization, as well as the detection of oscillation profiles of the proteins of interests. These methods should be applicable for both bioluminescence and fluorescence measurements originating from diverse cell lines and tissue types.

In vivo imaging is an efficient tool to study clock-driven biological processes non-invasively in living animals without sacrificing them. Remarkable features such as imaging of clock-controlled biological rhythms in real time, low costs, and a high resolution of the data and the availability of ultrasensitive devices will promote the use of these techniques. The utilization of luciferase expressing animals allows for the imaging of biological processes in healthy versus diseased animals. By monitoring the intensity and the location of clock-controlled luciferase expression, scientists can study the influence of diseases as well as the effect of different drug treatments, such as chemotherapeutics, on the circadian system. Since it has been shown that the deregulation of the circadian system correlates with many diseases, drug treatment effects could be measured in a daytime-dependent manner to optimize therapeutic strategies such as the emerging field of chronotherapy that aims to increase the efficacy of drugs and reduce toxic side effects. *In vivo* imaging might also be beneficial to monitor changes in tumor phenotype over time.

In addition to *in vitro* and *in vivo* methodologies, systems biology approaches have been increasingly employed to quantify and visualize clock-associated components resulting mainly in the development of deterministic models focusing on the core-clock of model organisms such as drosophila and mice. The advance of next-generation sequencing (NGS) techniques has led to a massive increase of publicly available datasets of molecular measurements for a wide range of cell types and species. By integrating this growing set of existing data in the parameterization process of quantitative modeling, it may be possible to create specific clock models that are able to simulate circadian behavior for different organisms in a tissue-specific manner. Furthermore, alternative modeling approaches such as stochastic modeling and the use of hybrid models that incorporate the intrinsic molecular noise of the system might prove beneficial for the robust simulation of circadian rhythms and the pathways of clock-controlled genes.

Taken together, the emerging developments of new experimental and theoretical models for the visualization and quantification of circadian rhythms can potentially provide novel insights into the molecular mechanisms of the clock and clock-controlled pathways. The hypotheses and findings generated by such models will continue to contribute meaningful data to the biological and biomedical domain including but not limited to health research.

Acknowledgments

The authors are grateful to the following funding agencies: R.E., J.M., L.F., M.A, and A.R. were funded by the German Federal Ministry of Education and Research (BMBF)—eBio-CIRSPLICE—FKZ031A316. L.F. and N.G. were additionally funded by the Berlin School of Integrative Oncology (BSIO) of the Charité Universitätsmedizin Berlin.

References

1. Bollinger, T. and U. Schibler, Circadian rhythms—from genes to physiology and disease. *Swiss Med Wkly*, 2014. **144**: (w13984).
2. Bell-Pedersen, D. et al., Circadian rhythms from multiple oscillators: Lessons from diverse organisms. *Nat Rev Genet*, 2005. **6**(7): 544–556.
3. Relógio, A. et al., Tuning the mammalian circadian clock: Robust synergy of two loops. *PLoS Comput Biol*, 2011. **7**(12): e1002309.
4. Fuhr, L. et al., Circadian systems biology: When time matters. *Comput Struct Biotechnol J*, 2015. **13**: 417–426.
5. Roenneberg, T. and M. Merrow, The circadian clock and human health. *Curr Biol*, 2016. **26(10)**: R432–R443.
6. Izumo, M. et al., Differential effects of light and feeding on circadian organization of peripheral clocks in a forebrain Bmal1 mutant. *Elife*, 2015. **3**: e04617.
7. Yamazaki, S. and J.S. Takahashi, Real-time luminescence reporting of circadian gene expression in mammals. *Methods Enzymol*, 2005. **393**: 288–301.
8. Liu, A.C. et al., Intercellular coupling confers robustness against mutations in the SCN circadian clock network. *Cell*, 2007. **129**(3): 605–616.
9. Zhang, E.E. et al., A genome-wide RNAi screen for modifiers of the circadian clock in human cells. *Cell*, 2009. **139**(1): 199–210.
10. Baggs, J.E. et al., Network features of the mammalian circadian clock. *PLoS Biol*, 2009. **7**(3): e52.
11. Tiscornia, G., O. Singer, and I.M. Verma, Production and purification of lentiviral vectors. *Nat Protoc*, 2006. **1**(1): 241–245.

12. Chen, Z. et al., Identification of diverse modulators of central and peripheral circadian clocks by high-throughput chemical screening. *Proc Natl Acad Sci USA*, 2012. **109**(1): 101–106.
13. Hirota, T. et al., High-throughput chemical screen identifies a novel potent modulator of cellular circadian rhythms and reveals CKIalpha as a clock regulatory kinase. *PLoS Biol*, 2010. **8**(12): e1000559.
14. Hirota, T. et al., A chemical biology approach reveals period shortening of the mammalian circadian clock by specific inhibition of GSK-3beta. *Proc Natl Acad Sci USA*, 2008. **105**(52): 20746–20751.
15. Zhang, J.D., R. Biczok, and Ruschhaupt, M., *The ddCt Algorithm for the Analysis of Quantitative Real-Time PCR (qRT-PCR)*. R package version 1.32.0 .2015.
16. Trayhurn, P., Northern blotting. *Proc Nutr Soc*, 1996. **55**(1B): 583–589.
17. Zhao, W.N. et al., CIPC is a mammalian circadian clock protein without invertebrate homologues. *Nat Cell Biol*, 2007. **9**(3): 268–275.
18. Emery, P., RNase protection assay. *Methods Mol Biol*, 2007. **362**: 343–348.
19. Balsalobre, A., F. Damiola, and U. Schibler, A serum shock induces circadian gene expression in mammalian tissue culture cells. *Cell*, 1998. **93**(6): 929–937.
20. de la Iglesia, H.O., In situ hybridization of suprachiasmatic nucleus slices. *Methods Mol Biol*, 2007. **362**: 513–531.
21. Watanabe, N. et al., Circadian pacemaker in the suprachiasmatic nuclei of teleost fish revealed by rhythmic period2 expression. *Gen Comp Endocrinol*, 2012. **178**(2): 400–407.
22. Mastropietro, A. et al., Proof of concept of an automatic tool for bioluminescence imaging data analysis. *Conf Proc IEEE Eng Med Biol Soc*, 2015. **2015**: 6269–6272.
23. Wulbeck, C. and C. Helfrich-Forster, RNA in situ hybridizations on Drosophila whole mounts. *Methods Mol Biol*, 2007. **362**: 495–511.
24. Arunakirinathan, K. and A. Heindl, *FISHalyseR: FISHalyseR a package for automated FISH quantification*. 2015: R package version 1.6.2.
25. Robinson, M.D. and T.P. Speed, A comparison of Affymetrix gene expression arrays. *BMC Bioinformatics*, 2007. **8**: 449.
26. Team, R.D.C., *R: A Language and Environment for Statistical Computing*. 2016, Vienna, Austria: R Foundation for Statistical Computing.
27. Team, R., *RStudio: Integrated Development Environment for R*. 2015, Boston, MA: RStudio.
28. Huber, W. et al., Orchestrating high-throughput genomic analysis with Bioconductor. *Nat Methods*, 2015. **12**(2): 115–121.
29. Carvalho, B.S. and R.A. Irizarry, A framework for oligonucleotide microarray preprocessing. *Bioinformatics*, 2010. **26**(19): 2363–2367.
30. Irizarry, R.A. et al., Exploration, normalization, and summaries of high density oligonucleotide array probe level data. *Biostatistics*, 2003. **4**(2): 249–264.
31. Kauffmann, A., R. Gentleman, and W. Huber, ArrayQualityMetrics–a bioconductor package for quality assessment of microarray data. *Bioinformatics*, 2009. **25**(3): 415–416.
32. Ritchie, M.E. et al., Limma powers differential expression analyses for RNA-sequencing and microarray studies. *Nucleic Acids Res*, 2015. **43**(7): e47.
33. Thaben, P.F. and P.O. Westermark, Detecting rhythms in time series with RAIN. *J Biol Rhythms*, 2014. **29**(6): 391–400.
34. Wu, G. et al., MetaCycle: An integrated R package to evaluate periodicity in large scale data. *BioRxiv*, 2016: 040345.
35. Cho, R.J. et al., Transcriptional regulation and function during the human cell cycle. *Nat Genet*, 2001. **27**(1): 48–54.
36. Kamphuis, W. et al., Circadian expression of clock genes and clock-controlled genes in the rat retina. *Biochem Biophys Res Commun*, 2005. **330**(1): 18–26.
37. Mortazavi, A. et al., Mapping and quantifying mammalian transcriptomes by RNA-Seq. *Nat Methods*, 2008. **5**(7): 621–628.

38. Scotton, C. et al., Deep RNA profiling identified CLOCK and molecular clock genes as pathophysiological signatures in collagen VI myopathy. *J Cell Sci*, 2016. **129**(8): 1671–1684.
39. Cohen, S.E. et al., Best practices for fluorescence microscopy of the cyanobacterial circadian clock. *Methods in enzymology*, 2015. **551**(2): 211–221.
40. Ettinger, A. and T. Wittmann, Fluorescence Live Cell Imaging. *Methods in cell biology*. 2014, Elsevier, pp. 77–94.
41. Waters, J.C., *Live-Cell Fluorescence Imaging*. 2013, Elsevier, pp. 125–150.
42. Bauer, C., *Bioluminescence-microscopy-New Avenues in Live Cell Imaging. G.I.T. Imaging & Microscopy*. 2013. **4/2013**: 32–34.
43. Enoki, R. et al., Single-cell resolution fluorescence imaging of circadian rhythms detected with a Nipkow spinning disk confocal system. *J Neurosci Methods*, 2012. **207**(1): 72–79.
44. Ciruela, F., Fluorescence-based methods in the study of protein-protein interactions in living cells. *Curr Opin Biotechnol*, 2008. **19**(4): 338–343.
45. Xu, Y., D.W. Piston, and C.H. Johnson, A bioluminescence resonance energy transfer (BRET) system: Application to interacting circadian clock proteins. *Proc Natl Acad Sci USA*, 1999. **96**(1): 151–156.
46. Ma, L. and R. Ranganathan, Quantifying the rhythm of KaiB-C interaction for in vitro cyanobacterial circadian clock. *PLoS One*, 2012. **7**(8): 1–10.
47. Meyer, P., L. Saez, and M.W. Young, PER-TIM interactions in living Drosophila cells: An interval timer for the circadian clock. *Science*, 2006. **311**(5758): 226–229.
48. Takanishi, C.L. et al., GFP-based FRET analysis in live cells. *Brain Res*, 2006. **1091**(1): 132–139.
49. James, J.R. et al., A rigorous experimental framework for detecting protein oligomerization using bioluminescence resonance energy transfer. *Nat methods*, 2006. **3**(12): 1001–1006.
50. Wang, Y., J.Y.J. Shyy, and S. Chien, Fluorescence proteins, live-cell imaging, and mechanobiology: Seeing is believing. Annu Rev Biomed Eng, 2008. **10**(1): 1–38.
51. Rapsomaniki, M.A. et al., easyFRAP: An interactive, easy-to-use tool for qualitative and quantitative analysis of FRAP data. *Bioinformatics*, 2012. **28**(13): 1800–1801.
52. Ollinger, R. et al., Dynamics of the circadian clock protein PERIOD2 in living cells. *J Cell Sci*, 2014. **127**(19): 4322–4328.
53. Minors, D.S. and J.M. Waterhouse. Mathematical and statistical analysis of circadian rhythms. *Psychoneuroendocrinology*, 1988. **13**(6): 443–464.
54. Sage, D. et al., A software solution for recording circadian oscillator features in time-lapse live cell microscopy. *Cell division*, 2010. **5**(1): 1.
55. Bieler, J. et al., Robust synchronization of coupled circadian and cell cycle oscillators in single mammalian cells. *Mol Sys Biol*, 2014. **10**(7): 739.
56. Muzzey, D. and A. van Oudenaarden, Quantitative time-lapse fluorescence microscopy in single cells. *Annu Rev Cell Develop Biol*, 2009. **25**: 301.
57. Blanchoud, S. et al., CAST: An automated segmentation and tracking tool for the analysis of transcriptional kinetics from single-cell time-lapse recordings. *Methods*, 2015. **85**: 3–11.
58. Lande-Diner, L. et al., Single-cell analysis of circadian dynamics in tissue explants. *Mol Biol Cell*, 2015. **26**(22): 3940–3945.
59. Karlebach, G. and R. Shamir, Modelling and analysis of gene regulatory networks. *Nat Rev Mol Cell Biol*, 2008. **9**(10): 770–780.
60. Klipp, E. et al., *Systems Biology in Practice: Concepts, Implementation and Application*. 2008, John Wiley & Sons.
61. Roenneberg, T. et al., Modelling biological rhythms. *Curr Biol*, 2008. **18**(17): R826–R835.
62. Smolen, P., D.A. Baxter, and J.H. Byrne, Modeling circadian oscillations with interlocking positive and negative feedback loops. *J Neurosci*, 2001. **21**(17): 6644–6656.

63. Goldbeter, A., A model for circadian oscillations in the Drosophila period protein (PER). *Proc R Soc Lond B: Biol Sci*, 1995. **261**(1362): 319–324.
64. Forger, D.B. and C.S. Peskin, A detailed predictive model of the mammalian circadian clock. *Proc Natl Acad Sci*, 2003. **100**(25): 14806–14811.
65. Gallego, M. et al., An opposite role for tau in circadian rhythms revealed by mathematical modeling. *Proc Natl Acad Sci*, 2006. **103**(28): 10618–10623.
66. Relógio, A. et al., Ras-mediated deregulation of the circadian clock in cancer. *PLoS Genet*, 2014. **10**(5): e1004338.
67. Mittag, M., S. Kiaulehn, and C.H. Johnson, The circadian clock in Chlamydomonas reinhardtii. What is it for? What is it similar to? *Plant Physiol*, 2005. **137**(2): 399–409.
68. Baker, C.L., J.J. Loros, and J.C. Dunlap, The circadian clock of Neurospora crassa. *FEMS Microbiol Rev*, 2012. **36**(1): 95–110.
69. Cahill, G.M., M.W. Hurd, and M.M. Batchelor, Circadian rhythmicity in the locomotor activity of larval zebrafish. *Neuroreport*, 1998. **9**(15): 3445–3449.
70. Lopez-Olmeda, J.F. and F.J. Sanchez-Vazquez, Zebrafish temperature selection and synchronization of locomotor activity circadian rhythm to ahemeral cycles of light and temperature. *Chronobiol Int*, 2009. **26**(2): 200–218.
71. Chen, W. et al., Regulation of Drosophila circadian rhythms by miRNA let-7 is mediated by a regulatory cycle. *Nat Commun*, 2014. **5**: 5549.
72. Jud, C. et al., A guideline for analyzing circadian wheel-running behavior in rodents under different lighting conditions. *Biol Proced Online*, 2005. **7**: 101–116.
73. Siepka, S.M. and J.S. Takahashi, Methods to record circadian rhythm wheel running activity in mice. *Methods Enzymol*, 2005. **393**: 230–239.
74. Xu, Y., T. Mori, and C.H. Johnson, Cyanobacterial circadian clockwork: Roles of KaiA, KaiB and the kaiBC promoter in regulating KaiC. *EMBO J*, 2003. **22**(9): 2117–2126.
75. Onai, K. et al., Circadian rhythms in the thermophilic cyanobacterium Thermosynechococcus elongatus: Compensation of period length over a wide temperature range. *J Bacteriol*, 2004. **186**(15): 4972–4977.
76. Kondo, T. et al., Circadian clock mutants of cyanobacteria. *Science*, 1994. **266**(5188): 1233–1236.
77. Relogio, A. et al., Tuning the mammalian circadian clock: Robust synergy of two loops. *PLoS Comput Biol*, 2011. **7**(12): e1002309.
78. Vatine, G. et al., Light directs zebrafish period2 expression via conserved D and E boxes. *PLoS Biol*, 2009. **7**(10): e1000223.
79. Weger, M. et al., Real-time in vivo monitoring of circadian E-box enhancer activity: A robust and sensitive zebrafish reporter line for developmental, chemical and neural biology of the circadian clock. *Dev Biol*, 2013. **380**(2): 259–273.
80. Kaneko, M. and G.M. Cahill, Light-dependent development of circadian gene expression in transgenic zebrafish. *PLoS Biol*, 2005. **3**(2): e34.
81. Yamazaki, S. et al., Resetting central and peripheral circadian oscillators in transgenic rats. *Science*, 2000. **288**(5466): 682–685.
82. Abraham, U. et al., Independent circadian oscillations of Period1 in specific brain areas in vivo and in vitro. *J Neurosci*, 2005. **25**(38): 8620–8626.
83. Yoo, S.H. et al., PERIOD2::LUCIFERASE real-time reporting of circadian dynamics reveals persistent circadian oscillations in mouse peripheral tissues. *Proc Natl Acad Sci USA*, 2004. **101**(15): 5339–5346.
84. Tahara, Y. et al., In vivo monitoring of peripheral circadian clocks in the mouse. *Curr Biol*, 2012. **22**(11): 1029–1034.

85. Curie, T. et al., In vivo imaging of the central and peripheral effects of sleep deprivation and suprachiasmatic nuclei lesion on PERIOD-2 protein in mice. *Sleep*, 2015. **38**(9): 1381–1394.
86. Tahara, Y. et al., Entrainment of the mouse circadian clock by sub-acute physical and psychological stress. *Sci Rep*, 2015. **5**: 11417.
87. Saini, C. et al., Real-time recording of circadian liver gene expression in freely moving mice reveals the phase-setting behavior of hepatocyte clocks. *Genes Dev*, 2013. **27**(13): 1526–1536.
88. Noguchi, T. et al., Dual-color luciferase mouse directly demonstrates coupled expression of two clock genes. *Biochemistry*, 2010. **49**(37): 8053–8061.
89. Nakamura, W. et al., In vivo monitoring of circadian timing in freely moving mice. *Curr Biol*, 2008. **18**(5): 381–385.
90. Jones, J.R., M.C. Tackenberg, and D.G. McMahon, Manipulating circadian clock neuron firing rate resets molecular circadian rhythms and behavior. *Nat Neurosci*, 2015. **18**(3): 373–375.
91. Mihara, T. et al., Day or night administration of ketamine and pentobarbital differentially affect circadian rhythms of pineal melatonin secretion and locomotor activity in rats. *Anesth Analg*, 2012. **115**(4): 805–813.

Chapter 14 Measurement of lysosomal ion homeostasis by fluorescence microscopy

Benjamin König,
Lisa von Kleist,
and Tobias Stauber

Contents

Introduction

Lysosomes are the final compartments of the animal cell's degradative endocytic and autophagic pathways. By degrading various molecules delivered to these organelles and providing the products as metabolites for new anabolic pathways, and by their additional function as signaling platform, lysosomes play a pivotal role in general cellular homeostasis [1,2]. The establishment and maintenance of various ion gradients between the interior of these compartments and the cytosol is crucial to

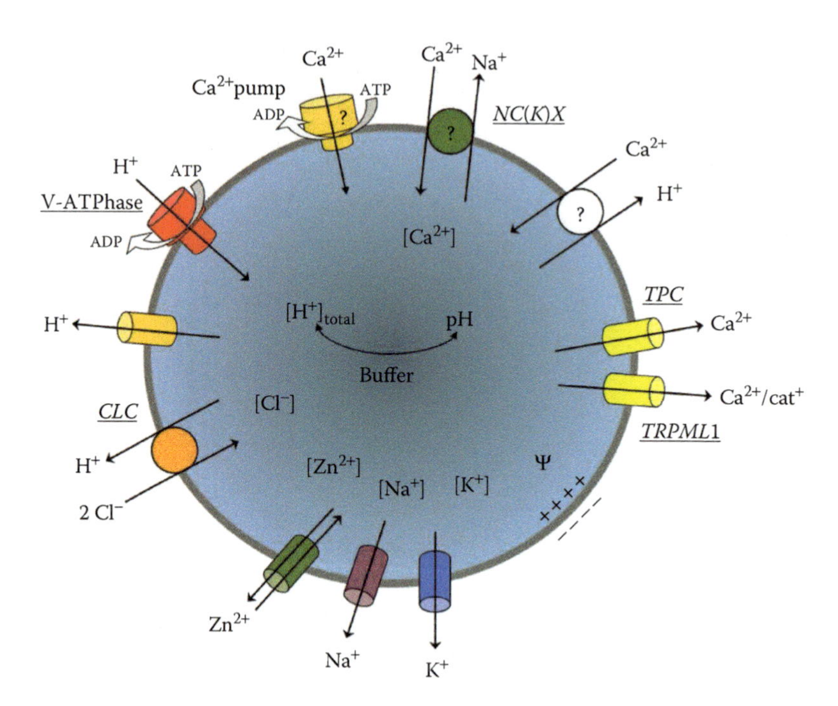

Figure 14.1 Ion transporters involved in lysosomal ion homeostasis. The energy-consuming V-ATPase pumps H^+ into the lumen. The luminal proton buffering capacity determines the pH change. H^+ can exit the organelle through an unspecific H^+ leak or its gradient can be converted into other ion gradients by symporters or antiporters, such as the Cl^-/H^+ exchanger ClC-7. Cation channels allow the flux of sodium and potassium, and there are transport systems for many more inorganic ions, such as zinc. Calcium is accumulated by an unknown mechanism and TPC or TRPML channels mediate its release from the lysosome. The total ionic composition of the organelle, including *trapped* charges determines the electrical potential Ψ relative to the surrounding cytosol.

facilitate the various physiological processes of lysosomes, and dysregulation of their ion homeostasis is involved in numerous diseases. A plethora of ion transport proteins ensures that the luminal concentrations of the different ions fit the needs of this organelle (Figure 14.1). The vacuolar-type ATPase (V-ATPase) provides for the acidic luminal pH ($\leq$5) by pumping protons (H^+) under the consumption of energy [3]. The low pH is important for the activity of lysosomal enzymes and for various membrane trafficking steps. Moreover, the pH gradient is used to drive secondary transport of metabolites and other ions against their electrochemical gradient by symporters and antiporters. The electrogenic proton pumping requires a parallel electrical shunt to prevent a rapid build up of an inside-positive potential that would inhibit further pumping by the V-ATPase. This could be the influx of anions such as chloride (Cl^-) and/or efflux of cations such as sodium (Na^+) or potassium (K^+) [4–6]. Besides their potential role in supporting the acidification, these ions may play further roles in lysosomal physiology [7,8]. Another ion that has gained much attention for its importance in lysosomal function is calcium (Ca^{2+}), whose luminal concentration may be up to 10,000-fold higher inside these organelles than in the cytosol. Its release is involved in signaling events and it can trigger the fusion of docked organelles with the plasma membrane and with other compartments [9,10]. Ca^{2+} release channels have been identified, whereas its uptake mechanism has remained enigmatic. In addition to the above-mentioned ions, many further inorganic ions—such as zinc (Zn^{2+}), iron, phosphate, magnesium, and copper—can be found in lysosomes and lysosomes play an important role in their cellular homeostasis.

For an understanding of the complex system of lysosomal ion concentrations, which are all linked at least by the lysosomal transmembrane potential, an accurate measurement of the concentrations is obviously required. There has been much progress in the development of lysosomal electrophysiology that led to tremendous insight about the function and regulation of lysosomal ion transporters within the past years [11]. Nonetheless, optical methods using ion-sensitive fluorescent molecules remain the first choice to monitor the respective ion concentrations and other parameters involved in their regulation. These do not rely on the artificial enlargement of lysosomes (in contrast to electrophysiological methods) and can be applied on isolated lysosomes as well as *in situ* in living cells, facilitating even the examination of individual lysosomes. In this chapter, we will give an overview about the optical measurement of the most important ion concentrations and of the transmembrane potential of lysosomes. We will discuss the choice, which dye to use, how to locate it to lysosomes, and how to determine the respective parameter.

General principles in fluorescence measurement of ion concentrations

There are vast possibilities to measure ion concentrations using fluorescent ion sensors. Some methods are useful to observe qualitative differences, whereas others are suitable for quantitative determinations, depending on the experimental setup and the used sensors. The sensors can be distinguished by the chemical nature of the sensor, which can be a small molecule or a genetically encoded sensor expressed by the cell. The sensitivity of the dye can be caused by fluorescence quenching or by a shift in the excitation/emission spectrum when the ion binds to the sensor, or binding of the ion can cause an accumulation of the dye within the compartment (e.g., in the case of protonation, which renders the dye membrane-impermeable). The sensor can reach lysosomes by simple diffusion, possibly in combination with an accumulation within the compartment. On the other hand, membrane-impermeable sensors can be delivered via the endocytic pathway, often coupled to macromolecular structures such as dextrans. Alternatively, lysosomal ion concentrations can be estimated indirectly by cytosolic measurement after lysosomal release, which can be triggered specifically or by disruption of lysosomal integrity.

Fluorescence measurement using a single wavelength will yield values that obviously do not only depend on the ion concentration of interest, but also on further parameters, such as importantly the amount or concentration of the fluorescent sensor and the size and possibly the number of the organelles, and the environment within these compartments. Hence, this kind of measurement is rather suited for qualitative estimations. For a more quantitative method, concentration-independent parameters such as the ion-sensitive fluorescence lifetime of the fluorophore can be measured. Most commonly, ratiometric imaging is performed to overcome the influence of parameters other than that to be measured. For this purpose, ion-sensitive dyes can be used in *pseudoratiometric* combination with ion-insensitive dyes, often cocoupled to one macromolecule, for normalization of the fluorescence intensities. In a superior approach single, intrinsically ratiometric fluorophores, such as fluorescein and derivatives for H^+ or Fura dyes for Ca^{2+}, are used [12]. Depending on the ion sensor, ratiometric fluorescence imaging requires a microscope setup that allows for fast dual-wavelength excitation or emission (Figure 14.2). By generating calibration curves with ionophores and buffers of fixed ion concentrations, ratiometric measurements are well suited to determine absolute values.

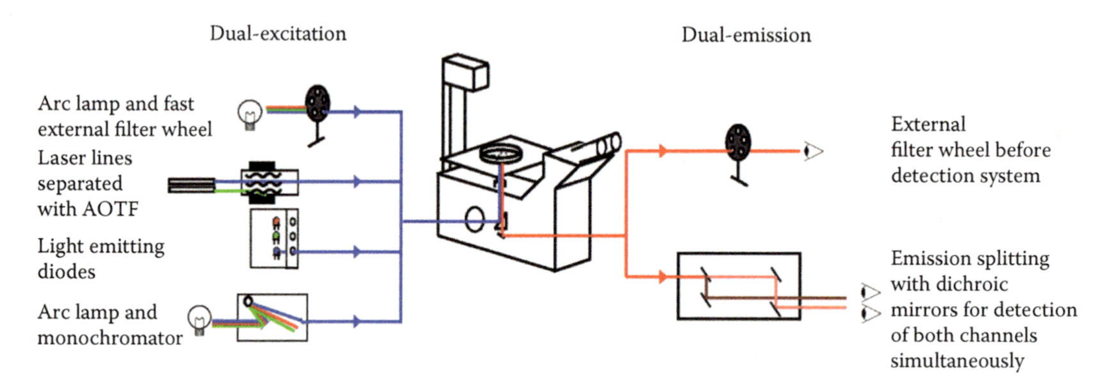

Figure 14.2 Technical setup for ratiometric microscopy measurements. Different technical setups allow for dual-excitation experiments that require excitation of the specimen with light of different wavelengths between which can be switched very quickly (left side). Fast external filter wheels can replace filter cubes used in conventional epifluorescence microscopy when using an arc lamp. An acousto-optical tunable filter (AOTF) enables fast transition between laser beams by changing the optical properties due to radiofrequency waves. Light-emitting diodes (LEDs) emit light of a specific wavelength very fast after activation. Finally, a monochromator splits the white light of an arc lamp with a diffraction grid and separates it by a fast transitioning mirror. For dual-emission experiments, the emitted light has to be detected separately at different wavelengths (right side). This can be achieved again by a fast filter wheel or by splitting of the respective wavelengths by dichroic mirrors and their simultaneous detection.

Lysosomal pH measurement

The acidic pH in lysosomes is central to their physiology [1,6]. Its alteration can be involved in diseases, in the regulation of the cell's metabolic state and it appears to be heterogeneous between individual lysosomes within a single cell [2,13,14]. To better understand its regulation and its role in (patho)-physiology, there is a need for techniques to measure the pH of lysosomes. Here, we will present methods to determine the pH by optical means with pH sensitive fluorophores. The low pH and degradative environment prevent the use of otherwise widely used fluorescent proteins, so instead synthetic dyes that are not digested are utilized.

Selection of the pH sensor and method

A wide range of pH-sensitive dyes exists (Figure 14.3), with different mechanisms underlying their sensitivity and absorption spectra, so that

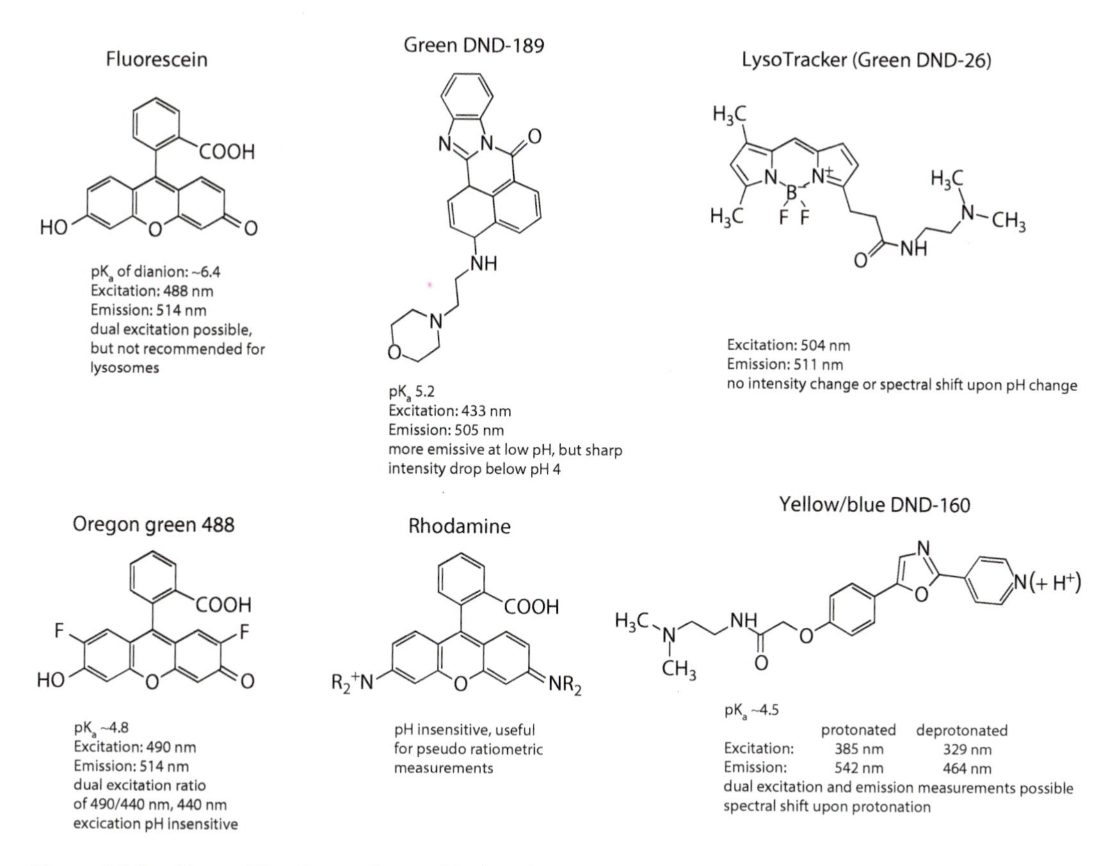

Figure 14.3 pH-sensitive fluorophores. Their pK$_a$ and excitation and emission maxima are given.

experiments can be planned according to the specific questions as well as to the available technical equipment [15,16].

For a qualitative analysis of lysosomal pH, one can use commercially available dyes such as PhrodoTM or the LysoTracker® series. LysoTracker® dyes are bound to a weak base and are membrane permeable. They accumulate in acidic compartments, such as lysosomes, and seem to be retained there by a not completely clarified mechanism. Due to its weak basic nature, LysoTracker® can cause a partial alkalization of lysosomes. Therefore, care must be taken not to incubate cells too long with these dyes, and time-lapse experiments are prone to errors caused by an artificially increased lysosomal pH. The dye is useful to detect acidic organelles through its accumulation in these compartments; however, there is no change in fluorescence due to pH alterations. Phrodo™, a rhodamine-based dye, on the other hand shows very weak fluorescence in near-neutral pH of the cytosol and extracellular medium. With decreasing pH, however, the fluorescence increases dramatically. Therefore, acidic organelles can be identified. This is advantageous compared to often used fluorescein isothiocyanate (FITC) and some derivatives, that show reduced fluorescence at lower pH. A recently described dye seems to overcome problems most of the above dyes shown in long-term experiments [17]: an iridium (III) complex (IR-Lyso) showed solid phosphorescence in acidic compartments when probed with two-photon microscopy. Two-photon microscopy makes use of the excitation of fluorophores (or in this case a phosphorophore) with two photons that are absorbed in a very short time, that have roughly double the wavelength of a photon needed for single-photon excitation. Accordingly, the used light is in the infrared (IR) range, light that is less deleterious to living cells than the visible/UV light used in single-photon excitation. In addition, it can penetrate relatively deep into tissue. Another advantage of two-photon microscopy is reduced out-of-focus light on account of the low probability that two photons excite fluorophores outside of the focal plane. IR-Lyso itself is not toxic to cells and it is very photostable, so it is possible to track lysosomes in living cells for up to four days [17].

Qualitative approaches are useful to track lysosomes and to obtain first indications of possible pH variations. On the other hand, ratiometric measurements enable precise determination of absolute pH values and are therefore the most commonly used methods [18]. They rely on fluorophores that show different sensitivity to pH in either their excitation or emission spectrum. An example is the fluorescein derivative Oregon Green 488. The emission of Oregon Green 488 at 514 nm is insensitive to pH variations when it is excited with 450 nm wavelength light. By contrast, when excited at 490 nm, there is a high dependency on the pH, with a reduction in fluorescence intensity correlating with a reduction in pH. By calculating the ratio of the emission after excitation with the two wavelengths, it is possible to generate a pH-dependent value that is unaffected by disturbing factors as cell thickness, volume and number of lysosomes, dye concentration, and photo bleaching, since these would affect the intensities in both channels in the same way. If the pH-sensitive dye itself is not ratiometric, it can be combined with a pH-insensitive dye for *pseudoratiometric* measurements. For this approach, it is possible to choose from the wide range of pH-sensitive dyes. Some advantages of single-fluorophore ratiometric measurement are lost with this method. The ratio does not correct for different photobleaching events occurring independently for the two fluorophores, leading to wrong ratios if the sample is illuminated too long and/or strongly [12].

The intensity is not the only property of fluorescence that can be measured and exploited for pH determination. Another one is the time that a population of fluorophores are emitting light, the so-called fluorescence lifetime [19]. It represents the time a fluorophore remains in the excited state before emitting light. The overall population of fluorophores displays a defined exponentially decaying intensity curve, from which the lifetime can be calculated. With fluorescence lifetime imaging microscopy (FLIM), a pseudocolored image displays the respective lifetimes of fluorophores in the sample. The lifetimes depend strongly on the environment of the fluorophore and can therefore reflect your parameter in question. LysoSensor Yellow/Blue was shown to have a well-detectable pH-dependent lifetime [19]. Since the lifetime of the fluorophores is measured, factors that would produce errors in intensity-based methods (with some exceptions in ratiometric methods) such as dye concentration and the scattering of light in the sample are irrelevant as long as enough photons are detected for calculating the exponential decay function. FLIM measurements can be performed similar to confocal microscopy by scanning every pixel one after the other or in widefield mode. Especially the widefield mode is a fast method and suitable for following the pH in living cells.

Another important feature of the dye that needs to be considered is its pK_a. When the pH equals the pK_a, half of the population is in its deprotonated and the other half in the protonated state (the pH sensitivity arises mostly from photo-induced electron transfers [PET] in either state quenching the fluorescence). Under this condition, enough fluorophores can be influenced by the surrounding environment in either direction for the detection of intensity differences. Trying to measure pH values far away from the pK_a of the used dye will lead to saturated fluorophores. A good example is fluorescein with a pK_a of ~6.4. It yields only very weak fluorescence in lysosomes, but is widely used. Newer derivatives of fluorescein, whose modified pK_as are closer to the pH of lysosomes, are better suited to measure the lysosmal pH, such as the Oregon Green series and 5(6)-carboxydichlorofluorescein (CDCF). Unfortunately, these dyes suffer from relatively fast photo bleaching and low quantum yields at low pH values. The intensity of these dyes is lower the more acidic the environment becomes, so measurements in lysosomes can suffer from high background noise. From the vast number of alternative dyes available, the LysoSensor™ probes are commonly used. LysoSensor™ Yellow/Blue can be used for dual emission or excitation imaging and is therefore attractive for different technical setups. With a pK_a of ~4.5 and good quantum yields in both protonated and deprotonated forms, it is suitable for ratiometric lysosomal pH measurements. A drawback is the required short excitation wavelengths, which are more prone to damage cells than longer ones (a problem that can be overcome by two-photon excitation). Furthermore, the free dye (called DND-160 or PDMPO) can alkalize lysosomes due to the weak base side chains, which reduces feasible incubation times [20]. Hence, it is not advisable to use free dyes to measure the pH of lysosomes. Instead, they should be linked to dextran to overcome problems with falsely stained organelles.

Dye delivery to lysosomes

The measurement of the pH in lysosomes by fluorescence microscopy requires the accumulation of the pH-sensitive dyes in these organelles. Acidotropic dyes, such as the widely used LysoTracker®, accumulate in acidic organelles due to their nature of being weak bases. In their deprotonated state, they are membrane permeable and simply enter the cell and lysosomes by diffusion. In acidic organelles, their protonation renders them membrane impermeable, so that they are trapped. This is a very straightforward way of dye delivery but is mostly restricted to qualitative pH assessments. Being weak bases, acidotropic dyes can impinge on the lysosomal pH and, in addition, they can stain other acidic organelles (e.g., late endosomes and autolysosomes). Another drawback is the increased osmotic potential of lysosomes, leading to water influx and altered luminal concentrations of enzymes, ions, and osmolytes in general [21].

The most common way to overcome the aforementioned problems is to deliver nonmembrane permeable dyes into lysosomes by coupling the fluorophores to dextran [18]. Dextran is a polysaccharide of α-linked D-glucopyranosyl units with varying molecular weights (10-kDa or 40-kDa dextrans are often used for lysosomal experiments) that is not digested by lysosomes. Conveniently, dextran is taken up by cells into endosomes and travels through the endocytic pathway to end in fully maturated lysosomes. Therefore, during a standard pulse-chase protocol, cells are incubated with fluorophore-coupled dextran for a certain time (pulse time) and the dye is chased into lysosomes to avoid measuring the pH of prelysosomal compartments of the endocytic pathway. The correct localization can be verified by colocalization of the dextran-coupled fluorophore with lysosome-specific marker proteins such as LAMP-1. Most pH-sensitive dyes named here are commercially available as dextran conjugates. One dextran molecule is linked to multiple fluorophores, increasing the observable brightness (and therefore the signal-to-noise ratio) and at the same time decreasing the osmotic stress for many fluorophores being delivered as a single osmolyte. The pH sensitivity of most dyes is not changed when linked to dextran, there is virtually no leakage of dyes, and unwanted interactions with proteins or other macromolecules are reduced. When performing pseudoratiometric measurements with different fluorophores coupled to dextran, it is of tremendous importance to ensure that all experiments and calibrations are done with the same solution of mixed dextran conjugates, since differences in the coupling ratio would lead to different calculated pH values.

From relative to absolute values: Calibration

To convert (pseudo-)ratiometrically measured intensities or lifetimes of the lysosomal dyes to absolute pH values, different calibration methods can be used. Measuring the respective fluorophores *in vitro* in buffers with defined pH values to create a calibration curve would neglect the important cellular context of the experiment. By contrast, by using the potassium/nigericin approach [18,22], calibration can be performed *in situ*. Nigericin is an ionophore that exchanges K^+ and H^+ across cellular membranes according to their concentration gradients. By bathing cells in a solution with a cytosolic potassium concentration, the pH inside the cell and lysosomes will adjust to the pH of the added solution through H^+ exchange. The cells are incubated in different solutions of varying pH values in the presence of nigericin, and the respective intensity ratios or lifetimes of the dye are measured. The resulting calibration curve serves to calculate the absolute pH value of measured ratios/lifetimes. The accuracy of calibration depends on the correct potassium concentration of the calibration buffer, so the overall potassium concentration of the cells in use needs to be known or measured as it can vary between cell types (and even due to experimental conditions) [23]. Therefore, additional ionophores, such as monensin, are often used [4,24].

An alternative calibration method is the null-point measurement [25]. It relies on the ability of weak acids and bases to move across cell membranes. Depending on their concentrations, they will alter the pH inside the compartment and the pH-sensitive dye will display changes in its behavior reflecting this alteration. A mixture of acids and bases with a pH equal to the lysosomal pH will lead to no net change in pH and therefore no difference in fluorescence. To determine the pH value of a given experiment cells are incubated with different acid/base mixtures and the condition with no change in fluorescence represents the pH of interest. In a less labor-intensive variant of this method, the exact null point is not directly measured, but rather interpolated [26]. Both methods avoid the drawback of the high potassium/nigericin approach that a certain cell-specific concentration needs to be known. However, they do assume that only the uncharged states of the acids and bases move freely through the cell membrane.

Measurement of lysosomal calcium concentrations

Lysosomes are directly involved in Ca^{2+} signaling both as the source compartment for Ca^{2+} and by their responsiveness to intracellular Ca^{2+} signals [10,27]. Nicotinic acid adenine dinucleotide phosphate (NAADP) is a second messenger that evokes calcium release from endosomes and lysosomes through the two-pore channel (TPC) family [28,29]. This Ca^{2+} release triggers further Ca^{2+} release from the endoplasmic reticulum (ER), which is of high importance for intracellular Ca^{2+} signaling. Moreover, endocytic transport and fusion of endosomes and lysosomes are dependent on the Ca^{2+} release from lysosomes [9]. In the lysosomal lumen, Ca^{2+} is bound by large to the luminal matrix, reducing the electrochemical gradient and preventing precipitation at high concentrations [10]. Depending on the stage of the endolysosomal pathway, the free luminal Ca^{2+} concentration is by a factor of 10 to 10,000 higher than the resting cytosolic Ca^{2+} concentration of approximately 100 nM [30,31]. After endocytosis, the luminal Ca^{2+} level decreases from the extracellular 1 mM to 4–40 μM in early endosomes. Potentially by H^+/Ca^{2+} exchange, late endosomes/lysosomes again accumulate Ca^{2+} to reach a free concentration of around 500 μM [27,30]. In the mammalian endolysosomal system, the TPC and transient receptor potential mucolipin (TPRML) protein families reportedly constitute the predominant Ca^{2+} release channels, whereas the transport protein(s) for Ca^{2+} uptake is unknown [10,32]. In agreement with the essential role of Ca^{2+} in regulating lysosomal function and trafficking, defects in lysosomal Ca^{2+} homeostasis can result in lysosomal storage disorders (LSDs) and are linked to further pathologies [10,27,31].

Measuring global cytosolic Ca^{2+} signals is technically straightforward using ratiometric Ca^{2+}-sensitive fluorophores such as Fura-2 [33]. By stark contrast, measuring Ca^{2+} in acidic compartments is generally more challenging due to the pH-dependency of available fluorescent Ca^{2+} indicators [34–36]. The most commonly used approach to circumvent this problem is to trigger the release of Ca^{2+} from a given Ca^{2+} store followed by measurement of Ca^{2+} in the cytosol. Assuming that higher levels of stored Ca^{2+} lead to a bigger response, this method serves to estimate relative Ca^{2+} contents qualitatively. Another common practice is to use pharmacological or genetic inhibition of a given Ca^{2+} source to elucidate the contribution of the targeted store. The main drawback of these indirect methods is the risk of misinterpreting the results. Depleting a Ca^{2+}

store may often lead to downstream effects or it may affect the Ca^{2+} response of neighboring organelles, for example, by cross talk between lysosomes and the ER [37,38]. An emerging approach to address this issue is to perform local measurements in the direct vicinity of the lysosome itself. This opens the possibility to record small or local Ca^{2+} signals that would be below the detection limit when measuring global cytosolic Ca^{2+} levels. One example of such a methodology is the use of genetically encoded Ca^{2+} indicators (GECIs), which are tethered to the membrane surface of lysosomes as fusion proteins with lysosome-specific membrane-associated proteins such as LAMP1 or TRPML1 [39,40]. However, the used indicator is required to display low Ca^{2+} affinity to measure local Ca^{2+} concentrations selectively.

Cytosolic measurement of released calcium

To measure Ca^{2+} released into the cytosol, cells can be loaded with a Ca^{2+} indicator as an acetoxymethyl ester (AM)-coupled variant of the fluorophore that enables the probe to cross cellular membranes. Once inside the cell, the AM group is cleaved by cellular esterases thus trapping the dye inside the cell [33]. A variety of fluorophores suitable for live-cell measurements that display different excitation/emission wavelengths and varying affinity to Ca^{2+} are available (Table 14.1), among them the Fluo family and the ratiometric Fura-2. The affinity to Ca^{2+} is important to consider for the selection of a fluorophore, since the reporter must operate over the concentration range appropriate for the compartment of interest. As an example, Fluo-4 or Fluo-8 would be suitable for recording cytosolic Ca^{2+} but not in Ca^{2+}-storing organelles such as the lysosome where Ca^{2+} levels are in the submillimolar range—this would saturate the fluorophore before starting the experiment. Furthermore, if the goal is to obtain quantitative results, using a ratiometric fluorophore, such as Fura-2 should be considered. As discussed earlier, ratiometric measurements are virtually insensitive to variables such as fluorophore concentration, photobleaching, and cell thickness [34].

Table 14.1 Calcium-Sensitive Fluorophores and their Binding Affinities for Ca^{2+}. Note that Fura-2 and Indo-1 can be used Ratiometrically by Dual-Excitation and Dual-Emission Microscopy, Respectively

	K_d of Ca^{2+} Binding (nM)	Excitation Maximum (nm)	Emission Maximum (nm)
Calcium Green-1	190	505	530
Fluo-4	345	490	520
Fluo-8	390	490	520
Fura-2	145	340/380	505
Indo-1	230	350	405/485
Oregon Green 488 BAPTA-1	170	500	525
Rhod-1	570	550	580

A commonly used way to release Ca^{2+} from lysosomes to measure its content in the cytosol is the treatment of the cells with glycyl-L-phenylalanine (GPN), a membrane-permeable tripeptide. GPN enters lysosomes by diffusion, where it is cleaved by the lysosomal enzyme cathepsin C. The cleavage products remain inside the lysosome, inducing osmotic swelling and eventually resulting in the permeabilization of the lysosomal membrane, so that Ca^{2+} and small solutes are released [36,41,42]. Another way to trigger lysosomal Ca^{2+} release is the use of V-ATPase inhibitors such as bafilomycin A1 or concanamycin A. This occurs as a consequence of disrupting the pH gradient, which leads to a loss of ions including Ca^{2+} from the vesicle due to the disrupted equilibrium between uptake and leak. In addition, the application of protonophores or of the ionophores nigericin and monensin that dissipate the lysosomal pH gradient leads to a Ca^{2+} release. More specifically, Ca^{2+} release can be triggered by cytosolic application of NAADP or by a TRPML1 agonist [40,43].

Measurement of luminal Ca^{2+} concentrations

The main concern when measuring luminal Ca^{2+} directly within lysosomes is that the acidic pH results mostly in the quenching of the chromophore motif or, even more problematic, affect the Ca^{2+}-binding site. Protons can compete with Ca^{2+} for binding to the sensor and hence cause a significant shift in Ca^{2+} affinity [30,36]. Accurate determination of the Ca^{2+} concentration within the lumen of lysosomes therefore requires additional pH calibration of the indicator, which means determining both the luminal pH and the K_d of the indicator at that particular pH. To accurately determine the K_d of the Ca^{2+}-sensitive dye at a given pH is, however, only possible *in vitro* as it requires strict control over the pH and Ca^{2+} concentration. To do this, the sensor fluorescence is determined as a function of Ca^{2+} concentration at the particular pH. The same experiment is to be performed at neutral pH in order to compare the K_d to literature as a control. Once a pH-specific K_d for the fluorophore has been determined, this information can be used to analyze results from ratiometric measurements in living cells. Ca^{2+} concentrations can then be calculated using the Equation 14.1

$$[Ca^{2+}] = K_d \times \left(\frac{S_{f2}}{S_{b2}}\right) \times \left(\frac{(R - R_{min})}{(R_{max} - R)}\right) \tag{14.1}$$

where:

R is the fluorescence ratio

R_{min} is the ratio in the absence of Ca^{2+}

R_{max} is the ratio at Ca^{2+} saturation

S_{f2} and S_{b2} are the fluorescence values of the denominator in the absence and presence of Ca^{2+}, respectively [44]

This approach has proven successful in shedding light on the role of lysosomal calcium homeostasis in Niemann–Pick type C1 (NPC1) disease by applying ratiometric dyes in combination with low-affinity rhod-dextran calcium probes for indirect and direct measurements, respectively [31].

Taken together, direct methods (luminal measurements) can minimize misinterpretation of results in comparison to indirect methods (cytosolic measurements). However, this approach is technically more challenging and often expensive. Therefore, the development of new, pH-insensitive Ca^{2+}-sensitive fluorophores would be of high value for future studies of lysosomal Ca^{2+} concentrations.

Further exemplary ions: Chloride and zinc

Measurements of lysosomal chloride concentrations

Chloride ions were thought for long time to merely provide the electrical shunt current to prevent the build up of an inhibitory membrane potential during the active pumping of H^+ into the lumen, thereby supporting the acidification of lysosomes. However, the requirement of lysosomal H^+/Cl^- exchange by ClC-7 that mediates in pH gradient-driven lysosomal Cl^- accumulation [24] suggests a role of luminal Cl^-. Reduced lysosomal Cl^- correlates with lysosomal pathology on loss of ClC-7 or its conversion into a pure Cl^- conductance [24]. Lysosomal acidification on the other hand can be supported by cation efflux and does not absolutely require Cl^- [4,24].

To measure the chloride concentration of lysosomes in cells by microscopy one cannot use common genetically encoded cytosolic Cl^- sensors such as Clomeleon or ClopHensor since these proteins would be quickly degraded in lysosomes and because of their pH sensitivity [45]. Another often-used sensor group is quinolinium derivatives (e.g., 6-methoxy-N-ethylquinolinium, MEQ) that reflect Cl^- concentrations by collisional quenching. As a free salt, MEQ diffuses slowly into cells and can be used to measure the cytosolic Cl^- concentration. For lysosomal Cl^- measurement, this dye needs to be targeted to lysosomes as a dextran conjugate by pulse-chase as described earlier. By additional coupling of a Cl^--insensitive fluorophore such as tetramethylrhodamine (TMR) to the MEQ–dextran, it is possible to determine the chloride concentration in lysosomes in a *pseudoratiometric* manner [24,46]. These complexes can show varying performance, since the exact localization of the fluorophores on dextran is unpredictable, and careful control experiments *in vivo* need to be performed. A newly developed, uniformly performing, localizable and pH-independent chloride sensor is the DNA nanodevice Clensor [47]. It consists of three DNA modules representing the sensing module based on the dye 10-10′-bis[3-carboxypropyl]-9,9′-biacridinium dinitrate (BAC) (which was also used as part of the BAC–TMR–dextran sensor in [46]), the normalization module bearing the Cl^--insensitive fluorophore Alexa 647 for pseudoratiometric measurement and the targeting module. The targeting module can be used to localize Clensor to different organelles. For the endosomal pathway (and therefore lysosomes), it consists of simple dsDNA that leads to lysosomal localization [47]. Clensor appears to be a very versatile scaffold that may be used to generate Cl^- sensors fitting the specific needs of different technical setups and experimental questions.

Monitoring lysosomal zinc

Zinc is one of the most abundant transition metals in eukaryotic cells and it exerts important functions as cofactor for various enzymes and as apoptosis signal [48]. Lysosomes were found to be Zn^{2+} storage organelles involved in the strict regulation of the cytosolic Zn^{2+} homeostasis [49]. During the past years, many fluorophores were developed that are sensitive to Zn^{2+} and enable monitoring of zinc levels and changes in cells by different microscopy techniques.

Various obstacles need to be considered when designing Zn^{2+}-sensitive probes for the measurement of lysosomal Zn^{2+}. Besides Zn^{2+}, there is also a substantial amount of Ca^{2+} in lysosomes (as discussed earlier). As both ions carry two positive charges, the fluorophore has to be very specific for Zn^{2+} and in the ideal case insensitive for Ca^{2+}. Furthermore, as for Ca^{2+} sensors, the low pH of lysosomes has impinges on most fluorophores by protonation that quenches, enhances, or alters the fluorescence in a way that is hard to discriminate from changes occurring due to interaction with Zn^{2+}. The fluorophore should therefore have a very low pKa below the pH of lysosomes to be insensitive to physiological pH changes or be specifically active at these low pH values. To our knowledge, there is no publication showing absolute quantification of Zn^{2+} concentration in lysosomes by fluorescence microscopy, so the described fluorophores monitor Zn^{2+} in a qualitative way. Control experiments include the incubation of cells in high-zinc buffer containing the zinc ionophore pyrithione, for measuring the fluorescence when the dye is saturated with Zn^{2+}, and incubation with a membrane-permeable Zn^{2+} chelator, such as N,N,N′,N′-tetrakis(2-pyridylmethyl)ethylenediamine (TPEN), to complex free Zn^{2+} and measure the fluorescence virtually in absence of Zn^{2+} [50–52].

Two commercially available dyes for lysosomal Zn^{2+} measurements are N-(6-methoxy-8-quinolyl)-p-toluenesulfonamide (TSQ) [53] and FluoZin-3TM. TSQ is a membrane-permeable dye that fluoresces when bound to Zn^{2+}—even to Zn^{2+} that is complexed by proteins [54].

This creates a cytosolic background signal deriving from chelated Zn^{2+}, since the dye spreads equally over the whole cell. FluoZin-3TM is a more sensitive Zn^{2+} probe that is furthermore insensitive to chelated Zn^{2+} and only displays increased fluorescence when bound to free Zn^{2+}. The unmodified dye is not membrane permeant, but an acetoxy-methyl ester (AM) variant is available that diffuses across cellular membranes. When using either dye, TSQ and FluoZin-3TM, lysosomes need to be identified by additional specific staining. This can be achieved, for example, by staining lysosomes with acidotropic dyes such as LysoTracker that accumulate in acidic organelles, or by dextran-coupled dyes (care has to be taken that the wavelength does not interfere with the Zn^{2+} measurement) that are delivered into lysosomes through the endocytic pathway. In direct comparison, FluoZin-3TM seems to be better suited for the detection of free Zn^{2+} in lysosomes because the use of TSQ gives substantial background signal owing to chelated Zn^{2+} [49].

The above-mentioned Zn^{2+} probes were not originally intended for measuring Zn^{2+} specifically in lysosomes. By contrast, the following two recent examples were designed specifically for this purpose. However, they are not commercially available. The first dye is a Förster resonance energy transfer (FRET)-based conjugate of a Zn^{2+}-sensitive arylvinyl–bipyridyl fluorophore and a BODIPY derivative [51]. The Zn^{2+}-sensing fluorophore possesses a very broad emission spectrum and it is easily photobleached. However, when covalently linked to BODIPY, whose absorption spectrum overlaps with the emission spectrum of the arylvinyl–bipyridyl fluorophore, the resulting FRET pair displays Zn^{2+}-dependent excitation leading to the narrow emission band of BODIPY. Furthermore, the problem of photobleaching of the arylvinyl–bipyridyl group can be overcome, since FRET is a very fast process usually occurring before photobleaching. The FRET-conjugate was shown to localize to lysosomes due to the aliphatic amines contained in its structure [51]. The second example of a lysosomal Zn^{2+} sensor is based on a napthalimide dye linked to a morpholine and a N,N-di-(2-picolyl)ethylenediamine (DPEN) group [52]. The napthalimide dye is the fluorescent unit of the probe that is controlled by the morpholine and DPEN group. DPEN contains nitrogens, whose free electrons quench the fluorescence of the naphthalimide dye by PET. These particular nitrogens with their electrons on the other hand chelate the Zn^{2+} ions, preventing further PET and thereby the naphthalimide dye fluoresces in the presence of Zn^{2+}. The probe is membrane permeable and distributes over the whole cell. However, its morpholine group has a pKa of ~5 and PET occurs also at more alkaline pH, therefore this sensor only fluoresces in an acidic environment with the pH of lysosomes and if Zn^{2+} is present [52].

Imaging approaches to assess the lysosomal membrane voltage

The driving forces for the transport of the diverse ions across the lysosomal membrane do obviously not only depend on their concentration gradients, but also on the transmembrane voltage ψ, the electrical potential difference across the lysosomal membrane. Logically, measurements of this central parameter in lysosomal ion homeostasis are required for a deeper understanding of this complex system. Native lysosomes are too small to be examined electrophysiologically by, for example, patch clamp techniques. Artificially enlarged lysosomes on the other hand are now approachable by electrophysiological techniques [11,55,56]. However, in these experiments, the enlarged lysosomes are measured outside of their physiological cellular context, so the resting potential of lysosomes and the overall influence of ψ cannot be clearly elucidated. To monitor the transmembrane potential of phagosomes *in situ*, a FRET-based method was established that relies on potentiometric dyes [57]. This method was subsequently refined for the lysosomal transmembrane potential [58].

There are two major classes of potentiometric dyes (i.e., dyes displaying changes on differing voltages): fast and slowly responding probes. Fast responding probes display different fluorescence properties as a direct consequence of an altered electrical field, making it possible to follow membrane potential alterations in a fast way. However, the magnitude of the change is comparably small and is not suitable for minor potential differences. Slowly responding probes reflect potential differences by alterations of their localization. They enter or leave organelles/cells according to the membrane potential, thereby displaying voltage as increased or decreased intensity. The probe used for the measurement of the lysosomal membrane potential [58] is the oxonol dye $DiBAC_4(3)$. This slowly responding probe enters depolarized cells and therefore shows increased fluorescence in depolarizing and decreased fluorescence in hyperpolarizing cells. It distributes over the whole cell volume when applied alone, leaving lysosomes indiscernible from the rest of the cytosol. To identify lysosomes, a rhodamine dye coupled to L-α-phosphatidylethanolamine (Rh-PE) was introduced. The insertion of PE (and fluorescently labeled derivatives) first into the plasma membrane and its subsequent trafficking to lysosomes [59] ensures lysosomal localization of Rh-PE. As $DiBAC_4(3)$ and rhodamine represent a very efficient FRET pair, it is possible to measure the membrane potential through the potentiometric dye $DiBAC_4(3)$ directly at lysosomes where Rh-PE is specifically localized (Figure 14.4).

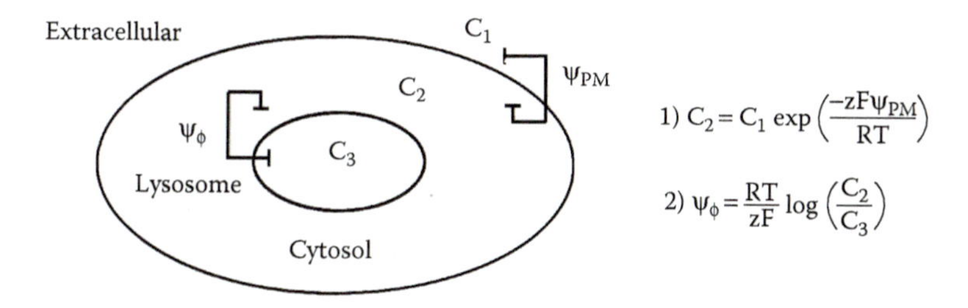

C_1 = Extracellular concentration of $DiBAC_4(3)$
C_3 = Lysosomal concentration of $DiBAC_4(3)$
ψ_{PM} = Membrane potential between cytosol/extracellular space
T = Temperature
ψ_ϕ = Membrane potential between lysosome/cytosol space
C_2 = Cytosolic concentration of $DiBAC_4(3)$
F = Faraday constant
R = Gas constant
z = Charge of $DiBAC_4(3) = -1$

Figure 14.4 Schematic overview of all parameters needed to calculate the lysosomal transmembrane potential ψ_ϕ by the method used in [58]. Although some values are known *a priori* (C_1, which is the applied concentration of DiBAC4[3]), others need to be measured: C_2 depends on C_1 and the plasma membrane potential ψ_{PM}, which needs to be measured or extracted from the literature for each specific cell line (Equation 1 in Figure 14.4). C_3 needs to be determined in each experiment. It can be calculated from the FRET signal of $DiBAC_4(3)$ and Rh-PE, which depends on the concentrations of both fluorophores and can therefore be calculated from a calibration curve of varying $DiBAC_4(3)$ concentrations and the corresponding FRET-signal. Although the absolute concentration of Rh-PE is irrelevant, it must be the same for the calibration and the experiment itself. For calibration, FRET-signals are measured after depolarizing the cells in rich K^+ buffers containing gramicidin A (facilitating the equilibration of cellular and extracellular K^+), the V-ATPase inhibitor concanamycin A, and varying concentrations of $DiBAC_4(3)$. Due to the dissipated membrane potential, $DiBAC_4(3)$ can freely diffuse, leading to equal concentrations in the compartments. With the calibration curve, C_3 can be determined, which allows to calculate the lysosomal membrane potential ψ_ϕ (Equation 2 in the Figure 14.4). C1 = extracellular concentration of $DiBAC_4(3)$; C2 = cytosolic concentration of $DiBAC_4(3)$; C3 = lysosomal concentration of $DiBAC_4(3)$; F = Faraday constant; ψ_{PM} = membrane potential between cytosol/extracellular space; R = gas constant; T = temperature; z = charge of $DiBAC_4(3) = -1$.

Acknowledgments

We apologize to those whose work was omitted owing to space and reference limitations. We are grateful for financial support from the German Federal Ministry of Education and Research (BMBF), e:Bio grant no. 031A314.

References

1. Xu, H. and D. Ren, Lysosomal physiology. *Annu Rev Physiol*, 2015. **77**: 57–80.
2. Settembre, C. et al., Signals from the lysosome: A control centre for cellular clearance and energy metabolism. *Nat Rev Mol Cell Biol*, 2013. **14**(5): 283–296.
3. Marshansky, V. and M. Futai, The V-type H^+-ATPase in vesicular trafficking: Targeting, regulation and function. *Curr Opin Cell Biol*, 2008. **20**(4): 415–426.
4. Steinberg, B.E. et al., A cation counterflux supports lysosomal acidification. *J Cell Biol*, 2010. **189**(7): 1171–1186.
5. Van Dyke, R.W., Acidification of rat liver lysosomes: quantitation and comparison with endosomes. *Am J Physiol*, 1993. **265**(4 Pt 1): C901–C917.
6. Mindell, J.A., Lysosomal acidification mechanisms. *Annu Rev Physiol*, 2012. **74**: 69–86.
7. Stauber, T. and T.J. Jentsch, Chloride in vesicular trafficking and function. *Annu Rev Physiol*, 2013. **75**: 453–477.
8. Scott, C.C. and J. Gruenberg, Lon flux and the function of endosomes and lysosomes: pH is just the start: The flux of ions across endosomal membranes influences endosome function not only through regulation of the luminal pH. *Bioessays*, 2011. **33**(2): 103–110.
9. Luzio, J.P., N.A. Bright, and P.R. Pryor, The role of calcium and other ions in sorting and delivery in the late endocytic pathway. *Biochem Soc Trans*, 2007. **35**(Pt 5): 1088–1091.
10. Morgan, A.J. et al., Molecular mechanisms of endolysosomal Ca^{2+} signalling in health and disease. *Biochem J*, 2011. **439**(3): 349–374.
11. Zhong, X.Z. and X.P. Dong, Lysosome electrophysiology. *Methods Cell Biol*, 2015. **126**: 197–215.
12. DiCiccio, J.E. and B.E. Steinberg, Lysosomal pH and analysis of the counter ion pathways that support acidification. *J Gen Physiol*, 2011. **137**(4): 385–390.
13. Johnson, D.E. et al., The position of lysosomes within the cell determines their luminal pH. *J Cell Biol*, 2016. **212**(6): 677–692.
14. Settembre, C. and A. Ballabio, Lysosomal adaptation: How the lysosome responds to external cues. *Cold Spring Harb Perspect Biol*, 2014. **6**(6): 1–15.
15. Han, J. and K. Burgess, Fluorescent indicators for intracellular pH. *Chem Rev*, 2010. **110**(5): 2709–2728.
16. Johnson, I. and M.T.Z. Spence, Molecular Probes Handbook, *A Guide to Fluorescent Probes and Labeling Technologies*, 11th Edition.
17. Qiu, K. et al., Long-term lysosomes tracking with a water-soluble two-photon phosphorescent iridium(III) complex. *ACS Appl Mater Interfaces*, 2016. 8(20): 12702–12710.
18. Canton, J. and S. Grinstein, Measuring lysosomal pH by fluorescence microscopy. *Methods Cell Biol*, 2015. **126**: 85–99.
19. Lin, H.J., P. Herman, and J.R. Lakowicz, Fluorescence lifetime-resolved pH imaging of living cells. *Cytometry A*, 2003. **52**(2): 77–89.
20. Wolfe, D.M. et al., Autophagy failure in Alzheimer's disease and the role of defective lysosomal acidification. *Eur J Neurosci*, 2013. **37**(12): 1949–1961.
21. Ohkuma, S. and B. Poole, Cytoplasmic vacuolation of mouse peritoneal macrophages and the uptake into lysosomes of weakly basic substances. *J Cell Biol*, 1981. **90**(3): 656–664.
22. Thomas, J.A. et al., Intracellular pH measurements in Ehrlich ascites tumor cells utilizing spectroscopic probes generated in situ. *Biochemistry*, 1979. **18**(11): 2210–2218.
23. Boyarsky, G., C. Hanssen, and L.A. Clyne, Inadequacy of high K+/nigericin for calibrating BCECF I. Estimating steady-state intracellular pH. *Am J Physiol—Cell Physiol*, 1996. **271**(4 40–4): C1131–C1145.

24. Weinert, S. et al., Lysosomal pathology and osteopetrosis upon loss of H^+-driven lysosomal Cl^- accumulation. *Science*, 2010. **328**(5984): 1401–1403.
25. Eisner, D.A. et al., A novel method for absolute calibration of intracellular pH indicators. *Pflügers Archiv*. **413**(5): 553–558.
26. Chow, S., D. Hedley, and I. Tannock, Flow cytometric calibration of intracellular pH measurements in viable cells using mixtures of weak acids and bases. *Cytometry*, 1996. **24**(4): 360–367.
27. Lloyd-Evans, E. et al., Endolysosomal calcium regulation and disease. *Biochem Soc Trans*, 2010. **38**(6): 1458–1464.
28. Morgan, A.J. and A. Galione, Two-pore channels (TPCs): current controversies. *Bioessays*, 2014. **36**(2): 173–183.
29. Ruas, M. et al., Expression of Ca(2)(+)-permeable two-pore channels rescues NAADP signalling in TPC-deficient cells. *EMBO J*, 2015. **34**(13): 1743–1758.
30. Christensen, K.A., J.T. Myers, and J.A. Swanson, pH-dependent regulation of lysosomal calcium in macrophages. *J Cell Sci*, 2002. **115**(Pt 3): 599–607.
31. Lloyd-Evans, E. et al., Niemann-Pick disease type C1 is a sphingosine storage disease that causes deregulation of lysosomal calcium. *Nat Med*, 2008. **14**(11): 1247–1255.
32. Patel, S. and X. Cai, Evolution of acidic Ca(2)(+) stores and their resident Ca(2)(+)-permeable channels. *Cell Calcium*, 2015. **57**(3): 222–230.
33. Takahashi, A. et al., Measurement of intracellular calcium. *Physiol Rev*, 1999. **79**(4): 1089–1125.
34. O'Connor, N. and R.B. Silver, Ratio imaging: Practical considerations for measuring intracellular Ca2+ and pH in living cells. *Methods Cell Biol*, 2007. **81**: 415–433.
35. Thomas, D. et al., A comparison of fluorescent Ca2+ indicator properties and their use in measuring elementary and global Ca2+ signals. *Cell Calcium*, 2000. **28**(4): 213–223.
36. Morgan, A.J., L.C. Davis, and A. Galione, Imaging approaches to measuring lysosomal calcium. *Methods Cell Biol*, 2015. **126**: 159–195.
37. Garrity, A.G. et al., The endoplasmic reticulum, not the pH gradient, drives calcium refilling of lysosomes. *Elife*, 2016. **5**: e15887.
38. Morgan, A.J. et al., Bidirectional Ca(2)(+) signaling occurs between the endoplasmic reticulum and acidic organelles. *J Cell Biol*, 2013. **200**(6): 789–805.
39. McCue, H.V. et al., Generation and characterization of a lysosomally targeted, genetically encoded Ca(2+)-sensor. *Biochem J*, 2013. **449**(2): 449–457.
40. Shen, D. et al., Lipid storage disorders block lysosomal trafficking by inhibiting a TRP channel and lysosomal calcium release. *Nat Commun*, 2012. **3**: 731.
41. Berg, T.O. et al., Use of glycyl-L-phenylalanine 2-naphthylamide, a lysosome-disrupting cathepsin C substrate, to distinguish between lysosomes and prelysosomal endocytic vacuoles. *Biochem J*, 1994. **300** (Pt 1): 229–236.
42. Jadot, M. et al., Intralysosomal hydrolysis of glycyl-L-phenylalanine 2-naphthylamide. *Biochem J*, 1984. **219**(3): 965–970.
43. Parkesh, R. et al., Cell-permeant NAADP: A novel chemical tool enabling the study of Ca2+ signalling in intact cells. *Cell Calcium*, 2008. **43**(6): 531–58.
44. Grynkiewicz, G., M. Poenie, and R.Y. Tsien, A new generation of Ca2+ indicators with greatly improved fluorescence properties. *J Biol Chem*, 1985. **260**(6): 3440–3450.
45. Arosio, D. and G.M. Ratto, Twenty years of fluorescence imaging of intracellular chloride. *Front Cell Neurosci*, 2014. **8**: 258.
46. Sonawane, N.D., J.R. Thiagarajah, and A.S. Verkman, Chloride concentration in endosomes measured using a ratioable fluorescent Cl^- indicator: Evidence for chloride accumulation during acidification. *J Biol Chem*, 2002. **277**(7): 5506–5513.

47. Saha, S. et al., A pH-independent DNA nanodevice for quantifying chloride transport in organelles of living cells. *Nat Nano*, 2015. **10**(7): 645–651.
48. Lee, S.J. and J.Y. Koh, Roles of zinc and metallothionein-3 in oxidative stress-induced lysosomal dysfunction, cell death, and autophagy in neurons and astrocytes. *Mol Brain*, 2010. **3**(1): 30.
49. Kukic, I. et al., Zinc-dependent lysosomal enlargement in TRPML1-deficient cells involves MTF-1 transcription factor and ZnT4 (Slc30a4) transporter. *Biochem J*, 2013. **451**(2): 155–163.
50. Xue, L. et al., Rational design of a ratiometric and targetable fluorescent probe for imaging lysosomal zinc ions. *Inorg Chem*, 2012. **51**(20): 10842–10849.
51. Sreenath, K. et al., A Fluorescent Indicator for Imaging Lysosomal Zinc(II) with Forster Resonance Energy Transfer (FRET)-Enhanced Photostability and a Narrow Band of Emission. *Chem Eur J*, 2014. **21**(2): 867–874.
52. Lee, H.-J. et al., A two-photon fluorescent probe for lysosomal zinc ions. *Chem Commun*, 2016. **52**(1): 124–127.
53. Frederickson, C.J. et al., A quinoline fluorescence method for visualizing and assaying the histochemically reactive zinc (bouton zinc) in the brain. *J Neurosci Methods*, 1987. **20**(2): 91–103.
54. Meeusen, J.W. et al., TSQ, a common fluorescent sensor for cellular zinc, images zinc proteins. *Inorg Chem*, 2011. **50**(16): 7563–7573.
55. Saito, M., P.I. Hanson, and P. Schlesinger, Luminal chloride-dependent activation of endosome calcium channels: Patch clamp study of enlarged endosomes. *J Biol Chem*, 2007. **282**(37): 27327–27333.
56. Dong, X.P. et al., The type IV mucolipidosis-associated protein TRPML1 is an endolysosomal iron release channel. *Nature*, 2008. **455**(7215): 992–996.
57. Steinberg, B.E. et al., In situ measurement of the electrical potential across the phagosomal membrane using FRET and its contribution to the proton-motive force. *Proc Natl Acad Sci USA*, 2007. **104**(22): 9523–9528.
58. Koivusalo, M. et al., In situ measurement of the electrical potential across the lysosomal membrane using FRET. *Traffic*, 2011. **12**(8): 972–982.
59. Willem, J. et al., A non-exchangeable fluorescent phospholipid analog as a membrane traffic marker of the endocytic pathway. *Eur J Cell Biol*, 1990. **53**(1): 173–184.

Chapter 15 Capturing quantitative features of protein expression from *in situ* fluorescence microscopic images of cancer cell populations

Joana Figueiredo, Ana Sofia Ribeiro, Tânia Mestre, Sofia Esménio, Martina Fonseca, Joana Paredes, Raquel Seruca, and João M. Sanches

Contents

Introduction

The most common method used for measuring protein expression is quantitative immunoblot or western blot. However, in such approach, the cellular/tissue visualization of a specific protein is lost, since the protocol requires destructive cellular/tissue steps [1,2]. Further, cellular heterogeneity is not taken into consideration and the overall cell population is analyzed as a homogenous model [2]. Indeed, protein expression levels can be misestimated due to the presence of other cell types that *contaminate* the user's interest. Actually, for generating quantitative information concerning protein expression in particular cell types within heterogeneous samples, flow cytometry is the gold standard technology [3,4]. Though, flow cytometry is also a nonvisual approach and cell subpopulations with very different protein localizations, but with similar total levels of expression, are classified as similar events [4,5]. Another clear limitation of this technique is related to its inability to preserve tissue architecture and cell–cell interactions, since it requires cells in suspension. In analyses of solid tissues, this limitation is particularly relevant, as samples need to be subjected to multiple disruptive steps to isolate the cells of interest, before being stained with fluorescent antibodies [6]. These steps, namely enzymatic degradation, centrifugation, and/or filtration, may alter biophysical and biochemical properties of the sample and, consequently, the level of protein expression.

Microscopic imaging is, thus, the only methodology that overcomes the above limitations and allows detailed visual protein information at subcellular level, maintaining tissue architecture [7]. In several microscope image modalities, such as time-lapse, confocal laser-scanning microscopy (CLSM), and spinning disk microscopy, immunofluorescence (IF) offers a qualitative analysis of the target protein [8–10]. Nonetheless, those strategies imply a potential unconscious bias or subjectivity in data selection, accounting for their major drawback. In fact, imaging analyses are strongly operator-dependent, which is likely to influence the biological and clinical evaluation of the overall data. Moreover, quantification of the results lies on software that only measure total intensity of the fluorochrome, neglecting the expression profile of the target along the distinct subcellular/cellular/tissue components and missing possible deregulation events [11,12]. In addition, other central limitations in imaging quantification are the discrepancies of parameters occurring during image acquisition, and cell heterogeneity in terms of morphology (size and shape).

In this work, we develop a novel bioimaging strategy to map and quantify specific protein signatures in images of single cells or populations of cells. Noteworthy, our algorithm includes a geometric compensation strategy to deal with cell morphological variability. With this innovative approach, we open a new window of intervention to identify protein patterns associated to disease, as well as to disclose their molecular related mechanisms.

Experimental procedures

Herein, we described a procedure to characterize protein expression regarding its level of expression and distribution in intra- and intercellular space. By computing 1D internuclear (IN) and radial (RD) intensity profiles, we could achieve a rigorous quantification of proteins in specific cellular compartments and a representative virtual cell that mimics the typical distribution of the molecule in a cell population.

Our algorithm is composed of the following steps: (1) nuclei selection; (2) nuclei segmentation and centroid detection; (3) fluorescence profile extraction from selected single cells (in case of RD) or pairs of cells (in case of IN); (4) image map building; (5) intensity-preserved denoising of map image; (6) geometric compensation of each 1D column profile; and (7) computation of average and standard deviation profiles from compensated maps (Figure 15.1).

First, nuclei selection was performed, in each image, using a graphical user interface (*IDNuclei* interface) that allows manual selection of pairs of adjacent nuclei by the biologist and, thus, the exclusion of cells/situations that may represent pitfalls inherent to immunofluorescence or transfection techniques (Figure 15.1a) [13].

The segmentation of nuclei and calculation of centroid coordinates were subsequently performed to ensure the accuracy of results and independence of the point selected by the user in the graphical interface (Figure 15.1b). Segmentation was achieved by combination of the Otsu thresholding method with the watershed algorithm [14,15], whereas centroid detection was implemented using the *Regionprops* function of the image processing toolbox from MATLAB®. To convert watershed results in a binary image and simplify centroid detection, we applied the Canny edge detector algorithm [16].

Upon this processing, IN and RD profiles might be extracted. IN profiles report the average expression level of a protein between two contiguous cells and translate its distribution pattern along the medial axis of cell pairs (Figure 15.1c). Indeed, the software registered all fluorescence intensity values from the line joining two paired cells and also from a number of juxtaposed parallel segments, improving the accuracy of the data obtained and decreasing misleading effects [13]. RD profiles were generated to capture protein expression located outside of the internuclear axis and, consequently, that is not possible to be acquired by IN profiles (Figure 15.1d). Using this method, we were able to represent

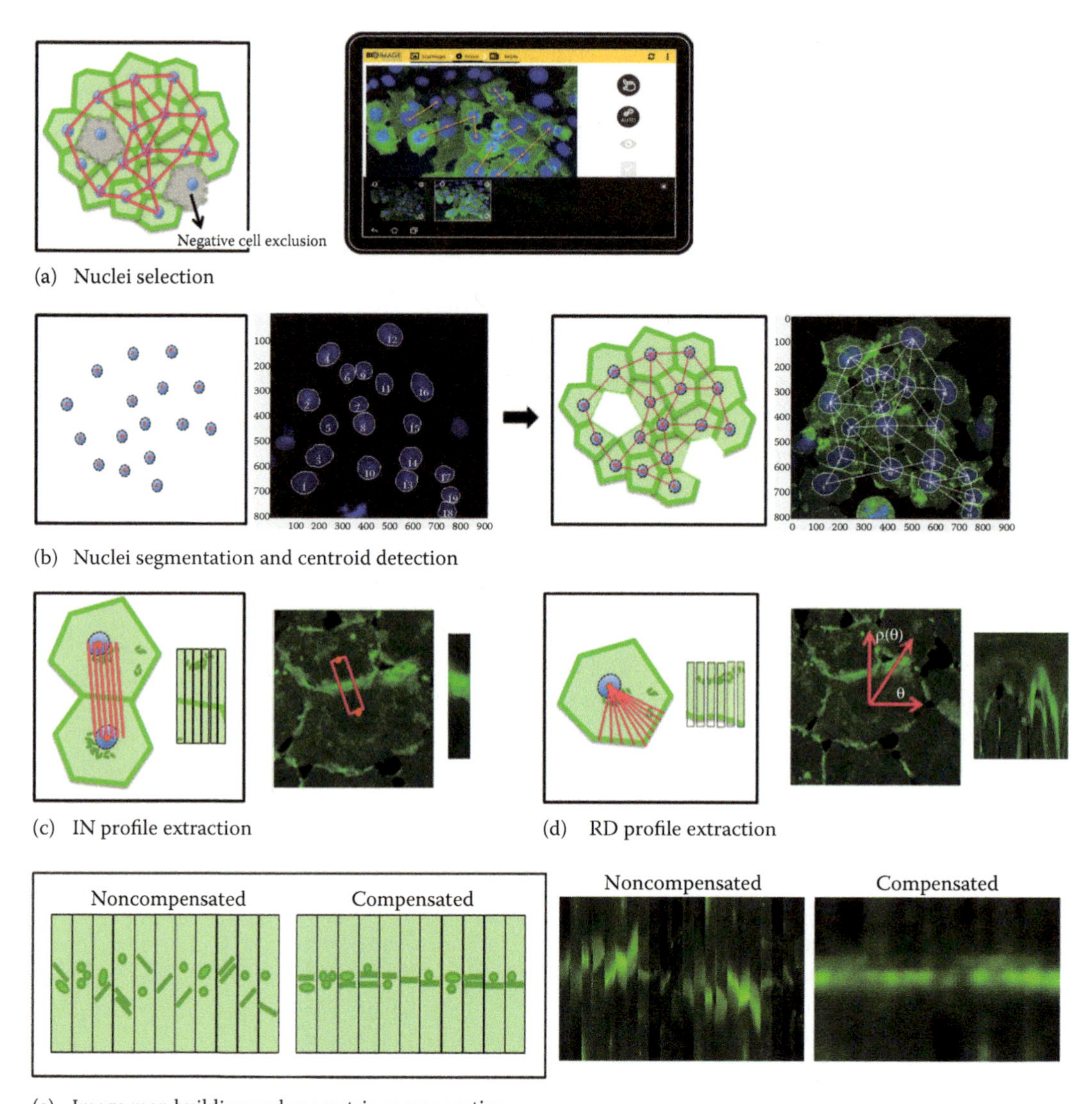

Figure 15.1 Schematic representation of the bioimaging pipeline: (a) Manual selection of pairs of adjacent nuclei was performed using a graphical user interface (IDNuclei interface); (b) nuclei segmentation and calculation of centroid coordinates were subsequently performed, ensuring the accuracy of results; (c) IN profiles were extracted to quantify the expression level and represent the protein distribution along two contiguous cells; (d) on the other hand, RD profiles reported total protein signature—located outside of the internuclear axis—of a single cell; (e) IN and RD profiles were assembled into maps that undergo a geometric compensation step to avoid cell morphological variability; and *(Continued)*

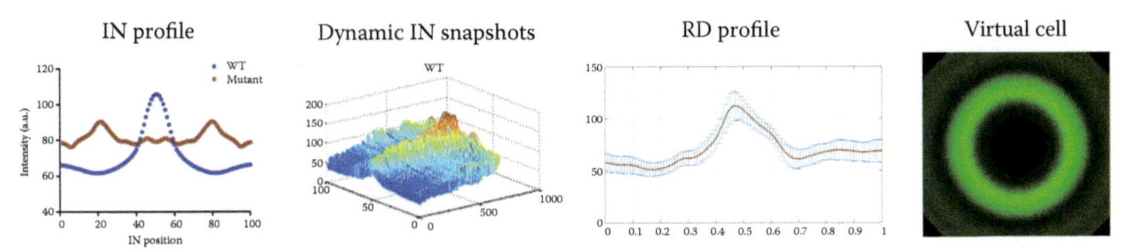

(f) Extraction of quantitative and qualitative data

Figure 15.1 (Continued) Schematic representation of the bioimaging pipeline: (f) compensated maps were then used to achieve a detailed dataset of fluorescence intensities and their respective locations.

the location and the expression level of a protein (at the membrane or cytoplasm) in a virtual cell, illustrating the typical protein signature of a large cell population. This output can be very advantageous to rapidly recognize abnormal protein patterns and detect possible trafficking defects (Figure 15.1e).

Extraction of RD profiles was performed using a polar coordinate representation [13]. All $\theta = [0, 360]^{\circ}$ orientations were considered at 1° intervals, and each pixel value was extracted as $\rho = [0, \rho_{max}]$, where ρ_{max} was given by the average of cell–cell distances of the sample.

IN and RD profiles obtained from the same image were integrated and normalized to a constant length in order to form IN or RD maps, respectively (Figure 15.1e). At this step, a robust denoising algorithm was applied to the resulting maps, assuming a Poissonian distribution and multiplicative noise for pixel intensity, as described by Rodrigues et al. [17]. Upon this procedure, noise and blur were reduced and image quality was greatly improved, which was of crucial importance to extract accurate image information.

Finally, the pipeline comprised a geometric compensation method to circumvent cell's morphological variability. This effect is particularly relevant in situations where the target molecule affects cell shape and induces epithelia reorganization. Geometric compensation is a common procedure in several image modalities and typically consists on the estimation of rigid or nonrigid transformations to make the objects under alignment, as similar as possible in terms of shape and size [18–21]. For that purpose, we used a Bayesian algorithm to model each column profile as a finite and continuous field, estimated from the initial observations. Similarity between adjacent columns was imposed and allowed

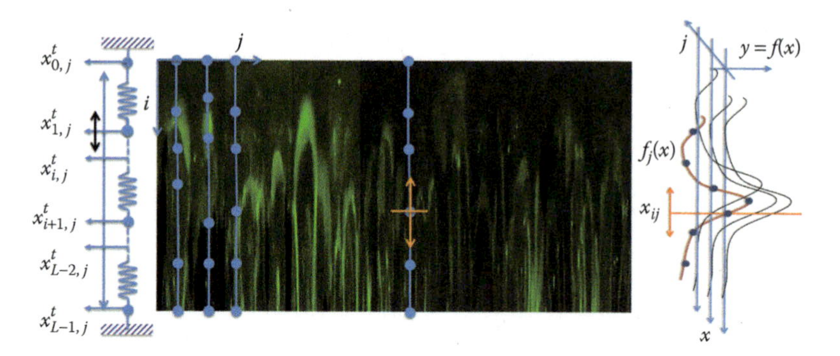

Figure 15.2 Geometric compensation model. Each profile/column composing the map was subjected to an iteration process of observation's adjustment. The location of the registered intensities (observations) was adjusted vertically by a tension regularization term and, simultaneously, by a similarity-driven force, imposing similitude between neighboring columns. At the end, the ideal profile and compensated locations of the intensity profile points were achieved.

the adjustment of each column position by smoothing the solution, as illustrated in Figure 15.2.

Based on the estimation of a continuous field in R^2, our method was divided in three major steps: (1) initialization, (2) similarity interpolation function, and (3) observation's position adjustments.

For initialization, a function $y = f(x_j, c_j)$, depending on a set of parameters, c_j, was estimated from the set of intensities, y_j, and the corresponding locations x_j. Minimization of an energy function was then used to achieve the ideal profile (c_j) and compensated locations of the intensity profile points (x_j):

$$c_j^t = \arg {}^{\min}_{c} E(y_j, x_j^t, c_j) \tag{15.1}$$

$$x_j^{t+1} = \arg {}^{\min}_{x} E(y_j, x_j, c_j^t) \tag{15.2}$$

The adjustment was performed through optimization of the energy function enforcing similarity between adjacent columns, and the final image was built by moving the set of intensities (y_i) to the new corresponding locations (x_j). Empty pixels were obtained by an interpolation process (see the description at Mathematical Formulation, section 15.3 and 15.4). In the end, the average and standard deviation were computed using the compensated maps.

Mathematical formulation

Let $Y = \{y_{i,j}\}$ be the $N \times M$ map of intensity profiles (IN or RD) obtained from M different N length normalized profiles, and $X = \{x_{i,j}\}$ be the corresponding locations along the jth profile. The initial locations of the noncompensated maps were evenly distributed in the interval $[0,1]$, defining that $x_{i,j}{}^0 = i/(N-1)$ with $0 \le i \le N-1$.

Let $f_j(x_j,c_j) = \sum_{k=0}^{L-1} c_{k,j}\phi_{k,j}(x)$ be a 1D finite continuous function, depending on a column vector of parameters, $c_j = [c_{0,j}, c_{1,j}, \ldots, c_{L-1,j}]^T$, to describe the underlying ideal j^{th} column intensity profile (Figure 15.2). $y_{i,j}$ is the ith intensity observation from the jth profile taken at the $x_{i,j}$ location, and the $\phi_{k,j}(x)$ is the kth basis function. These functions are, as well, evenly distributed in the interval $[0,1]$, with $0 \le k \le L-1$. Further, the basis/interpolation functions are shifted versions of a mother basis function in accordance with

$$\phi_{k,j}(x) = \phi\left(\frac{x}{\Delta} - k\right) \tag{15.3}$$

where $\phi(x) = \text{sinc}(x)$ is the sinc function and $\Delta = 1/L-1$. We assumed that the locations of the observations, $x_{i,j}$, are geometrically distorted and, thus, corrupted by position noise,

$$x_{i,j} = x^*_{i,j} + \epsilon_{i,j} \tag{15.4}$$

where $\epsilon_{i,j}$ is an unknown translation error vector.

Let $C = \{c_{i,j}\}$ be a $L \times M$ matrix of coefficients, where c_j represents the coefficients defining each jth column of the continuous map profile. Each ideal profile function can be defined as $f_j(x) = \Phi^T(x)c_j$, where $\Phi(x) = [\phi_0(x), \phi_1(x), \ldots, \phi_{L-1}(x)]^T$ is a column vector containing the values of the L basis functions computed at location x. The optimal coefficients c_j and the positions of the observations, x_j are estimated by solving the following optimization problem

$$[c_j, x_j]^* = arg \min_{c_j, x_j} E(x_j, y_j, c_j) \tag{15.5}$$

Here, the energy function minimized is

$$E(x_j, y_j, c_j) = E_y(x_j, y_j, c_j) + E_p(c_j) + E_c(c_j) + E_x(x_j) \tag{15.6}$$

In this equation, the energy function is composed by one data fidelity term and three prior terms. The data fidelity term,

$$E_y(x_j, y_j, c_j) = \sum_{i}^{N-1} (f_j(x_{i,j}) - y_{i,j})^2 \tag{15.7}$$

pushes the solution toward the data. The first prior term,

$$E_p(c_j) = \alpha \sum_i (c_{i,j} - c_{i-1,j})^2 \tag{15.8}$$

is used to stabilize the iterative process, smoothing the solution by imposing similarity of neighboring coefficients of each jth profile, $c_{i,j}$ and $c_{i-1,j}$. The second prior term,

$$E_c(c_j) = \beta \sum_{i,j=0}^{L-1,M-1} (c_{i,j} - c_{i,j-1})^2 \tag{15.9}$$

smoothes the solution imposing similitude between homologous coefficients on neighboring columns, $c_{i,j}$ and $c_{i,j-1}$, in order to force the similarity of neighboring profiles, $f_j(x)$. The third prior term,

$$E_x(x_j) = \gamma \sum_{i,j}^{N-1,M-1} (x_{i,j} - x_{i,j-1})^2 \tag{15.10}$$

is a prior function to keep under control the displacement compensation adjustment of the intensity locations at each profile, and to prevent degenerated solutions. The ending locations, $x_{0,j}$ and $x_{N-1,j}$ are fixed with values 0 and 1, respectively. α, β, and γ are prior hyper parameters.

Using matrix notation, Equations 15.7 through 15.10 can be written as follows:

$$E_y(x_j, y_j, c_j) = \sum_j \left\| \Phi_j^T(x_j)c_j - y_j \right\|_2^2 \tag{15.11}$$

$$E_p(c_k) = \alpha \sum_j \left\| \theta c_j \right\|_2^2 \tag{15.12}$$

$$E_c(c_j) = \beta \sum_j \left\| c_j - c_{j-1} \right\|_2^2 \tag{15.13}$$

$$E_x(x_j) = \gamma \sum_j \left\| \theta x_j \right\|_2^2 \tag{15.14}$$

where $\Phi_j(x_j) = [\Phi(x_{0,j}), \Phi(x_{1,j}), \ldots, \Phi(x_{N-1,j})]$ is a $N \times L$ matrix computed for each jth column profile and is the following difference operator

$$\theta = \begin{bmatrix} 1 & -1 & 0 & \ldots & 0 \\ -1 & 1 & 0 & \ldots & 0 \\ 0 & -1 & 1 & \ldots & 0 \\ \vdots & \vdots & \vdots & \vdots & \vdots \\ 0 & 0 & \ldots & -1 & 1 \end{bmatrix} \tag{15.15}$$

The energy function defined in Equation 15.6 is minimized in three steps according with

$$c_j^0 = arg \min_{c_j} E_y(x_j, y_j, c_j) + E_p(c_j),\ 0 \le j \le M-1 \tag{15.16}$$

$$c_j^t = arg \min_{c_j} E_y(x_j^{t-1}, y_j, c_j) + E_c(c_j),\ 0 \le j \le M-1 \tag{15.17}$$

$$x_j^t = arg \min_{x_j} E_y(x_j, y_j, c_j^t) + E_x(x_j),\ 0 \le j \le M-1 \tag{15.18}$$

where t is the iteration index of the iterative optimization process and Equations 15.17 and 15.18 alternate until convergence be achieved.

Optimization

The minimization step in Equation 15.16, is performed by finding the stationary point of Equation 15.6 regarding c_j, $\nabla_{c_j} E(x_j, y_j, c_j) = 0$, which leads to

$$(\Phi_j^T c_j - y_j)^T (\Phi c_j - y_j) + \alpha (\theta c_j)^T (\theta c_j) = 0 \tag{15.19}$$

with the following solution,

$$c_j^0 = (\Phi_j^T \Phi_j + \alpha \Theta^T)^{-1} \Phi_j^T y_j \tag{15.20}$$

where $\Theta = \theta^T \theta$.

Similarly, to the initialization step, the minimization step in Equation 15.1 is obtained by solving $\nabla_{c_j} E(x_j^{t-1}, y_j, c_j) = 0$

$$\Phi_j (\Phi_j^T c_j - y_j) + \alpha \Theta^T c_j + 2\beta [c_j - \bar{c}_j] = 0 \tag{15.21}$$

where $\bar{c}_j = c_{j-1} + c_{j+1}/2$ is the average of the neighboring columns of c_j. Using the fixed-point method, the solution of this equation can be computed with the following recursion:

$$c_j^t = (\Phi_j^T \Phi_j + \alpha \Theta^T + \beta I_L)^{-1} (2\beta \bar{c}_j^{t-1} + \Phi_j^T y_j) \tag{15.22}$$

where:

I_L is an L dimension identity matrix
t is the recursion index

The minimization step in Equation 15.2 is obtained by computing the derivative roots of the energy function regarding the ith element of the jth profile with intensities y_j,

$$\frac{\partial E}{\partial x_{i,j}} = z_{i,j} + 2\gamma(x_{i,j} - \bar{x}_{i,j}) = 0 \tag{15.23}$$

Here, $z_{i,j} = (f_j(x_{i,j}) - y_{i,j}) \dot{f}_j(x_{i,j})$ and $\bar{x}_{i,j} = (x_{i-1,j} + x_{i+1,j})/2$ are the average values of the neighboring intensity locations. Using the same fixed-point approach, x_j can be obtained as follows:

$$x_j^{t+1} = \Omega x_j^t - \frac{1}{2\gamma} z_j \tag{15.24}$$

where

$$\Omega = \begin{bmatrix} 0 & -2 & 0 & 0 & \dots & 0 \\ -1 & 0 & -1 & 0 & \dots & 0 \\ 0 & -1 & 0 & -1 & \dots & 0 \\ \vdots & \vdots & \vdots & \vdots & \vdots & \vdots \\ 0 & 0 & 0 & \dots & -2 & 0 \end{bmatrix} \tag{15.25}$$

Applications

To illustrate the application of this methodology, we used IF real images from cancer cells stained for cell–cell adhesion proteins, such as E- and P-cadherins, and one of their molecular interactors, namely p120-catenin. It is well known that in homeostatic conditions, cadherins are concentrated at the cell membrane, establishing a homophilic binding with similar cadherin molecules present at neighboring cells [22,23]. This sticky function awards cadherins (E- and P-cadherin) a key role in tissue architecture and homeostasis [23,24]. Importantly, the regulation of cadherins is highly dependent of their cytoplasmic interactors—the catenins—including p120-, α- and β-catenins [25–28]. Catenins are crucial for the transport of newly synthesized cadherin molecules to the plasma membrane and, once there, they can act as cadherin's stabilizers and, simultaneously, as intermediate linkers between them and the actin cytoskeleton [28–31]. Hence, the assembly of the cadherin–catenin complex is required for normal cell–cell adhesion [22,25].

In cancer, E-cadherin deregulation is a common event, resulting, in most of the cases, in its absence or decreased expression at the cell membrane, as well as in aberrant expression at the cytoplasm, causing loss of protein function [23,32,33]. Contrarily, P-cadherin upregulation is frequently found in many tumor types and it is an effective marker of invasive capacity of cancer cells, since it interferes with the normal function of E-cadherin [34–39]. Further, altered expression of β- and p120-catenins have been reported to be associated with E-cadherin loss/decreased expression or function and, thus, their expression delocalization have already been proposed as prognostic factors in some carcinomas [40].

Based on this, we used cells transfected with wild-type E-cadherin and with a panel of cancer-related E-cadherin mutants, with proven dysfunction to mediate cell–cell adhesion and to suppress invasion [41–50]. IF images of the different cell lines immunostained for E-cadherin were subjected to our analytical pipeline. As shown in Figure 15.3a, cells expressing E-cadherin pathogenic variants displayed a distinct IN protein profile from that displayed by wild-type cells, with decreased fluorescence intensity at the membrane and/or aberrant intensity peaks in the perinuclear region [51]. Some E-cadherin variants did not present aberrant cytoplasmic accumulation of the protein, but still presented

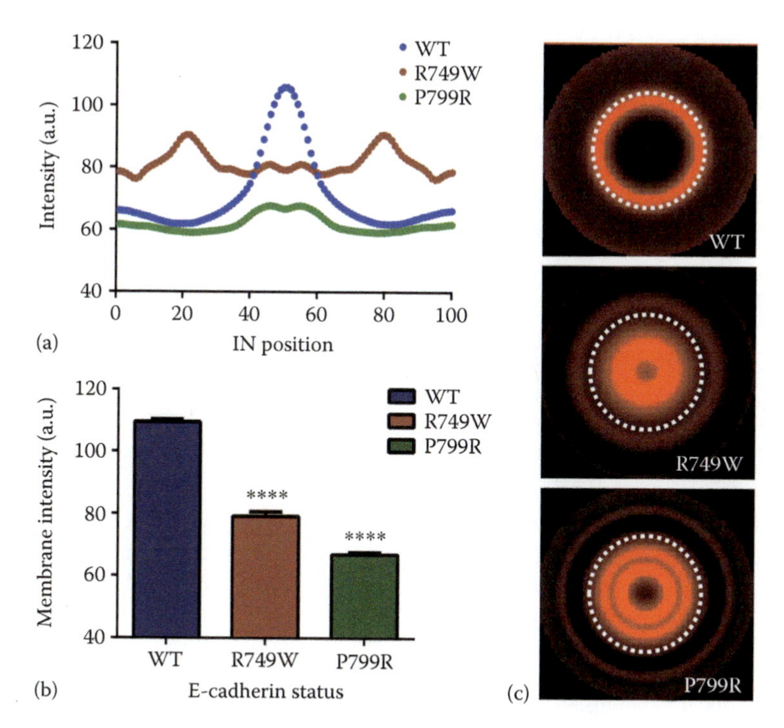

Figure 15.3 E-cadherin pathogenic variants displayed distinct IN and RD protein profiles from that of the wild-type cells: (a) Cells expressing mutant forms of E-cadherin showed decreased fluorescence intensity at the membrane and/or aberrant intensity peaks in the perinuclear region, when compared with the wild-type counterparts; average intensity of IN profiles of E-cadherin variants (red and green lines) are overlapped with that of the WT cells (blue line); (b) mean and standard error (SE) of the fluorescence intensity at the plasma membrane, which corresponds to the IN position 50, is represented, **** represents $p \leq 0.0001$; (c) virtual cells, generated from RD profiles, illustrated the particular signatures of E-cadherin mutants that are possibly due to trafficking deregulation and protein accumulation in distinct cell compartments.

less E-cadherin molecules at the membrane, supporting its pathogenic significance (Figure 15.3b) [51]. Clear differences were also observed in virtual cells, generated from RD profiles from each cellular condition (Figure 15.3c) [51]. We verified that E-cadherin mutants exhibited a particular protein signature due to different trafficking defects and consequent protein accumulation in distinct cell compartments. Therefore, our proposal is that this bioimaging tool could be applied as a complementary method to assess the clinical relevance of novel E-cadherin missense variants found in the context of hereditary diffuse gastric cancer [52,53].

Our algorithm was subsequently applied to infer the adhesion status of P-cadherin overexpressing cells, providing insightful information about its implication into the cancer cell invasion process. For that, we performed the quantification of intercellular networks and evaluated the IN profile of the E-cadherin/p120ctn complex [77]. An increase in the IN distances was observed in P-cadherin positive cells, greatly reflecting the loss of cell–cell adhesion and the increased invasive properties of breast cancers with P-cadherin overexpression [39,54,77]. These changes were also significantly associated with the activation of the E-cadherin/p120ctn signaling, through p120ctn delocalization. As analyzed by the fluorescent IN profile, we could observe a significant decrease in p120ctn membrane expression, without affecting E-cadherin profile [77]. We also applied the same bioimaging methodology in P-cadherin overexpressing cells treated with an inhibitor of Src, which is able to revert the invasive and tumorigenic effect of P-cadherin. Accordingly, IF images from treated cells showed a decrease in IN distances with an increase in p120ctn staining at the cell membrane, maintaining the expression and localization of membrane E-cadherin [77]. With these experiments, we demonstrated that this bioimaging methodology could be used as a screening platform to depict cellular effects and modifications in protein expression mediated by pharmacological drugs in a specific cancer context.

Deregulation of posttranslational mechanisms, such as that occurring during malignant transformation, are also detected by the application. Using gastric carcinoma cell lines, we observed the presence of an abnormal pattern of E-cadherin as a result of N-acetylglucosaminyltransferase V (GnT-V) glycosylation [55]. Cells expressing the Asn-554 mutant, which abrogates the availability of this site to be modified by GnT-V, exhibited a strong and well-defined protein staining restricted to the plasma membrane [55]. In contrast, mutants preserving this N-glycan specific site showed decreased E-cadherin membrane expression and a diffuse pattern of intensity throughout the cytoplasm [55]. This imaging assay proved to be crucial in demonstrating that aberrant glycan modifications on a precise site of E-cadherin can disturb E-cadherin stability and localization, affecting its tumor suppressor function and contributing for gastric cancer progression.

In addition to these, many other cellular effects are likely to be analyzed by our approach, either in basic or translational research. Protein trafficking studies, namely colocalization techniques involving immunofluorescence staining with simultaneous probes and fluorescent organelle markers, as well as time-lapse microscopy to follow and map newly synthesized proteins, would be greatly improved.

Physical transfer of membrane and cytoplasmic components through mechanisms of cell–cell communication will also benefit from this type of detailed analysis. Despite being a general and poorly understood mechanism of neighboring eukaryotic cells, the intercellular protein transfer (or ICT, intercellular transfer of cellular components) seems to be particularly relevant in the context of the immune system [56,57]. In fact, a large number of molecules/antigens are exchanged between inflammatory cells through internalization, dissociation-associated pathways, uptake of exosomes, trogocytosis, and membrane nanotube formation, orchestrating immune responses [58,59]. So, the characterization of that specific protein routes would be of major interest to unravel the paths for immune disorders. The same methodology could also be used to analyze and quantify the movement of organelles and proteins through tunneling nanotubes (TNTs). TNTs are F-actin-based membrane structures that form between cells in culture and in tissues, mediating intercellular communication ranging from electrical signaling to transfer of organelles [60]. Identical procedures can be applied to investigate signal transmission between neuronal cells and intercellular propagation of pathogens, such as Listeria, Legionella, and Mycobacterium [61–63].

Cytoskeletal rearrangements and protein flow, for example, during wound closure, are also suitable to be captured by our software and could be very indicative about mechanical, morphological, motile, and adhesive properties of the involved cells [64–66].

Notably, the applications of this tool are not restricted to *in vitro* cell cultures. *In vivo* models such as *Drosophila melanogaster*, *Caenorhabditis elegans*, or *Danio rerio* (Zebrafish) are excellent systems to state-of-the-art microscopy techniques, given their extended/thin epithelia and optical clarity [67–71]. These models have been widely used to study human diseases, due to a high similarity with man in basic machineries and signaling pathways [67–72].

In conclusion, the identification of aberrant protein signatures is currently a challenge in cancer and in the diagnosis of diseases characterized by aberrant protein expression or trafficking deregulation, including Alzheimer, Parkinson, Huntington, Creutzfeldt–Jakob, amyloidosis, cystic fibrosis, nephrogenic diabetes insipidus, Gaucher's disease, and many other misfolding and aggregation disorders [73–76]. Hence, automatic or semiautomatic inspection of protein level and localization constitutes a major advance in the field of proteinopathies.

Limitations

The advantages and applications of our algorithm open a complete new field in the quantification of biological processes. However, due to the fact that the mapping and quantification of protein expression are performed by computing 1D immunofluorescence images, some limitations still exist.

Indeed, one major drawback of this bioimaging tool is the fact that it can only be applied to cells in direct contact. To better characterize the behavior of a protein at the cell membrane, it is required that IN profiling would be extracted from neighboring parallel segments of two adjacent cells, depending on a tight cell–cell contact. If this premise is not accomplished, the output data could have no biological meaning or be wrongly interpreted. To overcome this limitation, RD profiling is the correct tool to be used in isolated cells, since it extracts the expression level and mapping at a single-cell level.

In addition, 1D IN and RD profiling are also not yet ideal for tissue samples. Tissues and organs are three-dimensional (3D) structures, with unique topological interactions and spatial organization of multilayers of cells. As so, using a single 1D IF image, cell nuclei or IN profiles of our protein of interest are not fully represented in that image, promoting again an incorrect data analysis.

A major challenge for this new bioimaging application is to perform an integrated analysis of the data, from multiple images from a single experimental condition. For example, to observe protein pattern variations along experimental time, time-lapse images from different time-points need to be quantified individually and the data must be combined manually to obtain a unique IN profile. An automated version addressing this issue is not available yet. The same happens for dynamic studies exploring the signaling of two or more proteins, colocalization assays, or 3D images, where it is not possible to have an integrated analysis of the IN profile variation due to the 3D positioning of cells. This integrative analysis would be a major feature, producing more precise, real, and informative results, and improving significantly the impact of this bioimaging tool on the investigation of complex biological processes.

Future

An integrated analysis combining the quantification and mapping of biomolecules with morphological parameters, namely cell area and cytoskeletal organization, would be of great value to unravel molecular paths or disease mechanisms in basic and clinical research. For example, in the analysis of drug-screening strategies, it will be possible to select compounds that rescue better protein scores and, simultaneously, normal cellular phenotypes. The design of geometric meshes representing intercellular organization, together with protein mapping and expression, can be a possible strategy in this type of analyses.

Quantitative immunohistochemistry in tissue samples provides an absorbing challenge concerning the development of accurate signal-processing algorithms. Indeed, a key limitation is to provide quantitative information of protein level and localization in 3D structures. Aside of the huge number of bioimaging methodologies proposed in the past years, tissue quantitative molecular analysis faces problematic factors, such as preprocessing, region-of-interest (ROI) identification, multicellular structure detection, tissue segmentation, and automated diagnosis. For instance, in cancer-automated diagnosis, many attempts have been made in order to establish grading scales based on size, shape, and texture of cellular components; nonetheless, most of the systems demonstrated lack of robustness and inability to solve these problems. Therefore, qualitative visual inspection of tissue immunohistochemistry with stable surface-functionalized probes remains the most common technique in cancer diagnosis. A clear opportunity to improve on signal emission and detection for an automated 3D analysis is, thus, the use of stable and highly sensitive light-emitting particles. The semiconductor quantum dots (QDs) are a new class of fluorescent labels, which have been regarded as a major breakthrough in molecular imaging domain. QDs have sharper emission profiles and are considerably brighter and more resistant to photobleaching than fluorescent proteins. With the use of QD, we will be able to develop more sensitive and efficient methods of signal detection and image acquisition, aiming to map and quantify molecular biomarkers.

In addition, modern machine learning and data mining techniques could also be applied to identify candidate proteins and model complex mechanisms involved in a large panel of diseases. In this perspective, our next goal is to develop an automated integration of heterogeneous, multiscale, and multimodal datasets, and increase exponentially the potential of this algorithm.

Acknowledgments

This work was supported by FEDER funds through the Operational Programme for Competitiveness Factors (COMPETE) and National Funds through the Portuguese Foundation for Science and Technology (FCT), under the projects PTDC/BIM-ONC/0171/2012, PTDC/BIM-ONC/0281/2014, PTDC/BBB-IMG/0283/2014, and postdoctoral grants SFRH/BPD/87705/2012-JF and SFRH/BPD/75705/2011-ASR. We acknowledge the Programa IFCT (FCT Investigator) for funding JP research. We also thank to the American Association of Patients with Hereditary Gastric Cancer *No Stomach for Cancer* for funding the project "Today's present, tomorrow's future on the study of germline E-cadherin missense mutations."

References

1. Towbin, H., T. Staehelin, and J. Gordon, Electrophoretic transfer of proteins from polyacrylamide gels to nitrocellulose sheets: Procedure and some applications. *Proc Natl Acad Sci USA*, 1979. **76**(9): 4350–4354.
2. Gallagher, S.R., One-dimensional SDS gel electrophoresis of proteins. *Curr Protoc Protein Sci*, 2001. **3**(10): 10.1.1–10.1.34.
3. Verschoor, C.P. et al., An introduction to automated flow cytometry gating tools and their implementation. *Front Immunol*, 2015. **6**: 380.
4. Chattopadhyay, P.K. and M. Roederer, Cytometry: today's technology and tomorrow's horizons. *Methods*, 2012. **57**(3): 251–258.
5. Tanner, S.D. et al., An introduction to mass cytometry: Fundamentals and applications. *Cancer Immunol Immunother*, 2013. **62**(5): 955–965.

6. van Beijnum, J.R. et al., Isolation of endothelial cells from fresh tissues. *Nat Protoc*, 2008. **3**(6): 1085–1091.
7. Lichtman, J.W. and J.A. Conchello, Fluorescence microscopy. *Nat Methods*, 2005. **2**(12): 910–919.
8. Muzzey, D. and A. van Oudenaarden, Quantitative time-lapse fluorescence microscopy in single cells. *Annu Rev Cell Dev Biol*, 2009. **25**: 301–327.
9. Sandison, D.R. et al., Quantitative fluorescence confocal laser scanning microscopy (CLSM). In *Handbook of Biological Confocal Microscopy*, J.B. Pawley (Ed.), 1995. Plenum, NY, pp. 39–53.
10. Nakano, A., Spinning-disk confocal microscopy—A cutting-edge tool for imaging of membrane traffic. *Cell Struct Funct*, 2002. **27**(5): 349–355.
11. Waters, J.C., Accuracy and precision in quantitative fluorescence microscopy. *J Cell Biol*, 2009. **185**(7): 1135–1148.
12. Hamilton, N., Quantification and its applications in fluorescent microscopy imaging. *Traffic*, 2009. **10**(8): 951–961.
13. Esménio, S. et al., E-cadherin radial distribution characterization for mutation detection purposes. Proceedings. Springer 2013 Lecture Notes in Computer Science, 2013 (ISBN 978-3-642-38627-5).
14. Otsu, N., A threshold selection method from gray-level histograms. *Sys, Man Cybernetics, IEEE Trans*, 1979. **9**(1): pp. 62–66.
15. Beucher, S. and C. Lantuéjoul, Use of watersheds in contour detection. *Proceedings of the International Workshop on Image Processing: Real Time Edge and Motion Detection/Estimation*, Rennes, France, 1979.
16. Wu, Q., F. Merchant, and K. Castleman, *Microscope Image Processing*. Academic press, NY, 2010.
17. Rodrigues, I.C. and J.M. Sanches, Convex total variation denoising of Poisson fluorescence confocal images with anisotropic filtering. *IEEE Trans Image Process*, 2011. **20**(1): 146–160.
18. Fonseca, L.M.G. and B.S. Manjunath, Registration techniques for multisensor remotely sensed imagery. *Photogramm Eng Remote Sensing*, 1996. **62**(9): 1049–1056.
19. Zitova, B. and J. Flusser, Image registration methods: A survey. *Image Vis Comput*, 2003. **21**(9): 977–1000.
20. Sanches, J.M. and J.S. Marques, Joint image registration and volume reconstruction for 3D ultrasound. *Pattern Recognit Lett*, 2003. **24**(4): 791–800.
21. Li, S., J. Wakefield, and J.A. Noble, Automated segmentation and alignment of mitotic nuclei for kymograph visualisation. *Biomedical Imaging: From Nano to Macro*, 2011 IEEE International Symposium on, pp. 622–625, 2011.
22. van Roy, F. and G. Berx, The cell-cell adhesion molecule E-cadherin. *Cell Mol Life Sci*, 2008. **65**(23): 3756–3788.
23. Paredes, J. et al., Epithelial E- and P-cadherins: Role and clinical significance in cancer. *Biochim Biophys Acta*, 2012. **1826**(2): 297–311.
24. Berx, G. and F. van Roy, Involvement of members of the cadherin superfamily in cancer. *Cold Spring Harb Perspect Biol*, 2009. **1**(6): a003129.
25. Aberle, H., H. Schwartz, and R. Kemler, Cadherin-catenin complex: Protein interactions and their implications for cadherin function. *J Cell Biochem*, 1996. **61**(4): 514–523.
26. Ozawa, M., H. Baribault, and R. Kemler, The cytoplasmic domain of the cell adhesion molecule uvomorulin associates with three independent proteins structurally related in different species. *EMBO J*, 1989. **8**(6): 1711–1717.
27. Yap, A.S., C.M. Niessen, and B.M. Gumbiner, The juxtamembrane region of the cadherin cytoplasmic tail supports lateral clustering, adhesive strengthening, and interaction with p120ctn. *J Cell Biol*, 1998. **141**(3): 779–789.
28. Bajpai, S. et al., α-Catenin mediates initial E-cadherin-dependent cell-cell recognition and subsequent bond strengthening. *Proc Natl Acad Sci USA*, 2008. **105**(47): 18331–18336.

29. Chen, Y.T., D.B. Stewart, and W.J. Nelson, Coupling assembly of the E-cadherin/beta-catenin complex to efficient endoplasmic reticulum exit and basal-lateral membrane targeting of E-cadherin in polarized MDCK cells. *J Cell Biol*, 1999. **144**(4): 687–699.
30. Davis, M.A., R.C. Ireton, and A.B. Reynolds, A core function for p120-catenin in cadherin turnover. *J Cell Biol*, 2003. **163**(3): 525–534.
31. Ireton, R.C. et al., A novel role for p120 catenin in E-cadherin function. *J Cell Biol*, 2002. **159**(3): 465–476.
32. Oliveira, C. et al., E-cadherin alterations in hereditary disorders with emphasis on hereditary diffuse gastric cancer. *Prog Mol Biol Transl Sci*, 2013. **116**: 337–359.
33. Carneiro, P. et al., E-cadherin dysfunction in gastric cancer–cellular consequences, clinical applications and open questions. *FEBS Lett*, 2012. **586**(18): 2981–2989.
34. Shimoyama, Y. and S. Hirohashi, Expression of E- and P-cadherin in gastric carcinomas. *Cancer Res*, 1991. **51**(8): 2185–2192.
35. Albergaria, A. et al., P-cadherin role in normal breast development and cancer. *Int J Dev Biol*, 2011. **55**(7–9): 811–822.
36. Paredes, J. et al., P-cadherin expression in breast cancer: A review. *Breast Cancer Res*, 2007. **9**(5): 214.
37. Van Marck, V. et al., P-cadherin promotes cell-cell adhesion and counteracts invasion in human melanoma. *Cancer Res*, 2005. **65**(19): 8774–8783.
38. Paredes, J. et al., P-cadherin overexpression is an indicator of clinical outcome in invasive breast carcinomas and is associated with CDH3 promoter hypomethylation. *Clin Cancer Res*, 2005. **11**(16): 5869–5877.
39. Ribeiro, A.S. et al., Extracellular cleavage and shedding of P-cadherin: A mechanism underlying the invasive behaviour of breast cancer cells. *Oncogene*, 2010. **29**(3): 392–402.
40. Paredes, J. et al., Breast carcinomas that co-express E- and P-cadherin are associated with p120-catenin cytoplasmic localisation and poor patient survival. *J Clin Pathol*, 2008. **61**(7): 856–862.
41. Suriano, G. et al., E-cadherin germline missense mutations and cell phenotype: Evidence for the independence of cell invasion on the motile capabilities of the cells. *Hum Mol Genet*, 2003. **12**(22): 3007–3016.
42. More, H. et al., Identification of seven novel germline mutations in the human E-cadherin (CDH1) gene. *Hum Mutat*, 2007. **28**(2): 203.
43. Kaurah, P. et al., Founder and recurrent CDH1 mutations in families with hereditary diffuse gastric cancer. *JAMA*, 2007. **297**(21): 2360–2372.
44. Simoes-Correia, J. et al., Endoplasmic reticulum quality control: A new mechanism of E-cadherin regulation and its implication in cancer. *Hum Mol Genet*, 2008. **17**(22): 3566–3576.
45. Oliveira, C. et al., Screening E-cadherin in gastric cancer families reveals germline mutations only in hereditary diffuse gastric cancer kindred. *Hum Mutat*, 2002. **19**(5): 510–517.
46. Zhang, Y. et al., Germline mutations and polymorphic variants in MMR, E-cadherin and MYH genes associated with familial gastric cancer in Jiangsu of China. *Int J Cancer*, 2006. **119**(11): 2592–2596.
47. Keller, G. et al., Germline mutations of the E-cadherin(CDH1) and TP53 genes, rather than of RUNX3 and HPP1, contribute to genetic predisposition in German gastric cancer patients. *J Med Genet*, 2004. **41**(6): e89.
48. Oliveira, C. et al., E-Cadherin (CDH1) and p53 rather than SMAD4 and Caspase-10 germline mutations contribute to genetic predisposition in Portuguese gastric cancer patients. *Eur J Cancer*, 2004. **40**(12): 1897–1903.
49. Yabuta, T. et al., E-cadherin gene variants in gastric cancer families whose probands are diagnosed with diffuse gastric cancer. *Int J Cancer*, 2002. **101**(5): 434–441.

50. Figueiredo, J. et al., The importance of E-cadherin binding partners to evaluate the pathogenicity of E-cadherin missense mutations associated to HDGC. *Eur J Hum Genet*, 2013. **21**(3): 301–309.
51. Sanches, J.M. et al., Quantification of mutant E-cadherin using bioimaging analysis of in situ fluorescence microscopy. A new approach to CDH1 missense variants. *Eur J Hum Genet*, 2015. **23**(8): 1072–1079.
52. Oliveira, C. et al., Familial gastric cancer: genetic susceptibility, pathology, and implications for management. *Lancet Oncol*, 2015. **16**(2): e60–e70.
53. van der Post, R.S. et al., Hereditary diffuse gastric cancer: Updated clinical guidelines with an emphasis on germline CDH1 mutation carriers. *J Med Genet*, 2015. **52**(6): 361–374.
54. Ribeiro, A.S. et al., P-cadherin functional role is dependent on E-cadherin cellular context: A proof of concept using the breast cancer model. *J Pathol*, 2013. **229**(5): 705–718.
55. Carvalho, S. et al., Preventing E-cadherin aberrant N-glycosylation at Asn-554 improves its critical function in gastric cancer. *Oncogene*, 2015. **35**(13): 1619–1631.
56. Niu, X. et al., Physical transfer of membrane and cytoplasmic components as a general mechanism of cell-cell communication. *J Cell Sci*, 2009. **122**(Pt 5): 600–610.
57. Ahmed, K.A. and J. Xiang, Mechanisms of cellular communication through intercellular protein transfer. *J Cell Mol Med*, 2011. **15**(7): 1458–1473.
58. Davis, D.M., Intercellular transfer of cell-surface proteins is common and can affect many stages of an immune response. *Nat Rev Immunol*, 2007. **7**(3): 238–243.
59. Rechavi, O., I. Goldstein, and Y. Kloog, Intercellular exchange of proteins: The immune cell habit of sharing. *FEBS Lett*, 2009. **583**(11): 1792–1799.
60. Wang, X. and H.H. Gerdes, Transfer of mitochondria via tunneling nanotubes rescues apoptotic PC12 cells. *Cell Death Differ*, 2015. **22**(7): 1181–1191.
61. Russell, D.G., The evolutionary pressures that have molded Mycobacterium tuberculosis into an infectious adjuvant. *Curr Opin Microbiol*, 2013. **16**(1): 78–84.
62. Lebreton, A., F. Stavru, and P. Cossart, Organelle targeting during bacterial infection: insights from Listeria. *Trends Cell Biol*, 2015. **25**(6): 330–338.
63. Ensminger, A.W., Legionella pneumophila, armed to the hilt: Justifying the largest arsenal of effectors in the bacterial world. *Curr Opin Microbiol*, 2015. **29**: 74–80.
64. Antunes, M. et al., Coordinated waves of actomyosin flow and apical cell constriction immediately after wounding. *J Cell Biol*, 2013. **202**(2): 365–379.
65. Cordeiro, J.V. and A. Jacinto, The role of transcription-independent damage signals in the initiation of epithelial wound healing. *Nat Rev Mol Cell Biol*, 2013. **14**(4): 249–262.
66. Danjo, Y. and I.K. Gipson, Actin "purse string" filaments are anchored by E-cadherin-mediated adherens junctions at the leading edge of the epithelial wound, providing coordinated cell movement. *J Cell Sci*, 1998. **111** (Pt 22): 3323–3332.
67. Brumby, A.M. and H.E. Richardson, Using Drosophila melanogaster to map human cancer pathways. *Nat Rev Cancer*, 2005. **5**(8): 626–639.
68. Potter, C.J., G.S. Turenchalk, and T. Xu, Drosophila in cancer research. An expanding role. *Trends Genet*, 2000. **16**(1): 33–39.
69. Berghmans, S. et al., Making waves in cancer research: new models in the zebrafish. *Biotechniques*, 2005. **39**(2): 227–237.
70. Gonzalez-Moragas, L., A. Roig, and A. Laromaine, C. elegans as a tool for in vivo nanoparticle assessment. *Adv Colloid Interface Sci*, 2015. **219**: 10–26.
71. O'Reilly, L.P. et al., C. elegans in high-throughput drug discovery. *Adv Drug Deliv Rev*, 2014. **69–70**: 247–253.

72. Calahorro, F. and M. Ruiz-Rubio, Caenorhabditis elegans as an experimental tool for the study of complex neurological diseases: Parkinson's disease, Alzheimer's disease and autism spectrum disorder. *Invert Neurosci*, 2011. **11**(2): 73–83.
73. Chaudhuri, T.K. and S. Paul, Protein-misfolding diseases and chaperone-based therapeutic approaches. *FEBS J*, 2006. **273**(7): 1331–1349.
74. Cohen, F.E. and J.W. Kelly, Therapeutic approaches to protein-misfolding diseases. *Nature*, 2003. **426**(6968): 905–909.
75. Muchowski, P.J., Protein misfolding, amyloid formation, and neurodegeneration: A critical role for molecular chaperones? *Neuron*, 2002. **35**(1): 9–12.
76. Bartolini, M. and V. Andrisano, Strategies for the inhibition of protein aggregation in human diseases. *Chembiochem*, 2010. **11**(8): 1018–1035.
77. Ribeiro, A.S., et al., *Atomic force microscopy and graph analysis to study the P-cadherin/SFK mechanotransduction signalling in breast cancer cells.* Nanoscale, 2016. **8**(46): p. 19390–19401.

Chapter 16 Cancer cell invadopodia

Visualization and quantification tools

Angela Margarida Costa
and Maria José Oliveira

Contents

Introduction

Cancer metastasis is the major cause of cancer high-rate mortality, although the mechanisms that regulate this process are not fully understood. In the case of solid tumors, the metastatic process consists on the displacement of cancer cells from the primary tumor site and the formation of secondary tumors in other organs. Metastasis is a multistep process that comprises sequential events such as local invasion, intravasation (entry into blood or lymph vessels), survival in circulation, extravasation (exit from blood or lymph vessels), and colonization (homing at secondary sites).

Invasion, the hallmark of malignancy, is the process that distinguish a locally constricted benign from a malignant tumor or cancer. It occurs when cancer cells gain the ability to cross tissue boundaries, reaching the underlying stroma. To invade adjacent tissues, tumor cells are dependent of intrinsic properties as migration, the ability to move through the tissue, and proteolysis, the ability to induce extracellular matrix (ECM) degradation and reorganization, and [1,2] of extrinsic factors as the ECM composition and stiffness, hypoxia, extracellular vesicles, and other cells from the microenvironment [2–8].

To invade, tumor cells form invadopodia, which are highly dynamic actin- and cortactin-enriched protrusions, with intense proteolytic activity. These structures were first described in Rous sarcoma virus transformed cells [9] and are associated with cancer cell invasion *in vitro* and *in vivo* [10–14]. Invadopodia, generally located at the invasive cell-basal membrane, elongate and guide cells to the stroma, being crucial to locally degrade ECM components, as collagen type-I and IV, laminin, and fibronectin [15–17]. The majority of the invadopodia effector proteins are associated with cytoskeleton regulation (Arp 2/3 complex, cortactin, N-WASP, cofilin, and RhoGTPases), or with proteolytic activity (matrix metalloproteases, MMPs) [18], being both the activation of cytoskeleton-associated proteins and proteolysis, necessary to accurately identify these structures.

Unraveling the components of invadopodia, how and where they are expressed, and in which manner they are regulated is of critical importance to understand the invasive process, and to develop novel therapeutic strategies. However, the quantification and understanding of invadopodia assembly and disassembly kinetics are still limited by technical constrains, namely *in vivo* visualization, by the nature of the invasive process that occurs deep in tissues, in a highly dynamic way. In this chapter, we will focus on the most common assays used to study invadopodia formation and regulation as well as on the most common and recent advances on invadopodia visualization and quantification.

Experimental procedure

In vitro assays

To visualize and quantify invadopodia, the classical approaches consist of fluorescently label ECM components, as collagen and fibronectin, or of other matrices such as gelatin or matrigel, on which the invasive cells are seeded and allowed to invade (Table 16.1). After cell-matrix adhesion, invadopodia are formed at the cell basal side, as protrusions enriched in classical markers (F-actin, cortactin, Tks5, and proteolytic enzymes), and appear as areas of fluorescence loss, due to fluorescent matrix components degradation. The quantification of invadopodia is usually done by analysis of at least 100 random cells, and by the combination of epifluorescence with widefield microscope images [15].

Table 16.1 Invadopodia Formation Assays

	Models	Microscopy Analysis	Advantages	Disadvantages	Application	References
In vitro	• Cancer cells seeded on/within fluorescently-labeled ECM components • ECM-coated transwells • Cell culture in dense epidermal decellularized matrices	• Epifluorescence • Confocal • High-resolution time-lapse • Transmission electron microscopy (TEM) • High-resolution structured illumination microscopy (SIM) imaging • Total internal reflection fluorescence (TIRF) • Fluorescence resonance energy transfer (FRET)	• Easy to perform and image • Controlled environment • More exact colocalization analysis • Analysis in native basement membrane or isolated ECM components (2D or 3D) • Handling of matrices as coverslips	• Manual quantification of invadopodia • Artifacts resulting from contact of cells with hard substract beneath them • Static assays do not replicate the tumor microenvironment • Chemicals used in electron microscopy can interfere with actin filaments	• Invadopodia formation overtime driven by an invasive stimuli	[52] Sharma, 2013
					• Effect of molecular silencing on invadopodia dynamics	[19,48], Sharma, 2013
					• Patients samples analysis by DQ–gelatin–FITC to assess areas of degradation that colocalize with a protein of interest	[20]
					• Use of FRET to assess the role of Rac1 in invadopodia formation	[21]
					• Use of TEM to demonstrate that invadopodia are in close proximity to the nucleus and Golgi	[22–25]
					• Use of SIM to study invadopodia behavior in distinct matrices	[26]
					• TIRF used to study colocalization of molecules at invadopodia, and the spatial localization of invadopodia in the cell	[25,27]
					• Influence of matrix rigidity on invadopodia formation	[28,29]

(*Continued*)

Table 16.1 (*Continued*) Invadopodia Formation Assays

	Models	Microscopy Analysis	Advantages	Disadvantages	Application	References
In vivo	Avian chorioallantoic membrane (CAM) assay	• Widefield • Confocal • High-resolution time-lapse or multiphoton intravital microscopy	*In vivo* 3D system, which closely mimics the complex tumor microenvironment	• Assay in which mammalian cells are not in an human • Microenvironment Only labeled cells can be used • Lack of markers for ECM degradation that could be imaged overtime • Intravital imaging have to be optimized to high resolution in all planes, with rapid image acquisition, to follow several markers simultaneously	• Role of specific invadopodia molecules on cancer cell extravasation	[30]
	Cancer cell lines xenografts in mice				• Intravital live imaging to analyze invadopodia formation during cancer-cell intravasation, and its dependence on the microenvironment	[6,31]
	C. elegans				• Genome-wide RNAi screening for genes promoting invadopodia	[32]
	Zebrafish				• Assess the need of specific molecules or pathways to the invadopodia formation	[33]

The ECM-digested areas and the invadopodia morphology are generally measured using different softwares as Threshold command of Metamorph, Analyze Particles command of ImageJ, Velocity 5.5, or iVision software [15,34–36]. Despite recent software advances for invadopodia analysis and quantification [37], the majority of the works published in the field includes manual analysis, in a time-consuming and partially bias way.

A variant of this technique is the use of a noncross-linking substrate with an excessive number of fluorescent dyes attached that due to their close proximity lead to signal quenching. During invasion, substrate enzymatic degradation induces the separation of the dye molecules, resulting in the emission of a fluorescent signal. On this way, invadopodia matrix degradation sites are visualized as fluorescent dots in a nondigested black background [20]. An example is the fluorogenic DQ™ collagen assay [21], which in comparison with other invadopodia formation assays, avoids high fluorescent background and potential 2D cross-linking artifacts, permitting their exact localization.

Since invadopodia extension is impaired by the resistance of the glass coverslip, a new chemoinvasion assay was developed to overcome this limitation (Table 16.1). Such assay comprises the use of a transwell system where the upper and lower compartments are separated through 1 μm pore size filter. These filters, coated with commercially available or native matrix components, allow invadopodia extension and matrix degradation, without permitting matrix cancer cell invasion, as it happens when using matrigel filters with 8 μm pore size [15]. This assay has the advantage of enabling the use of native basement membranes, isolated directly from human or animal models, instead of just commercially available recombinant ECM components or artificial gel-based matrices, allowing an accurate study of native cancer cells–ECM interactions [15,38].

To better mimic native matrices, invadopodia analysis was also established using cells cultured over dense pig dermal decellularized matrices (XeDerma®), a three-dimensional mesh of collagen and elastic fibers (Table 16.1). These studies revealed that over a complex 3D environment, the molecular distribution of major invadopodia components differs from classical 2D assays. Over 3D matrices, thin filopodia-like protrusions, emerging from a central invadopodia and lacking cortactin and MMP vesicles, are visible. These decellularized native matrices have the advantage of enabling direct handling to electron microscopy and immunofluorescence analysis as a classical coverslip [22].

The use of confocal microscopy, in combination with distinct fluorescent markers directed to ECM and invadopodia components, enables the detailed assessment of invadopodia localization and structure. However, to study invadopodia assembly/disassembly kinetics a more dynamic analysis is required. It is reported that invasive cells are able to form invadopodia protrusions with 1 to 7 μm long and 0.5–2 μm diameter [15], just few hours after seeding. The invadopodia kinetics and its formation dynamics can be monitored and evaluated through high-resolution time-lapse microscopy, by tracing cells stably transfected with specific fluorescent invadopodia markers. Under these circumstances, invadopodia assembly is considered to occur when spots of colocalized markers appear, whereas its disassembly ensues when those spots disappear [21].

Transmission electron microscopy (TEM), although providing static observations, is also a valuable method to understand the relative location of invadopodia within the cell and to evaluate invadopodia cytoskeleton ultrastructure and vesicle trafficking [15,22,34]. The most advanced microscope systems are also useful tools for invadopodia studies, as evidenced by the use of high-resolution structured illumination microscopy (SIM) imaging [26], and of total internal reflection fluorescence microscopy [26,39]. Other recent techniques are proposed to complement the above-described methods, such as (1) fluorescence resonance energy transfer (FRET)-based sensors, which enable, after photoactivation, the direct subcellular visualization of the molecule of interest [34]; and the (2) study of invadopodia dynamics, through barbed ends assay, once free barbed ends are essential to actin polymerization [34].

Most interestingly, the majority of these visualization techniques are compatible, enriched, and complemented with advanced cellular and molecular analysis tools. After image acquisition, the differential expression of invadopodia-specific markers along the invasive cell body, or of its active protrusions, is possible. Through a differential extraction method, by scraping cells from the top of artificial or native matrices, or through the recovering of the entrapped invadopodia arrested within these matrices, it is possible to dissect in more detail the molecular mechanisms underlying invadopodia formation and kinetics [18,20].

In vivo assays

In an attempt to study *in vivo*, the dynamics of invadopodia formation along the extravasation process, the chick chorioallantoic membrane (CAM) assay was used in association with widefield, confocal fluorescence, or high-resolution time-lapse multiphoton intravital microscopy (Table 16.1). Therefore, cancer cells were injected intravenously in the CAM, a highly organized capillary network supported by arteries, veins, and stromal cells. The combination of fluorescent labeling of lectins (stain the luminal surface of endothelial cells), dextrans (stain the vessels lumen), avian endothelium, and cancer cells, with the accessibility, transparency, and high CAM vascularization, allowed 3D visualization of invadopodia formation along the circulation and extravasation of tumor cells into the CAM stroma [30,40]. These studies evidenced that invadopodia facilitates tumor cell extravasation, a crucial step for metastasis.

Through the combination of human breast carcinoma xenografts in SCID mice with high-resolution multiphoton microscopy, it was also possible to identify invadopodia *in vivo*, through tracing of cortactin, Tks5, and MMP-enriched vesicles (Table 16.1) [6,31]. Data supporting the role of invadopodia on basement membrane degradation was also accomplished using the *Caenorhabditis elegans* (*C. elegans*) model, in combination with high-resolution confocal and time-lapse microscopy. During the uterine-vulval development, the anchor cells, specialize uterine cells, invade the uterine basement membrane, projecting invadopodia [41,42]. Using *C. elegans* mutants to the proteins of interest, it is possible to address *in vivo* their role on the invadopodia-formation process.

Zebrafish can also be used as a promising model to study invadopodia formation and kinetics, given the similarity of the mechanisms that support mammalian cancer-cell invasion and migration of fish neural crest or of intestinal epithelial cells during development, that are also known to form actin-rich protrusions (Table 16.1) [33]. Detailed analysis revealed that these basal side protrusions are enriched in actin, cortactin, Tks5, and MT1-MMP, in close proximity with mammalian cancer-cell invadopodia [43].

Altogether, the combination of the most recent advances in artificial and native matrices, animal models of invasion and metastasis, and high-resolution microscopy and intravital animal imaging highlighted the role of specific invadopodia markers on invadopodia formation, kinetics, and therapeutics [44].

Application

The majority of our knowledge on invasion has been achieved through *in vitro* assays established with ECM components, as Matrigel, 3D collagen gels, and spheroid assays, or through *in vivo* animal models, as the CAM assay, and the evaluation of tumor growth, invasion, and metastasis in flies, fishes, nematodes, or mice [45,46]. The advances in cellular and molecular techniques, the use of transgenic animal models, and of intravital microscopy imaging improved our ability to dissect invasion-associated mechanisms and to develop novel therapeutic strategies [47].

Invadopodia formation assays are valuable and complemental tools for the assessment of ECM degradation and cytoskeleton alterations, driven by invasive stimulus, throughout the distinct steps of the metastatic process, as primary tumor invasion, vascular intravasation, vascular extravasation, and distant metastasis [47]. The majority of these assays benefits from the advances in high-resolution microscopy and brings together two elements of the tumor microenvironment essential to disease progression: (1) the cancer cells themselves and (2) the extracellular matrix components. The effect of distinct pharmacological inhibitors or of silencing experiments on the impairment of invadopodia formation dynamics exploits driven mechanisms and possible routes of therapeutic intervention [19,48–51].

The use of advanced microscopy techniques, in combination with fluorescent markers, directed to artificial or native ECM components and to distinct invadopodia-associated molecules as cortactin and actin, enables the detailed analysis of invadopodia formation kinetics [52]. High resolution time-lapse microscopy studies performed on gelatin fluorescently labeled substrates, combined with immunofluorescence analysis, revealed that invadopodia precursors already comprise the same markers as mature invadopodia (cortactin, cofilin, N-WASp, Arp2/3 complex, F-actin, Tks5, and the membrane type-I matrix metalloprotease MT1-MMP), although without ECM proteolytic activity [53] Similarly, using the fluorescent dye-quenched (DQ)-collagen type-IV it was possible to track the colagenolytic activity of colorectal cancer cells and to dissect the role of vasodilator-stimulated phosphoprotein (VASP) on invadopodia formation [51]. The results indicated that phosphorylation of VASP (Ser239), an actin-binding protein implicated in membrane protrusion dynamics, reduced the number and length of filopodia and invadopodia and impaired cancer-cell collagenolytic activity. Moreover, these achievements suggested that VASP activation may constitute a putative target to control colorectal cancer invasion and metastasis [51].

Invadopodia analysis may also combine microscopy with *in situ* zymography, a technique that allows detection of MMP activity directly in histological sections. Therefore, frozen tissue sections, derived from animal models or directly from cancer patients' surgical resections, are placed on glass coverslips and coated with fluorescently labeled ECM proteins. After incubation with a proper MMP activity buffer, and using an inverted epifluorescence microscope, tissue areas of proteolytic activity are visible as black digested areas over a fluorescently-labeled undigested matrix. At these focalized areas of proteolysis, cells can be even extracted for further invadopodia molecular analysis [20]. Applying this strategy to distinct tumor cryosections, other researchers evidenced the presence of cortactin and MMP-containing vesicles in cells adjacent to the tumor edge and to tumor blood vessels [6,31,54,55].

FRET was probably the technique that most contributed to understand invadopodia assembly mechanisms and kinetics [53]. Based on this knowledge, we know that at the beginning of invadopodia formation, cortactin, N-WASp, and Arp2/3 proteins are recruited to areas of invadopodia precursors and establish an actin nucleation complex, to which other proteins are recruited. Cortactin binds then to cofilin, sequestering it and inhibiting its severing activity [53]. When cortactin is phosphorylated, cofilin is released and reestablishes its severing activity, which generates free barbed ends for Arp2/3 complex F-actin nucleation and polymerization, permitting invadopodia elongation [41]. Cortactin dephosphorylation blocks cofilin severing and subsequent actin nucleation, stabilizing invadopodia. At the tip of this elongating invadopodia, MMP-enriched vesicles are then recruited and released to permit extracellular matrix degradation and cellular invasion [53]. Since the formation of distinct cell motility structures (lamellipodia, fillopodia, and stress fibers) has been associated with enhanced smallGTPases activity [56], the involvement of these proteins on invadopodia formation was also intensively investigated. Therefore, using a specific Racl-FRET biosensor along invadopodia formation, Moshfegh and collaborators evidenced that Racl is not required during these motility structures formation but only for their stability and disassembly [21].

Microscopy techniques are absolutely crucial and valuable tools in invadopodia studies. TEM was useful to evidence that invadopodia are membrane protrusions found in close proximity to the nucleus and to the Golgi apparatus [22,23–25]. Based on this technique, researchers demonstrated that the Golgi complex was normally oriented towards the invadopodia [23,24]. Invadopodia apical end interacts with the nuclear envelope, whereas its body extends through the cytoplasmic microtubule network, ending at basal integrin-enriched areas [25]. In addition, using TEM analysis on a cell-free dermis-based matrix as a new model to study invasion-associated structures, Tolde and collaborators evidenced that matrix proteolytic areas are localized at the base of invadopodia and not at the tip of their matrix-penetrating protrusions [22].

Notably, intravital animal imaging, using cells labeled toward invadopodia markers, was used to demonstrate that invadopodia-like structures are formed in cells invading blood vessels and that its formation is dependent from tumor microenvironment factors [6,44]. In fact, using high-resolution multiphoton microscopy in a breast cancer mouse model, Gligorijevic and collaborators demonstrated that only slowly migrating cells exhibit invadopodia and found that only tumor cells in the invadopodia-rich microenvironment exert ECM proteolysis and disseminated. These findings might be extremely important to explore future therapeutic applications, as also speculated by the authors [6].

High-resolution structured illumination microscopy (SIM), which allows images with a higher resolution than the diffraction limit, brought optical microscopy into the nanodimension and was very useful in invadopodia dynamics analysis. It demonstrated that in breast cancer cells seeded on Matrigel, NaV1.5 Na^+ channels colocalize with Na^+/H^+ exchanger type-I (NHE-1) and caveolin-1 at intense spots of matrix degradation and remodeling. These invadopodia were surrounded by F-actin rings, forming projections below the cell body with high penetration into the ECM (>5 μm depth) [26].

On its turn, total internal reflection fluorescence (TIRF) microscopy, by using an evanescent wave to selectively illuminate and excite fluorophores in a well-defined region of the specimen, enables high-resolution visualization of surface regions, as the basal plasma membrane. Using this technique, researchers demonstrated that sorting nexin 9 (SNX9) colocalizes with the invadopodia marker Tsk5 and negatively regulates invadopodia formation and function [27]. Notably, SNX9 depletion increased invadopodia incidence and local matrix degradation, by recruiting MT1-MMP at these invasion-specialized structures [27]. The reduced expression of SNX9 in breast tumors and in more aggressive non-small cell lung cancers further validated these results [27].

Taking advantage of animal models it was possible to further gain insights in the *in vivo* process of invadopodia formation along the distinct steps of carcinogenesis (invasion, intravasation, extravasation, and metastasis formation). Using the *ex ovo* CAM assay, a mouse model of lung metastasis and 3D time-lapse intravital imaging, researchers visualized in detail the molecular interactions established between cancer and endothelial cells during extravasation [30]. To escape from blood vessels and initiate new organ colonization cancer cells extend, through the endothelial junctions, invadopodia enriched in cortactin, Tsk4, Tsk5, and MT1-MMP1. Importantly, the silencing of these molecules, both by pharmacological inhibitors or small interference RNA (siRNA), decreased extravasation efficiency, and impaired metastasis formation [30].

Along the past decades, zebrafish (*Danio rerio*) became an excellent model to study transformed cell invasion and to test the application of potential therapeutic agents. Notably, a mutation that leads to constitutive activity of zebrafish smooth muscle myosin heavy chain gene, enhanced: smooth muscle cell contractile tone, the production of reactive oxygen species and stimulation of intestinal epithelium invasive remodeling [43]. Such invasive phenotype is characterized by the formation of actin-enriched invadopodia structures, which extend through the intestinal epithelium basement membrane. Similarly, to mammalian cells, the cortactin, Src, and Tks5 signaling pathway is required for the formation of these zebrafish actin-rich protrusions [33]. The zebrafish studies highlighted invadopodia dynamics and signaling in non-mammalian cells and provided further evidences that invadopodia are present *in vivo* and that may take part of the natural epithelium remodeling.

The nematode worm *Caenorhabditis elegans* (*C. elegans*) has been considered as a valuable tool in research, and was used in genome-wide RNAi screening when searching for genes promoting invadopodia function *in vivo*. From this high-throughput genomic analysis, 13 novel putative invadopodia regulators were identified [32].

Along *C. elegans* uterine-vulval development, the anchor cell, a specialized uterine cell, breaches the adjacent vulval epithelium basement membrane to initiate uterine-vulval attachment [44]. The transparency of the worm and its easy laboratory manipulation allowed the performance of the first time-lapse high-resolution imaging of cell-basement membrane interactions during invasion *in vivo*. These images provided the evidences that *C. elegans* anchor cells establish dynamic invadopodia structures during the invasion of the vulval epithelium basement membrane [41,42]. Those F-actin-enriched protrusions are extremely similar to cancer-cell invadopodia sharing similar molecular players as integrins, smallGTPases, the phospholipid PI(4,5)P2, and the actin regulator Ena/VASP. Other important cancer-cell invadopodia regulators as cortactin, the Tks4/5 adaptor proteins and the MT1-MMP metalloprotease are absent in anchor cells protrusions. However, their function may be compensated by other molecules. Notably, *C. elegans* Abp1 actin-binding protein also activates the Arp2/3 complex similarly to mammalian cell cortactin; or the function of the mammalian MT1-MMP may be replaced by other matrix metalloproteases. Interestingly, while in cancer cells invadopodia may persist over an hour, the anchor cell protrusions generally last only few minutes. These differences may rely on distinct kinetics between the molecular players, or simply reflect microenvironment variations. Once again, these studies highlighted the molecular mechanisms underlying invadopodia formation and kinetics, and elucidated that invadopodia are not exclusive from cancer cells but also conserved subcellular structures, used during the normal development by cells penetrating basement membrane barriers.

To explore the interplay between invasive cells and ECM, invadopodia formation was also used as a method to access the influence of matrix rigidity on cell behavior and invasion. Through integrins and focal adhesion complexes cells sense modifications on ECM composition, organization, or biomechanical properties. Alterations detected by these external sensors are then transmitted to their cytoplasmic interacting partners, leading to cytoskeleton reorganization, altered actomyosin contractility, MMP activity, or gene expression. Interestingly, the use of polyacrilamide gels with different stiffness allowed the establishment of the relation between cellular traction stress and invadopodia activity. Notably, enhanced ECM rigidity during tumorigenesis is associated with increased cellular contractility, matrix proteolysis, and invadopodia formation. These results highlighted the relevance of ECM biomechanical properties for the regulation of cellular actin dynamics, invadopodia kinetics, and cancer-cell invasion, and metastasis [28,57].

Limitations

Despite the advances in the area, our knowledge on invasion modulation is still insufficient, and more efforts have to be engaged to understand the temporal regulation of protein–protein interactions during invadopodia establishment. Highlighting the molecular mechanisms underlying these structures will provide important breakthroughs on the development of more efficient anticancer therapies.

Classical invadopodia assays enable the use of sophisticated staining and visualization techniques, but care must be taken about possible artifacts resulting from samples preparation or from the contact of cells with the hard substratum beneath them (Table 16.1). The use of fluorescently labeled proteins and substrates overcame our limitations of detection and visualization, improving considerably our ability to study invadopodia regulatory mechanisms and induced proteolysis. Nevertheless, it is important to consider that these molecules may, even marginally, affect cell behavior, the assembly/disassembly of cell cytoskeleton molecules, or even invadopodia dynamics. As an example, some chemicals used in electron microscopy samples preparation as OsO4, can interfere with actin filaments preservation and dynamics.

Another evident limitation is the use of static *in vitro* assays to evaluate invadopodia kinetics and regulatory mechanisms. The alternative of high-resolution time-lapse microscopy, in combination with fluorescently labeled cell proteins or substrates, permitted a more realistic observation of those highly dynamic structures.

Notably, the use of 3D artificial matrices has the advantage of permitting stiffness modulation to better understand the impact of substrate biomechanical properties on cancer-cell behavior. Simultaneously through their simplicity, artificial matrices may help to understand the contribution of single ECM components on invadopodia formation and kinetics. Nevertheless, due to their oversimplification, artificial matrices do not recapitulate the complex ECM network or the interactions present at the tumor microenvironment. More recently, the use of decellularized native matrices, obtained directly from animal models or from patients surgical resections, permit a more realistic approach.

The greatest limitation for tracking invadopodia assembly *in vivo* has been the difficulty of visualizing in detail such tiny structures, which occur deep in tissues, at the interface of invading cells with adjacent ECM components. Despite these limitations, the use of intravital microscopy in distinct animal models brought remarkable insights in understanding invadopodia dynamics during organogenesis or cancer invasion and metastasis (Application Section). A limitation inherent to these animal models, and certainly the most difficult to overcome, is the limited reproduction of the human tumor microenvironment. However, the knowledge provided by these applications, when transposed to *ex vivo* human tumors in controlled and advanced microfluidic conditions may improve significantly our understanding in this field.

Within the next decades, the optimization of current animal models, of more advanced microscopic techniques, of novel and more stable fluorescent proteins, of rapid image acquisition following simultaneously several markers, will certainly overcome the current limitations and improve our knowledge in invadopodia research.

Future

Despite the advances in understanding the molecular mechanisms that drive invadopodia formation and function, there is still a lack of knowledge regarding its subcellular structure, namely in understanding why functional similar proteins are present simultaneously at the invadopodia and how do they timely regulate each other. Highlighting these questions would improve our knowledge on invadopodia kinetics and regulation and open novel perspectives for therapeutic intervention.

In addition, the achievement of a consensus of specific invadopodia markers to be used among the scientific community, and the development of *in vitro* and *ex vivo* models that more accurately mimic the native microenvironment will be of capital importance. In parallel, some of the already available models or techniques can be further explored. For example, the emerging studies with decellularized native matrices derived from patient's surgical resections, and further repopulated for *ex vivo* evaluations, will constitute an excellent system to recapitulate the biomechanical and biochemical events of the native microenvironment. In addition, the previously described CAM assay can be used as a screening for the identification of pharmacological agents targeting invadopodia formation but also invadopodia-mediated cancer-cell invasion, intravasation, and extravasation.

The technological advances in high-resolution intravital microscopy will certainly improve our visualization and understanding of invadopodia *in vivo* formation and dynamics. This knowledge will corroborate the *in vivo* relevance of the *in vitro* findings and will contribute to clarify the role of these active actin-enriched structures in tissue homeostasis, organogenesis and cancer invasion, and metastasis.

Importantly, other techniques as atomic force microscopy (AFM), already explored to measure ECM biomechanical forces and the stiffness of other invadopodia similar protrusions, as macrophage podosomes [58,59], may provide an important contribution in this research field. This technique named protrusion force microscopy, used in the context of cancer invasion would be extremely valuable to (1) increase our insights into the mechanisms sustaining invasive cell–ECM interplay; (2) dissect invadopodia kinetics regulation; and (3) test the effect of distinct agents on cellular mechanosensing and invadopodia biophysics.

Even in *in vitro* studies, the already developed methodologies may improve our knowledge about the invasion. The work of Berginski et al. [37], when applied to high-throughput imaging technologies can contribute to simplified analysis of screening of inhibitors of invadopodia formation in cancer cells, and to quantify cell heterogeneity in the cells. Another tool that can help in the study on invadopodia–ECM relationship is the adaptation of the computational modeling tool developed by Kim et al. to study filopodia, in which it is possible to predict the behavior of filopodia that penetrate in a particular 3D ECM fiber network [60]. In addition, the work in podosomes of Proag and colleagues can be adapted to the study of invadopodia [61]. In this work, it was shown a technique that allows the simultaneous tracking of multiple podosomes, which is very useful to understand the dynamic of these protrusions organization, their mechanical characteristics, and to study its collective behavior overtime.

In the future, the combination of more advanced microscopic techniques, namely intravital animal imaging, with improved invasion and metastasis animal models and native extracellular matrices, which better mimic the tumor microenvironment, will improve our knowledge in invadopodia research and highlight novel and more efficient therapeutic applications.

Acknowledgments

This work was financially supported through National Funds and cofinanced by the FEDER via the PT2020 Partnership Agreement under the 4293 Unit I&D and the Program COMPETE FCOMP-01-0124-FEDER-010915. We also acknowledge the Investigator FCT program (MJ Oliveira) and the PostDoc fellowship (SFRH/BPD/109446/2015 to AM Costa), both supported by Fundação para a Ciência e Tecnologia (Portugal).

References

1. Friedl, P. and S. Alexander, Cancer invasion and the microenvironment: Plasticity and reciprocity. *Cell*, 2011. **147**(5): 992–1009.
2. Sabeh, F., R. Shimizu-Hirota, and S.J. Weiss, Protease-dependent versus -independent cancer cell invasion programs: Three-dimensional amoeboid movement revisited. *J Cell Biol*, 2009. **185**(1): 11–19.
3. Paszek, M.J. et al., Tensional homeostasis and the malignant phenotype. *Cancer Cell*, 2005. **8**(3): 241–254.
4. Hoshino, D. et al., Exosome secretion is enhanced by invadopodia and drives invasive behavior. *Cell Rep*, 2013. **5**(5): 1159–1168.
5. Diaz, B. et al., Notch increases the shedding of HB-EGF by ADAM12 to potentiate invadopodia formation in hypoxia. *J Cell Biol*, 2013. **201**(2): 279–292.
6. Gligorijevic, B., A. Bergman, and J. Condeelis, Multiparametric classification links tumor microenvironments with tumor cell phenotype. *PLoS Biol*, 2014. **12**(11): e1001995.
7. Condeelis, J. and J.W. Pollard, Macrophages: Obligate partners for tumor cell migration, invasion, and metastasis. *Cell*, 2006. **124**(2): 263–266.
8. Arsenault, D. et al., HDAC6 deacetylase activity is required for hypoxia-induced invadopodia formation and cell invasion. *PLoS One*, 2013. **8**(2): e55529.
9. Tarone, G. et al., Rous sarcoma virus-transformed fibroblasts adhere primarily at discrete protrusions of the ventral membrane called podosomes. *Exp Cell Res*, 1985. **159**(1): 141–157.

10. Seals, D.F. et al., The adaptor protein Tks5/Fish is required for podosome formation and function, and for the protease-driven invasion of cancer cells. *Cancer Cell*, 2005. **7**(2): 155–165.
11. Bowden, E.T. et al., An invasion-related complex of cortactin, paxillin and PKCmu associates with invadopodia at sites of extracellular matrix degradation. *Oncogene*, 1999. **18**(31): 4440–4449.
12. Coopman, P.J. et al., Phagocytosis of cross-linked gelatin matrix by human breast carcinoma cells correlates with their invasive capacity. *Clin Cancer Res*, 1998. **4**(2): 507–515.
13. Yamaguchi, H., J. Wyckoff, and J. Condeelis, Cell migration in tumors. *Curr Opin Cell Biol*, 2005. **17**(5): 559–564.
14. Bravo-Cordero, J.J., L. Hodgson, and J. Condeelis, Directed cell invasion and migration during metastasis. *Curr Opin Cell Biol*, 2012. **24**(2): 277–283.
15. Schoumacher, M. et al., Actin, microtubules, and vimentin intermediate filaments cooperate for elongation of invadopodia. *J Cell Biol*, 2010. **189**(3): 541–556.
16. Weaver, A.M., Invadopodia: Specialized cell structures for cancer invasion. *Clin Exp Metastasis*, 2006. **23**(2): 97–105.
17. Kelly, T. et al., Invadopodia promote proteolysis of a wide variety of extracellular matrix proteins. *J Cell Physiol*, 1994. **158**(2): 299–308.
18. Caldieri, G. et al., Cell and molecular biology of invadopodia. *Int Rev Cell Mol Biol*, 2009. **275**: 1–34.
19. Beaty, B.T. et al., Beta1 integrin regulates Arg to promote invadopodial maturation and matrix degradation. *Mol Biol Cell*, 2013. **24**(11): 1661–1675, S1–S11.
20. Busco, G. et al., NHE1 promotes invadopodial ECM proteolysis through acidification of the peri-invadopodial space. *FASEB J*, 2010. **24**(10): 3903–3915.
21. Moshfegh, Y. et al., A Trio-Rac1-Pak1 signalling axis drives invadopodia disassembly. *Nat Cell Biol*, 2014. **16**(6): 574–586.
22. Tolde, O. et al., The structure of invadopodia in a complex 3D environment. *Eur J Cell Biol*, 2010. **89**(9): 674–680.
23. Baldassarre, M. et al., Dynamin participates in focal extracellular matrix degradation by invasive cells. *Mol Biol Cell*, 2003. **14**(3): 1074–1084.
24. Buccione, R., G. Caldieri, and I. Ayala, Invadopodia: specialized tumor cell structures for the focal degradation of the extracellular matrix. *Cancer Metastasis Rev*, 2009. **28**(1–2): 137–149.
25. Revach, O.Y. et al., Mechanical interplay between invadopodia and the nucleus in cultured cancer cells. *Sci Rep*, 2015. **5**: 9466.
26. Brisson, L. et al., NaV1.5 Na(+) channels allosterically regulate the NHE-1 exchanger and promote the activity of breast cancer cell invadopodia. *J Cell Sci*, 2013. **126**(Pt 21): 4835–4842.
27. Bendris, N. et al., Sorting nexin 9 negatively regulates invadopodia formation and function in cancer cells. *J Cell Sci*, 2016. **126**: 2804–2816.
28. Jerrell, R.J. and A. Parekh, Cellular traction stresses mediate extracellular matrix degradation by invadopodia. *Acta Biomater*, 2014. **10**(5): 1886–1896.
29. Jerrell, R.J. and A. Parekh, Matrix rigidity differentially regulates invadopodia activity through ROCK1 and ROCK2. *Biomaterials*, 2016. **84**: 119–129.
30. Leong, H.S. et al., Invadopodia are required for cancer cell extravasation and are a therapeutic target for metastasis. *Cell Rep*, 2014. **8**(5): 1558–1570.
31. Gligorijevic, B. et al., N-WASP-mediated invadopodium formation is involved in intravasation and lung metastasis of mammary tumors. *J Cell Sci*, 2012. **125**(Pt 3): 724–734.
32. Lohmer, L.L. et al., A sensitized screen for genes promoting invadopodia function in vivo: CDC-42 and rab GDI-1 direct distinct aspects of invadopodia formation. *PLoS Genet*, 2016. **12**(1): e1005786.
33. Murphy, D.A. et al., A Src-Tks5 pathway is required for neural crest cell migration during embryonic development. *PLoS One*, 2011. **6**(7): e22499.

34. Magalhaes, M.A. et al., Cortactin phosphorylation regulates cell invasion through a pH-dependent pathway. *J Cell Biol*, 2011. **195**(5): 903–920.
35. Razidlo, G.L. et al., Vav1 as a central regulator of invadopodia assembly. *Curr Biol*, 2014. **24**(1): 86–93.
36. Valenzuela-Iglesias, A. et al., Profilin1 regulates invadopodium maturation in human breast cancer cells. *Eur J Cell Biol*, 2015. **94**(2): 78–89.
37. Berginski, M.E. et al., Automated analysis of invadopodia dynamics in live cells. *PeerJ*, 2014. **2**: e462.
38. Beaty, B.T. and J. Condeelis, Digging a little deeper: the stages of invadopodium formation and maturation. *Eur J Cell Biol*, 2014. **93**(10–12): 438–444.
39. Beaty, B.T. et al., Talin regulates moesin-NHE-1 recruitment to invadopodia and promotes mammary tumor metastasis. *J Cell Biol*, 2014. **205**(5): 737–751.
40. Kim, Y. et al., Quantification of cancer cell extravasation in vivo. *Nat Protoc*, 2016. **11**(5): 937–948.
41. Hagedorn, E.J. et al., The netrin receptor DCC focuses invadopodia-driven basement membrane transmigration in vivo. *J Cell Biol*, 2013. **201**(6): 903–913.
42. Hagedorn, E.J. et al., Integrin acts upstream of netrin signaling to regulate formation of the anchor cell's invasive membrane in C. elegans. *Dev Cell*, 2009. **17**(2): 187–198.
43. Seiler, C. et al., Smooth muscle tension induces invasive remodeling of the zebrafish intestine. *PLoS Biol*, 2012. **10**(9): e1001386.
44. Lohmer, L.L. et al., Invadopodia and basement membrane invasion in vivo. *Cell Adh Migr*, 2014. **8**(3): 246–255.
45. Katt, M.E. et al., In vitro tumor models: Advantages, disadvantages, variables, and selecting the right platform. *Front Bioeng Biotechnol*, 2016. **4**: 12.
46. Kramer, N. et al., In vitro cell migration and invasion assays. *Mutat Res*, 2013. **752**(1): 10–24.
47. Ellenbroek, S.I. and J. van Rheenen, Imaging hallmarks of cancer in living mice. *Nat Rev Cancer*, 2014. **14**(6): 406–418.
48. Sharma, V.P. et al., Tks5 and SHIP2 regulate invadopodium maturation, but not initiation, in breast carcinoma cells. *Curr Biol*, 2013. **23**(21): 2079–2089.
49. Sun, J. et al., STIM1- and Orai1-mediated Ca(2+) oscillation orchestrates invadopodium formation and melanoma invasion. *J Cell Biol*, 2014. **207**(4): 535–548.
50. Mader, C.C. et al., An EGFR-Src-Arg-cortactin pathway mediates functional maturation of invadopodia and breast cancer cell invasion. *Cancer Res*, 2011. **71**(5): 1730–1741.
51. Zuzga, D.S. et al., Phosphorylation of vasodilator-stimulated phosphoprotein Ser239 suppresses filopodia and invadopodia in colon cancer. *Int J Cancer*, 2012. **130**(11): 2539–2548.
52. Sharma, V.P., D. Entenberg, and J. Condeelis, High-resolution live-cell imaging and time-lapse microscopy of invadopodium dynamics and tracking analysis. *Methods Mol Biol*, 2013. **1046**: 343–357.
53. Oser, M. et al., Cortactin regulates cofilin and N-WASp activities to control the stages of invadopodium assembly and maturation. *J Cell Biol*, 2009. **186**(4): 571–587.
54. Sibony-Benyamini, H. and H. Gil-Henn, Invadopodia: the leading force. *Eur J Cell Biol*, 2012. **91**(11–12): 896–901.
55. Genot, E. and B. Gligorijevic, Invadosomes in their natural habitat. *Eur J Cell Biol*, 2014. **93**(10–12): 367–379.
56. Ridley, A.J., Rho GTPase signalling in cell migration. *Curr Opin Cell Biol*, 2015. **36**: 103–112.

57. Yang, H. et al., Mechanosensitive caveolin-1 activation-induced PI3K/Akt/mTOR signaling pathway promotes breast cancer motility, invadopodia formation and metastasis in vivo. *Oncotarget*, 2016. **7**(13): 16227–16247.
58. Labernadie, A. et al., Dynamics of podosome stiffness revealed by atomic force microscopy. *Proc Natl Acad Sci USA*, 2010. **107**(49): 21016–21021.
59. Labernadie, A. et al., Protrusion force microscopy reveals oscillatory force generation and mechanosensing activity of human macrophage podosomes. *Nat Commun*, 2014. **5**: 5343.
60. Kim, M.C. et al., Cell invasion dynamics into a three dimensional extracellular matrix fibre network. *PLoS Comput Biol*, 2015. **11**(10): e1004535.
61. Proag, A. et al., Evaluation of the force and spatial dynamics of macrophage podosomes by multi-particle tracking. *Methods*, 2016. **94**: 75–84.

Index

Note: Page numbers followed by f and t refer to figures and tables, respectively.

D

E

H

I

J

K

L

M

S